ALLE ZEIT WACH 1842

Archives of Toxicology, Supplement 16

Use of Mechanistic Information in Risk Assessment

Proceedings of the
1993 EUROTOX Congress
Meeting Held in Uppsala, Sweden, June 30 – July 3, 1993

Edited by the Publication Committee
H. M. Bolt, B. Hellman and L. Dencker

With 35 Figures and 31 Tables

Springer-Verlag
Berlin Heidelberg New York
London Paris Tokyo
Hongkong Barcelona
Budapest

Hermann M. Bolt
Institut für Arbeitsphysiologie an der Universität Dortmund
Ardeystr. 67
D-44139 Dortmund

Björn Hellman
Uppsala University Hospital
Department of Occupational Medicine
S-75185 Uppsala

Lennart Dencker
Uppsala University
Department of Pharmaceutical Biosciences:
Toxicology
Box 594, BMC
S-75124 Uppsala

ISBN-13:978-3-642-78642-6 e-ISBN-13:978-3-642-78640-2
DOI: 10.1007/978-3-642-78640-2

Softcover reprint of the hardcover 1st edition 1994

Typesetting: Camera ready by authors
27/3130-5 4 3 2 1 0 – Printed on acid-free paper

Contents

Developmental Toxicity

Role of Mechanistic Studies of Toxicity in Risk Assessment

Mechanisms of Metal Toxicity

Toxicokinetics of Drugs

Mechanisms of Carcinogenicity and Risk Assessment

Late Manuscripts

Young Scientists Poster Award

Baltic Sea Pollution

(Chair: A. Jernelöv, Sweden)

Ecology of the Baltic - Introduction

M. Pliński
Institute of Oceanography, Gdańsk University
46 Piłsudskiego Av. 81-378 Gdynia, Poland

The Baltic Sea with 366,000 km^2 is only 0.1% of the world's oceans but nevertheless, unique in many ways. The Baltic Sea is the largest area of brackish water in the world. As a formation of nature, it is a sea that in many ways resembles a lake. This makes its flora and fauna interesting and well adapted to the brackish water environment. The recent stage of biocoenosis composition is influenced by several features. Historically, during the last glaciation, when the Baltic was sometimes a huge marine bay and sometimes a large, freshwater lake, several ecosystems developed and were successively replaced. The flora and fauna composition of those previous biocoenosis was discovered from the geological layers corresponding to the Yoldic Sea, the Ancylus Lake and the Littorina Sea periods. Recently the marine influence has come from the North Sea water inflow, which greatly affects the salinity of the deep water. The freshwater influence comes from the fluvial water which gives the surface layer a lower salinity value.

Salinity in the Baltic regions varies greatly - Gulf of Bothnia 2-4°/oo, the central Baltic proper 7-13°/oo, the Kattegat 15-30°/oo in the surface water and 32-34°/oo in the bottom water, the Skagerrak 20-30°/oo in the surface and 32-35°/oo in the bottom water. The water column is stratified by a halocline at 60-80 meters depth in the Baltic proper, and at 10-20 meters in the Kattegat and eastern Skagerrak, in the summer season with the thermocline. The salinity gradient structures marine life, with a few marine species occurring in the Baltic and increasing numbers of this species towards the Skagerrak.

Disturbances in the Baltic environment came in the 1960s, when significant decreasing trends in oxygen concentrations were found in bottom waters in several deep basins (Fonselius 1969). An increasing concentration of nutrients had already been observed in the deep water. Since then, further sings of eutrophication have appeared (Pliński 1990). The first obvious effects of eutrophication in the Baltic were noted in the mid-1970s when filamentous algae were washed ashore in large quantities. In 1980 and 1981, fish mortality due to oxygen deficiency was reported by fishermen. Recently, since several years the toxic algae blooms have been observed in different areas of the Baltic.

Hydrology

The Baltic Sea has a meridional extension of more than 1500 km and a latitudinal extension of about 650 km. It consists of a series of basins of which the Bothnian Bay, the Bothnian Sea, the Gulfs of Finland, Riga and Gdańsk, and the Baltic Proper are the main bodies of water. The Belt Sea, including the Danish straits and the Sound, form together with the Kattegat the transition zone of the Baltic Sea with the Atlantic Ocean. The drainage basin covers an area which is 4.3 times as large as the area of the sea itself.

Due to the geographical and climatological differences in the drainage basin the system of water courses transporting fresh water to the sea is different in the northern and southern regions. Many rivers of variable sizes discharge water from the Scandinavian mountains and from the forest areas into the northern part, in the southeastern and southern areas, however, there are only a few main rivers flowing into the Baltic Sea. The largest rivers discharging into the Baltic Sea are the Neva flowing into the Gulf of Finland and the Vistula flowing into the Gulf of Gdańsk. There is a great seasonal variation in the river discharge into the Baltic Sea. The maximum run-off occurs in spring during the thawing period and the minimum during later summer or in midwinter.

The Baltic Sea has a positive water balance e.g. the inflow of fluvial water from the drainage area and the precipitation are higher than evaporation. This surplus amount of water flows out into the North Sea. However, water flows from the North Sea to the Baltic are exist. These inflows are mainly sudden and intensive but of short duration. The salinity of the inflowing water is distinctly lower than that of the oceanic water, due to the fact that the inflowing water partly consists of water that has flowed out of the Baltic Sea and become mixed with inflowing oceanic water in the Sounds and the Kattegat. The salinity usually varies within a range of 15-25‰. The water inflow is an element extremely variable in time, both in its annual course and in the multi-year one. The volume and intensity of the inflows from the North Sea depend on the actual anemobaric situation and the differences in water level and salinity. The salt-water inflow is made up of two principal components: a frequently occurring deep-water stream, generated by the horizontal salinity gradient between the Baltic Sea and the North Sea, and an episodic, much more intensive inflow connected with persistent westerly winds. The latter form usually occurs during autumn and winter. However, the interval between large successive inflows of this kind can be several years.

A dominating feature of the Baltic ecosystem is the marked permanent salinity stratification. This feature is of great importance for the chemical and biological conditions in the deep waters. The surface layer is separated from the deep water by the primary halocline. The thickness of the surface layer varies from basin to basin. In the large deep water reservoir of the Baltic Sea in the central Baltic Proper it varies from 60 to 70 m (Fonselius 1969). The surface layers are warmed up during the summer season with a mean temperature during the warmest month of 16°C, and a thermocline is formed at ca. 15 m depth. The water below, the winter water, has a stabile, low temperature usually below 7-8°C during the whole year and is rich in oxygen. The deep water below the halocline is warmer than the winter water. At the end of the year, when the thermocline is broken up, a vertical convection in the water mass above the primary halocline can take place. The accumulated nutrients in the water between the thermocline and halocline, to a great extent originating from diffusion through the halocline, now reaches the photic zone. Due to the shortage of light, however, the effects of this upwelling will not be seen until next spring. During winter large areas of the Baltic, especially the coastal regions, are covered by ice. The Bothnian Bay sometimes for more than 2/3 of the year. The presence of an ice cover cuts off the wind fetch and, therefore, prevents the normal turbulence of the water.

During prolonged periods of westerly winds, great injections of North Sea water may occur. This very dense water proceeds into the Baltic as a fluid layer on the bottom of the sea, forming a secondary halocline towards the deep water e.g. in the Gothland Basin at a depth ca. 150 m. The flow rate of this bottom water is a function of the amount injected and the density. The inflow in 1951 reached the Gothland Deep in 5-6 months, whereas the smaller inflow in 1969 spent 8-9 months on the same distance (Jansson 1972).

The inflowing North Sea water is the only oxygen source for aerobic organisms in the deep and bottom water. During periods of restricted inflow, the oxygen in the bottom layer is rapidly used up for the respiration of bacteria and higher organisms. Therefore, anaerobic processes predominate and as an end product of the activity of the sulphate-reducing bacteria hydrogensulphide is formed. Between 1950 and 1970, 13 inflows of oxygenated, saline water has been recorded, some of them followed by heavy stagnation periods with H_2S-formation (Fonselius and Rattanasen 1970). The next strong inflow from the North Sea was noticed during 1983 and after ten years period, in January 1993 a major salt water inflow to the Baltic occurred. However the nonliving conditions area for the bottom living animals during the last decades has been increased (Wiktor and Pliński 1992).

Apart from the oxygen depletion, the nutrients increase is another important ecological feature for the Baltic ecosystem. A recent estimation of external supplies of nutrients to the Baltic from rivers, point sources and atmospheric deposition, indicates levels of $980 \cdot 10^3$ tonnes of nitrogen and $50 \cdot 10^3$ tonnes of phosphorus (Rosenberg et al. 1990). Another estimation indicated $1190 \cdot 10^3$ tonnes N and $78 \cdot 10^3$ tonnes P (Elmgren 1989). Nutrients input to the Baltic at the turn of the century were much lower and have been estimated by Larsson et al. (1985) to have been about $300 \cdot 10^3$ tonnes N and $10 \cdot 10^3$ tonnes P. In the Baltic, most of the increase in input of nutrients is likely to have taken place since 1950s.

Biology

The brackish-water conditions exert influence on the composition of species in the Baltic Sea. Mainly limnic and marine organisms with a wide osmotic tolerance have been able to survive. The number of species is small. The phytoplankton is dominated by the diatoms Chaetoceros, Achnanthes, Sceletonema and Thalassiosira, the dinoflagellates Gymnodinium and Gonyaulax and the blue greens Aphanizomenon, Nodularia and Anabaena. The main zooplankton representatives are the rotiferan Synchaeta, the cladocerans Bosmina, Podon and Evadne and the copepods Acartia, Temora and Pseudocalanus.

Within both the phytoplankton and the zooplankton populations a typical succession takes places, different species replacing each other. At the early spring the phytoplankton bloom starts with diatoms and Gonyaulax. During the summer the dominance of the diatoms decreases but the dinoflagellates persist, accompanied by a dominance of blue greens in August. Towards the end of the year the diatoms increase again. Of the zooplankton species Acartia dominates in late winter and spring, benefiting from the rich supply of diatoms. In late spring Synchaeta is the dominating species. In July-August Podon, Temora and Eurytemora are numerous and during the late summer and autumn, when new diatom blooms appear, the populations of Acartia increase again. The peaks of potential food during the year, represented by phytoplankton and zooplankton are utilized by migrating fish, e.g. herring, which not only are able to utilize food where it is present in excess thanks to their great mobility but also act as living flows of matter tying different storages together and stabilizing the whole system.

The Baltic ecosystem is a typical detritus based system (Jansson 1972). The sea floor within the upper 50 m in the Baltic is very heterogeneous, ranging from bare rock to soft, highly organic sediments, but mostly dominated by sand, sandy silt and gravel, especially in the gulfs and bays. Below this depth the mud bottoms dominate. Here the remnants of the plankton blooms accumulate together with the feces from the pelagic organisms, the fecal pellets. At small depths the surface of the sediment has a brownish-greyish colour indicating a predominantly oxic condition. Deeper down the bottom material is blue to black and often has a low organic content, indicating a strongly reducing environment often with the formation of H_2S already at the sediment surface.

The oxidized surface layer of the mud bottoms has a very rich organic life and the microbes, the meiofauna and the macrofauna interact in the decomposition of the organic material. The dominating groups of meiofauna are: nematodes, polychaetes, oligochaetes, ostracods, herpacticoids and larvae of the clam Macoma baltica. The macrofauna is dominated by the crustaceans Pontoporeia and Mesidothea, the worms Halicryptus and Harmathoe and the clam Macoma baltica. The western and south parts of the Baltic have a richer variety, especially of polychaetes and clams and, therefore, act as important grazing areas for the migrating bottom fish, e.g. flounder and cod.

In the shallow soft bottom areas the influence from land run-off is great and the vegetation is, therefore, dominated by fresh water plants such as Phragmites, Potamogeton, Chara, Ruppia, Myriophyllum. A few marine algae, e.g. Enteromorpha and Chorda also occur, however. Vast areas of Phragmites extend especially in polluted bays, where the soft bottoms continuously expand due to heavy sedimentation from plankton blooms. The rocky shores show a typical stratification of the vegetation, especially during the summer: microscopic blue-green algae in the splash zone, green algae, mostly Cladophora glomerata between 0-1m, brown algae of the genus Fucus at a depth of 1-7 m, and below this a sparse vegetation of red algae.

The above presented pattern of the species distribution in the Baltic Sea has changed recently because of the eutrophication and pollution. During the last years considerable changes in the composition and distribution of the phytobenthos were noted. The number of species e.g. in the Gulf of Gdańsk diminished by 50% during eighty years, it can be assumed, however, that the dropping out of species took place in the last years (Pliński 1990). Green algae from genus Enteromorpha and Cladophora as well as brown algae Pilayella litoralis are the most numerous species. The presence of several species of red algae and vascular plants (Zostera, Ruppia, Zannichellia) is still found but very seldom. However the progressive reduction of the ranges of Zostera marina, Furcellaria fastigiata and Fucus vesiculosus has been observed for several years. This fact is reflected also in the biomass. The limitation of the bathymetric range of the macroflora is the other important symptom of biocoenotic changes in the Gulf of Gdańsk. At the beginning of the present century benthic algae appeared to the depth of 25 m, in the years 1979-1980 they were found at the depth of 9 m, and in 1985 the maximum range was found to be 6 m. (Pliński and Florczyk 1990).

Some changes which are directly connected with the deterioration of the physical and chemical conditions at the bottom have also been detected in the zoobenthos. The presence of Macoma - Pontoporeia community, contained a wet weigh value of 25.6 g/m^2 was noted in the Gdańsk Deep in 1925 (Hagemeier 1926). In 1951 the presence of Scoloplos - Macoma community with the average biomass of about 3.0 g/m^2 was detected (Demel and Mulicki 1954). However in 1980 in the same area of the Gdańsk Deep the presence of only Harmothoe sarsi with the average biomass of 2.11 g/m^2 was proved, while in 1981 the complete destruction of macrozoobenthos was stated; only Harmothoe sarsi was present at the depth of 80-90 m, producing the biomass of 0.07-0.14 g/m^2 (Okołotowicz 1985).

Phytoplankton, in the opposite way, pointed out the increase of biomass. Over a period of many years an upward trend has been observed in the primary production within the areas of the open waters of the southern Baltic, which amount's to 1-3% annually. An increase in the chlorophyll concentration in the water of the southern Baltic during the last several years has been noted as well. The mean trend in the increase of the concentration of chlorophyll-a in the Gulf of Gdańsk amounts to ca. 2% annually. The composition of phytoplankton has undergone a modification as well. Since mid 1970s the intensive blooming of Heterocapsa triquetra has been observed in the Gulf of Gdańsk, and since last year Gonyaulax catenata, two dinoflagellates species which are considered to be the toxin producers.

Conclusion

The location of the Baltic Sea in the northern high latitudes affects the structure and function of the Baltic ecosystem. Also the salinity conditions, which are very stable influenced the composition and distribution of biocoenosis. The double vertical stratification produces the specific features of the Baltic ecosystem. In recent years there has been stated a number of unfavourable environmental changes in the Baltic, resulting in biological changes which have led to the replacement of certain species by others more persistent, to an eutrphicated and contaminated environment.

References

Demel K, Mulicki Z, (1954) Studia ilościowe nad wydajnością dna południowego Bałtyku (Quantitative studies on the biological yield on the southern Baltic bottom). Prace MIR w Gdyni 7:48-57

Elmgren R, (1989) The eutrophication status of the Baltic Sea: input of nitrogen and phosphorus, their availability for plant production and some management implications. Baltic Sea Environment Proc Helsinki, 30:12-31

Fonselius SH (1969) Hydrology of the Baltic deep basins III. Fishery Bd Sweden ser Hydrography Report 23:1-97

Fonselius S, Rattanasen Ch (1970) On the water renewals in the Eastern Gothland Basin after world war II. Medd Harsfiskelab Lysekil 90:1-11

Hagemeier A (1926) Die Arbeiten mit dem Petersenschen Bodengreifer auf der Ostsee. Ber Deutsch Komm Meeresforsch NF 4:1-4

Jansson BO (1972) Ecosystem approach to the Baltic problem. Bull Ecol Res Comm Sweden 16:1-82

Larsson U, Elmgren R, Wulff F (1985) Eutrophication and the Baltic Sea: Causes and consequences. Ambio 14:9-14

Okołotowicz G (1985) Biomasa makrozoobentosu polskiej strefy Bałtyku wskaźnikiem jej zanieczyszczenia (Biomass of the macrozoobenthos of the Polish Baltic zone as the indicator of its pollution). Biuletyn MIR Gdynia 5/6:27-40

Pliński M (1990) Important ecological features of the Polish coastal zone of the Baltic Sea. Limnologia (Berlin) 20:39-45

Pliński M, Florczyk I (1990) Skład i rozmieszczenie fitobentosu w Zatoce Gdańskiej w latach 1984 i 1985 (Composition and distribution of the phytobenthos in the Gulf of Gdańsk in 1984 and 1985). Zesz Nauk UG ser Oceanografia 12:59-77

Rosenberg R, Elmgren R, Fleischer S, Jonsson P, Persson G, Dahlin H (1990) Marine Eutrophication Case Studies in Sweden. Ambio 19:102-108

Wiktor K, Pliński M (1992) Aktualny stan biocenozy Bałtyku południowego (Recent stage of the southern Baltic biocoenosis) in: Problemy zanieczyszczenia i ochrony wód powierzchniowych - dziś i jutro. UAM Poznań 49:351-359

Presence and Sources of Pollutants in the Baltic Sea

E. Leppäkoski

Department of Biology, Åbo Akademi University, SF-20520 Turku/Åbo, Finland

Environmental problems of the Baltic Sea have been among the major concerns of its coastal states during the past 25 years. Practically all the threats to the Baltic environment discussed during the First UN Conference on the Human Environment in Stockholm in 1972 are still actual topics in the environmental debate and objects of environmental research. At the same time new, severe problems have been identified, not least those caused by toxic substances. There are still good reasons to be worried about the presence and effects of PCBs, DDTs and heavy metals but for most of "new" contaminants we still have to deal with first-generation estimates only. For some of these substances concentrations in, e.g., sediments or selected organisms are known (even if fragmentarily) whereas both their sources and effects are unknown or badly understood: polychlorinated benzenes, naphthalenes, paraffins and terphenyls, polycyclic aromatic hydrocarbons, chlorinated camphenes, hexachlorocyclohexanes, dibenzofurans, dibenzodioxins, etc.

A challenge for ecotoxicology

The purpose of this paper is to point out some of the difficulties in mapping the sources and monitoring the presence of pollutants in the Baltic Sea. It is a Sea of Gradients, both physical, chemical and biological (surface-bottom; east-west; entrance-innermost bays (distance about 1,500 km, latitudinally from 54° to 66°N). The surface water salinity declines from >20 ‰ in Kattegat to <3 ‰ in the inner parts of the Gulfs. Gradients largely modify not only the target ecosystem but also the effects of chemicals on biological systems; consequently, results from ecotoxicological studies carried out at one point along a gradient cannot, without caution, be extrapolated and applied to the whole Baltic ecosystem (see Leppäkoski and Bonsdorff 1989 for further discussion).

The usefulness of the main gradients as baselines for ecotoxicological statements is restricted due to stochastic variations in the non-living environment. The system is subjected to huge variations between seasons; some properties (e.g. ice cover, annual benthic algae, migrating birds and fish) vary from abundant presence to total absence according to season. Water movements on different scales (from mm to hundreds of km) and of varying duration (from seconds to years) tend to redistribute not only nutrients and organisms but also chemicals, and influence exposure conditions. Currents (induced chiefly by winds and water-level fluctuations), upwelling-downwelling phenomena and turbulent mixing contribute to this hydrodynamic instability, resulting in unpredictable and intermittent exposure to point-source discharges.

Gradients are easy to describe on a basin-wide scale. When the scales are reduced to levels at which local populations of organisms live, no gradient factor or exposure to chemical substances is acute or chronic; they are generally intermittent, but even so most testing is acute (Rand & Petrocelli 1984). A pollutant may be pulse-dosed into the receiving water, e.g. due to the rhythmic nature of several industrial processes. The duration of exposure to a certain discharge may thus vary from minutes to several months, with regular (or most often completely irregular) intervals, and the actual concentrations of a toxicant change, within a short period of time, from 0 to 100 per cent of the potential maximum.

It follows from the ecosystem variability (in combination with the multidimensional network of gradients) that exposure conditions are difficult to predict. It is necessary to limit the number of subsystems to be monitored to those which are ecologically well-known and to areas from which relevant background documentation is available. In a hazard assessment process of any particular chemical substance always three components are involved: (1) exposure-related criteria, (2) effects-related criteria, and (3) characteristics of target organisms and ecosystems; thus, "a chemical substance is considered as hazardous to the environment if it (by itself or in combination with other substances), when released to the environment and because of its exposure-related properties (partitioning, persistence, bioaccumulation potential), may cause significant damage to ecosystems or their living components" (Landner 1989).

Keeping these triplicate demands in mind, for an exposure analysis in a Baltic Sea perspective following minimum-set of exposure criteria is needed (Landner 1989): (1) annual volume used in surrounding countries, (2) fraction released into the environment, (3) initial recipient (air, water, soil, sewage works), (4) estimated environmental partitioning, (5) transformation potential, degradability, (6) bioaccumulation potential, and (7) bioavailability.

A major influx (> 300 km^3) into the Baltic Sea of well-oxygenated Kattegat water took place in January 1993 for the first time since 1977. During these more than 15 years of stagnation, large amounts of metals (e.g., Hg, Cd, Pb, Zn) together with other harmful substances and nutrients have been deposited in the sediments in remobilable forms.The recent inflow may oxygenate and remobilize (not least by bioturbation, caused by mixing by bottomliving animals that rapidly recolonize the formerly lifeless bottoms) remarkable amounts of chemicals in the deep basins that have been virtually anoxic covering periodically about 100,000 km^2, i.e., one firth of the total bottom area of the Baltic. The highest metal concentrations were found in the centre of basins with very fine-grained material (local sites, contaminated by heavy metals from industrial sources, excluded; Hallberg 1991, Perttilä and Brügmann 1992). The contents of Zn in the uppermost sediment layer of the deepest parts of the Gotland Basin were 400 to 500 ppm, compared to values <100 ppm obtained in shallower areas. High correlation factors have been recorded between Zn and other trace metals (except Hg); therefore, Zn can be roughly used as a universal indicator of contamination of sediments by heavy metals.

This example illustrates one of the many factors that influence the spatial distribution pattern of contaminants in the Baltic Sea (Gaul 1992; Brügmann et al. 1992; modified): (1) background contamination levels (high in the Baltic due to the size of the drainage area in relation to the area of the sea surface; high proportion of airborne toxic substances), (2) the location of point and/or non-point sources of discharge, (3) the water exchange, (4) conditions for speciation (salinity, redox conditions), (5) degradation rate and environmental properties influencing it, (6) the partition of contaminants between dissolved and absorbed form, (7) sedimentation rate, and (8) uptake by biota.

There are a lot of data on sum emissions from the major point sources of pollutants in the Baltic Sea region compiled by The Baltic Marine Environment Protection Commission - Helsinki Commission (HELCOM 1993), whereas data on individual chemicals are insufficient. In 1981 the first thorough review of the pollution of the Baltic was published by HELCOM. Periodic assessments of the state of the marine environment of the Baltic have been published by HELCOM in 1987 and 1990; the third assessment is scheduled for 1995. First compilation of pollution load was available in 1987 and the second one in 1993 (based on input data of 1990), on airborne pollution load in 1991 and on the state of coastal waters in 1991.

The mere presence of any harmful chemical substance can be seen as a warning signal. However, marine pollution always involves also biological effetcs. Chemistry data give evidence of contamination, concurrent biological data (laboratory and field) are needed to measure pollution. Thus any assessment of the impact of chemicals on the health of the sea (Long and Chapman 1985, Leppäkoski and Bonsdorff 1989) must be based on information on at least (1) harmful substances (chemical species and possible derived products of them), (2) inputs and concentrations (actual or predicted) of them in the main compartments of the aquatic ecosystem (water, sediment, organisms), (3) toxicity of these substances (preferably under conditions prevailing *in situ*), and (4) evidence of changes in populations and communities actually living in environments from which concentration data have been obtained.

Toxic substances in an ecosystem under multi-source stress

The Baltic Sea is suffering from man-induced stresses, including man's physical impacts, disposal of pollutants, and biological impacts (Leppäkoski 1993). The Baltic is turning from its original clean-water, nutrient-poor (oligotrophic) conditions into a mesotrophic or eutrophic state. Marine eutrophication can be regarded as the most severe long-term change threatening the Baltic Sea as a whole; it is one of the enclosed seas damaged world-wide by over-fertilization. Several of the consequences of eutrophication are closely linked with the science of ecotoxicology. In nutrient-rich bodies of water there are naturally formed toxic substances (hydrogen sulphide formed in sediments and water in the anoxic deep basins of the Baltic; algal toxins in connection with blooms of planktonic cyanobacteria).

From an ecotoxicological viewpoint there are no principal differencies between naturally formed toxins (e.g. algal toxins), persistent substances and their compounds naturally present in the environment (heavy metals, hydrogen sulphide, nitrogen compounds) and man-made xenobiotics. All of them act as stressors; the last-mentioned group, however, is largely characterized by their persistence and tendency to bioaccumulate and biomagnify.

The fate in and effects on the Baltic ecosystem of chemical substances, its capacity to process them and resist harmful changes in ecosystem structure and functions are influenced by eutrophication. Toxic substances have became "diluted" in increased biomass of animals and plants and in particulate organic matter in the water and, especially, in sediments (on the deeper accumulation bottoms the organic matter in the surficial sediments doubled from 1930 to late 1980s; Jonsson 1992) and, subsequently, their concentrations in living animals have been in decrease.

The state of the art

The Baltic Sea covers no more than 0.1 per cent of the total area of the world ocean, but it is large and complicated enough to offer a good framework for a critical discussion of basic approaches in regional ecotoxicology (here particularly from the "eco" point of view). The basic physico-chemical features and biological properties of the Baltic Sea are relatively well-known. It is one of the best studied large aquatic ecosystems of the world, covering gradients of vital importance to the understanding of ecosystem structure and functions both in marine and freshwater environments, not least in those subjected to multisource stressors, induced or highly modified through human activities.

Second Periodic Assessment of the State of the Marine Environment of the Baltic Sea (HELCOM 1990) was chiefly based on results gained from years of surveys within the The Baltic Sea Monitoring Programme, started in 1979. The harmful substances included (in part as tentative determinands) are heavy metals (Cd, Cu, Hg, Pb and Zn), petroleum hydrocarbons and some chlorinated hydrocarbons in the seawater and sediment. Species to be sampled for analysis are the bivalve *Macoma balthica*, Baltic herring (*Clupea harengus*) and cod (*Gadus morhua*) and, as tentative species in coastal areas, blue mussel (*Mytilus edulis*), flounder (*Platichtys flesus*), pike (*Esox lucius*), yellow eel (*Anguilla anguilla*), the crustacean *Saduria entomon*, and common shrimp (*Crangon crangon*) (HELCOM 1983). The data processed provide information of concentration levels only; the potential biological effects of observed concentrations are largely unknown.

Data on production and use of toxic substances are lacking for most parts of the Baltic drainage area, their distribution and fate in the environment are not known and trends are uncertain because new sampling, storage and treatment techniques and improved analytical methods make it difficult to compare past and present results.

All pollutants are unevenly distributed in the Baltic ecosystem, not only by reasons of sources, hydrography and chemistry but also of biology: migration of birds, age-dependent migration behaviour of the fish species monitored for toxic substances (Baltic herring, salmon) and food-web interactions contribute to large-scale transports (in space and time) of contaminants.

Scrutinized against this background the number of samples and sampling frequency are still too low for drawing far-going conclusions; since even the joint resources of the Baltic Sea states are limited one has to critically examine both the strenghtnesses and weaknesses of a monitoring programme.

Multi-pollutant discharges from multiple sources

The manysideness of pollution of the Baltic Sea can be illustrated by two arbitrarily chosen examples, shipping and port operations and pulp and paper industry (Northern Europe's Seas - Northern Europe's Environment 1989; completed).

Pollutants from **shipping** may be released in connection with normal operation (spills during loading and unloading; intentional discharges, both legal and illegal, of waste water (nutrients, bacteria), bilge water, oil and solvents); emission of exhaust gases into the air (sulphur and nitrogen oxides, halogenated hydrocarbons, heavy metals); dumping and incineration of solid wastes at sea; treatment of ship hulls with anti-fouling paints (organotin compounds); release of "biological contaminats", i.e., exotic species, carried on ship hulls and in ballast tanks; discharges of "physical contaminats" (underwater noise and ship's waves) or as an immediate or delayed (ship-wrecks) result of accidents at sea. The list of potential pollutants from shipping covers all chemicals carried by ships, which may be released in connection with accidents at sea or as a result of careless handling.

Discharges from **forest-based** industries consist of chlorinated compounds including dioxines, fungicides (slimicides), dissolved and particulate organic material (hydrocarbons, organic acids, alcohols, lignin), sulphur and nitrogen oxides, solids, and nutrients.

Sources and trends

There is still major uncertainty as to the amounts of toxic substances that end up in the Baltic Sea. Very rough estimates only are presented below. Reliable data on trace elements and persistent organic compounds in the open Baltic Sea are still very limited. The Second Pollution Load Compilation of HELCOM (1993) covers a "ten-in-top" list (including Hg, Cd, Cu, Zn, and Pb) on total loads from major pollution sources only; small rivers (<5 m^3/sec), small settlements (<10,000 p.e.) and non-point loading from the coastal zone between rivers have not been included.

Much of this map has yet to be filled in; information on discharges of metals is not available from all countries around the Baltic (HELCOM 1991, 1993). Due to sampling, sample treatment and analytical problems, not all data available from different parts of the Baltic Sea are comparable. In addition to the problems and difficulties already mentioned, the Baltic Sea itself poses pecularities that greatly influence the actual environmental distribution and concentration of pollutants, e.g., (1) the low salinity of the water affects speciation and uptake of heavy metals, (2) the long residence time of the water (25-40 years) favours the accumulation of contaminants, (3) the occurrence of oxygen-free deep water and sediments influences the fate of pollutants, and (4) the relatively high content of dissolved organic matter reduces the fraction of free or labile-bound metals, which are potentially harmful to organisms, to 20-50 per cent of the total amount (Brügmann 1979).

In conclusion, even if the overall concentrations of trace metals and organic pollutants were known, the behaviour (speciation, dissolved/particulate ratio, retention of metals as sulphides in the sediment during anoxic stagnation periods, etc.), bioavailability, uptake and effects of any particular pollutant are not.

Metals

An overall description of metal loading and presence data seems not advisable; the amounts presented below of substances released into the Baltic are based on several reports published in 1980-1993 (Hallberg 1991, Perttilä and Brügmann 1992, Brügmann 1993, HELCOM 1993) and aimed at giving the orders of magnitude only.

Arsenic and its compounds	*sources*: mines, smelters, wood preservatives *amount per year*: unknown; *trend*: falling but still a major concern in the Gulf of Bothnia
Cadmium	*sources*: mining, metal industries, combustion of fossil fuels, burning of waste, fertilizers, pigments *amount per year*: 120-140 tons (50-85 per cent from the atmosphere); *trend*: constant or increasing (in fish in the N Bothnian Bay)
Copper	*sources*: mines, metal works *amount per year*: 2,000-3,000 (?) tons (35-90 per cent from the atmosphere); *trend*: constant
Lead	*sources*: car traffic, pigments, car batteries, hunting *amount per year*: >2,000 (?) tons (>90 per cent from the atmosphere); *trend*: constant or falling (in the water and biota) since the early 1980s (possibly as a result of the increased use of unleaded petrol; prior to that the share of traffic in lead emissions, e.g., in Finland was over 90 per cent
Mercury	*sources*: burning of fossil fuels, fungicides, slimicides, seed dressing pesticides (banned), chlorine-alkali industry, mining, dentistry, batteries, thermometers *amount per year*: <100 (?) tons (50 tons/year via rivers or from direct discharges from municipalities and industrial plants plus the same amount from the atmosphere); *trend*: constant or falling; concentrations in biota do not significantly differ from those in the North Sea and the NE Atlantic. Several coastal waters are still blacklisted because of too high mercury contents in fish
Organometallic compounds	*sources*: antifouling agent in paints (organotin compounds; recently banned in pleasure boats in many countries, e.g., Finland and Sweden), petrol additives *amount per year*: unknown; *trend*: decreasing
Zinc	*sources*: mines, metal works *amount per year*: 10,000-12,000 tons (50-65 per cent from the atmosphere); *trend*: constant, increasing in the N Gulf of Bothnia

Generally there seems to be an increasing trend from north to south for some metals (Pb, Hg) in the abiotic environment whereas concentrations of, e.g., Cd and Hg in organisms (phytoplankton, fish) are higher in the north; these conditions may reflect not only the influence of local sources from metal works but also increasing Cd uptake due to decreasing salinity of the water in the north. Zn loading is of geological reasons high in the Bothnian Sea. Loads of Cd, Pb and Hg are highest in the Baltic Proper, corresponding to 5-7 times the background level; these metals are the most urgent to deal with (Table 1). The inputs of Cu and Zn have been estimated at 2-3 times the background, whereas the increase of Co, Ni, Cr, and As are in the order of 10-40 per cent (Brügmann 1993). Residence times for metals in the water body are relatively short, months for Fe, Pb and Co, 2-3 years for Zn and Hg, 4-5 years for Cu, Ni and Cd (Brügmann 1986). Therefore, the use sediments and biota (especially bottom living animals and perennial algae) can be strongly recommended for follow-up studies on heavy metals in the Baltic Sea.

The coastal zone acts as a biofilter and sediment trap for compounds discharged from local point sources before they enter the open Baltic Sea. By mechanical trapping and bioaccumulation the phytal system forms an intermediate collector for harmful substances, from where they are transported to other subsystems; consequently, concentrations of harmful substances are generally higher in those areas than in the open sea (HELCOM 1992). Metal levels 5-10 times higher than the assumed background have been observed, e.g., the inner part of the Gulf of Finland, the Gulf of Riga, the Kurs'iu Bay, the Gulf of Gdansk and the Vistula Lagoon, from the Oder estuary and Pomeranian Bay, the Sound, and several localities along the Swedish and Finnish coasts, polluted by local industrial discharges (HELCOM 1991).

A reflux of toxic substances from the aquatic to the terrestrial ecosystem is brought about to a large extent through the littoral subsystem by fish-eating birds. The fish and invertebrates are therefore transformers and distributors of energy and matter between the subsystems, and subsequently also an important transport mechanism for chemical substances in the aquatic food web.

Table 1. Enrichment factors for metal concentrations in surface sediments compared to the levels prior to the 1850s (Perttilä and Brügmann 1992)

Metal	Kattegat	Belt Sea	W Baltic
Pb	3.2	1.3-5.1	1.3-4.1
Cd	-	1.1-3.7	2.0-6.9
Cu	1.3	0.7-1.6	1.1-1.9
Zn	1.4	0.8-2.9	1.5-7.0
Hg	4.5	3.4-12.8	3.2

In the open Baltic Proper a major part of heavy metals seems to be of atmospheric origin. This statement is still open to discussion and further studies are greatly needed, e.g., it is important but difficult to distinguish between anthropogenic and diagenetic influences on the observed contents of heavy metals at different depths in sediment cores (Hallberg 1991, Perttilä and Brügmann 1992).

Organic pollutants

Halogenated organic compounds have received a great deal of scientific and public interest. This group of pollutants consists of countless individual substances and their derivatives, formed both naturally and in industrial processes, combustion, incineration of wastes, chlorination of drinking water, etc. Inputs of organohalogens (brominated, chlorinated, fluorinated and iodinated organic compounds) into the Baltic Sea are impossible to estimate.

The body burdens of halogenated organic contaminants in herring are high in the Baltic Sea compared with data from the North Sea (Table 2); this fact, again, reflects the vulnerable situation of the Baltic due to its topography and poor water exchange.

Table 2. Concentrations of some halogenated organic substances in herring (ng/g fat) in 1988 (Anon. 1990)

Substance	Bothnian Sea	Baltic Proper	Skagerrak
Chlordane	190	180	35
Toxaphene	2200	2100	620
DDT	2000	4300	570
PCB	4700	6600	1900
HCB	120	140	41

The following attempt to give the approximate orders of magnitude is based mainly on Enell *et al.* (1991 with some additions) but it still lies, to the very letter, at the level of "intelligent guesses". As pointed out by Enell *et al.*, the only fairly reliable figures relate to water-borne inputs from Sweden and Finland; even these figures are expressed as a sum parameter only (absorbable chlorinated hydrocarbons, AOX). No data are available on non-point and point-source discharges for other countries in the Baltic Sea area.

Chlorinated organic matter

sources: pulp bleaching, PVC and metal industry, water-treatment plants, nature

amount per year: thousands of tons (some 30,000 tons/year of organic chlorine, expressed as TOCl, was discharged into the Baltic Sea in the early 1980s; Jonsson et al. 1993); *trend*: falling; chlorine has been substituted by chlorine dioxide and oxygen as bleaching agents; several internal measures have been included in the process and waste water treatment improved in order to reduce the concentrations of chlorinated organics in the waste water. The amounts of AOX in effluents have decreased from 5-7 to 1-2 kg/ton pulp. Some pulp mills have ceased to use chlorine-based bleaching chemicals. Concentrations of organochlorine residues in fish from the Baltic Proper are, however, still 3 to 10 times higher than in catches from the Northern Atlantic

(HELCOM 1992). These substances will be found in the Baltic Sea for several decades; about 80 per cent of the total amount of organochlorines (expressed as EOCl) is found in the sediments (Wulff *et al.* 1993)

Chlorinated paraffins	*sources*: flame retardants, plasticizers, cutting fluid additives;*amount per year*: unknown; *trend*: unknown
Σ DDT	*sources*: pesticide (banned in the Baltic Sea area) *amount per year*: tons?; *trend*: constant. A substantial decrease of DDT levels in the biota (e.g., up to 90 per cent in Baltic herring) has been recorded since the 1970s. Largely airborne, mainly from tropical countries
Dioxins	*sources*: combustion, metal works, pulp bleaching *amount per year*: kilograms; *trend*: constant or falling due to, e.g., improved burning techniques, gas cleaning systems and bleaching methods in pulp and paper industry
Nonylphenol	*sources*: degradation product of nonylphenol ethoxylates, used as a large volume surfactant *amount per year*: tons; *trend*: constant or falling
PAHs	*sources*: combustion, incl. car traffic *amount per year*: tons; *trend*: constant or rising
PBDEs	*sources*: flame retardants *amount per year*: unknown; *trend*: rising
PCBs	*sources*: technical uses in electrical equipment, hydraulic and heat-transfer fluids, sealants, plastics, paints; largely banned *amount per year*: tons; *trend*: constant, stabilized at a level of one third of the maximum levels in the 1970s
Phthalates	*sources*: plasticizer in paints, lubricants, PVC *amount per year*: tons; *trend*: rising
Toxaphene	*sources*: pesticide (chiefly a substitute for DDT) *amount per year*: unknown; *trend*: constant
Petroleum hydrocarbons	*sources*: shipping, accidents at sea, ship wrecks, oil refineries, petrochemical and other industries, oil heating, vehicles, metal industries *amount per year*: thousands (21,000 - 66,000) of tons, approximately a half of this amount via rivers; atmospheric input 1,000 -10,000 tons per year; *trend*: constant
Dumped war gas ammunition	*sources*: dumped at several places after the World War II *amount per year*: unknown (mustard gas as main potential pollutant); *trend*: increasing risk in the course of time (containers corroded by rust)

Discussion and Conclusions

It is obvious that our knowledge of the sources and presence of pollutants in the Baltic Sea still is very fragmentary. Even less is known about the interactions between indidual pollutants and the fate of contaminants in relation to elevated nutrient levels - an urgent matter to be considered in ongoing ecotoxicological research focused on the Baltic Sea. In the long run continuous monitoring of pollutant levels in the biota is of great importance: from the experiences gained, conclusions can be drawn regarding the real-world fate, both chemical and biological, as well as regarding the damage to biological systems. The Baltic Monitoring Programme will hopefully provide valuable information for estimates of bio-availability, bioaccumulation and biomagnification etc. of the most important chemicals in the Baltic environment in some indigenous species in their actual habitats and food web context and - under natural exposure conditions. But, unless coupled with observed or predictable effects on single populations and entire communities, even the most comprehensive chemical surveys will still be half-products (McIntyre 1986, Leppäkoski and Bonsdorff 1989).

In conclusion, the knowledge of rapid deterioration of the environmental situation of the Baltic Sea since the 1950s has made the public, politicians and, not least, scientists more aware of the pollution problems. The altered political and socio-economic situation in the Baltic region since the late 1980s has opened new possibilities for scientific cooperation. There is now great international interest not only in further assessment of the effects of man's impacts on the Baltic Sea but also in implementing environmental protection measures within its drainage area. Preventing continued contamination of the biota by harmful chemical substances is one of the most important tasks; testing and hazard evaluation of new and existing chemicals form a central part of these activities. Besides the goal of creating a scientific basis for universal test systems for chemical compounds, there are also some particular needs that may be identified in the brackish Baltic Sea and its coastal areas, i.e., the Baltic approach to regional ecotoxicology.

References

Anonymous (1990) Persistent organic contaminants. Ambio Spec Rept 7:13-14.

Brügmann L (1979) Spurenmetalle in marinen Proben - Formen des Auftretens und ihre analytische Bestimmung. Thesis, Univ. Rostock.

Brügmann L (1986) The influence of coastal zone processes on mass balances for trace metals in the Baltic Sea. Rapp. P.-v. Réun. Cons. Int. Explor. Mer 186:329-342.

Brügmann L (1993) How to use the sea: management interactions with special reference to the Baltic and its fisheries. ICES Coop Res Rept 186:1-19.

Enell M, J Lothigius, A Thurén (1991) Toxic organics have come a long way ... and all of it ends up in the Baltic. Enviro No. 11 (April, 1991):12-15.

Gaul H (1992) Temporal and spatial trends of organic micropollutants in the sea water of the Baltic Sea, the North Sea, and the Northeast Atlantic. ICES mar Sci Symp 195: 110-126.

Hallberg RO (1991) Environmental implications of metal distribution in Baltic Sea sediments. Ambio 20:309-316.

HELCOM (1983) Guidelines for the Baltic Monitoring Programme for the Second Stage. Baltic Sea Environ Proc 12:1-251.

HELCOM (1990) Second Periodic Assessment of the State of the Marine Environment of the Baltic Sea, 1984-1988. Baltic Sea Environ Proc 35B:1-432.

HELCOM (1991) Airborne Pollution Load to the Baltic Sea 1986-1990. Baltic Sea Environ Proc 39.

HELCOM (1992) The Baltic Sea Joint Comprehensive Environmental Action Programme. HELCOM ad hoc High Level Task Force, Helsinki, Finland, 9 April 1992.

HELCOM (1993) Second Pollution Load Compilation (in print).

Jonsson P (1992) Large-scale changes of contaminants in Baltic Sea sediments during the twentieth century. Acta Univ Upsaliensis, Compr Summ Uppsala Diss Fac Sci 407: 1-52.

Jonsson P, C Rappe, L-O Kjeller, A Kierkegaard, L Håkansson, B Jonsson (1993) Pulp-mill related polychlorinated organic compounds in Baltic Sea sediments. Ambio 22:37-43.

Landner L (1989) The ESTHER approach to environmental hazard assessment of chemicals. In: Landner L (ed) Chemicals in the aquatic environment. Spinger-Verlag, Berlin, Heidelberg, pp 59-72.

Leppäkoski E (1993) A biologist's view on the problems of the Baltic Sea. Eur Rev (in print).

Leppäkoski E, E Bonsdorff (1989) Ecosystem variability and gradients. Examples from the Baltic Sea as a background for hazard assessment. In: Landner L (ed) Chemicals in the aquatic environment. Spinger-Verlag, Berlin, Heidelberg, pp 6-58.

Long ER, PM Chapman (1985) A sediment quality triad: measures of sediment contamination, toxicity and infaunal community composition in Puget Sound. Mar Pollut Bull 16: 405-415.

McIntyre AD (1986) Biological effects and pollution assessment. Water Sci Tech 18:155-160.

Northern Europe's Seas - Northern Europe's Environment (1989). Report to the Nordic Council's International Conference on the Pollution of the Seas, 16-18 October 1989, 247 pp.

Perttilä M, L Brügmann, Eds. (1992) Review of contaminants in Baltic sediments. ICES Coop Res Rept 180:1-135.

Rand GM, SR Petrocelli (eds) (1984) Introduction. In: Fundamentals of aquatic toxicology. Hemisphere Publ, Washington, New York, London, pp 1-28.

Wulff F, L Rahm, P Jonsson, L Brydsten, T Ahl, Å Granmo (1993) A mass-balance model of chlorinated organic matter for the Baltic Sea - a challenge for ecotoxicology. Ambio 22:27-31.

Interrelationship between eutrophication and effects of pollutants

Per Larsson
Limnology, Department of Ecology, University of Lund,
Box 65, 221 00 Lund, Sweden

Uptake of persistent pollutants in aquatic biota has traditionally been explained by differing load of pollutants to the aquatic ecosystem, by chemical properties of substances and by position of the animal in the food web (van der Oost et al. 1988). These approaches have been summarized in models, which to a large extent has been focussed on the physical/chemical behaviour of the pollutants, e.g. the fugacity model developed by Gobas et al. (1988) or the equilibrium partitioning model first described by Hamelink et al. (1971). There is no doubt that the models have lead to a better understanding of pollutant transfer to aquatic biota, however, there is a lack of understanding of ecological processes that may be equally important. Such processes include studies of pollutant distribution in populations of fish where age, sex, physiological status, fat content and phase in the reproduction cycle could be of importance, shift of prey during the life cycle, importance of habitat and nische as well as importance of ecological processes such as eutrophication and the physical environment like humic content of the water.

Recently, various authors have explained differences in persistent pollutant uptake in fish by ecological factors. Borgmann and Whittle (1991) investigated PCBs, mirex and p,p-DDE in lake trout (*Salvelinus namaycush*) of Lake Ontario. They explained decreasing levels of pollutants in trout from 1977 to 1985 by the rate of growth of the alewife population (*Alosa pseudoharengus*), the principal food of lake trout. When the alewife population is increasing it should be composed to a greater part of young individuals, and presumably less contaminated. When the population is declining, older age classes dominate that have higher concentrations of contaminants. PCB concentrations in trout decreased in 1979 to 1985, one to two years after a large winter kill of alewife (whereafter a high recruitment of young alewife occurred). The authors also concluded that all contaminant concentration increased by age and bodysize, but the bodysize-effect was greater across age-classes than within. In another paper, Borgmann and Whittle (1992) pointed out growth as an important factor for pollutant levels in lake trout, faster growing fish "diluted" pollutants in the faster growing biomass and, consequently, had lower contaminant levels than slow growing fish.

The variable trophic structure of pelagic communities was the major explanation to the large differences in PCB levels in lake trout found in lakes of the glacial regions of Canada (Rasmussen et al. 1990). PCB concentrations increased in trout in lakes that contained intermediate trophic levels, such as *Mysis relicta*, smelt (*Osmerus mordax*) and coregonids (e.g. *Coregonus clupeaformis*). That *Mysis relicta* could act as a mediator for persistent pollutants in the sediment and *Mysis* predators in the pelagic zone by it's vertical migration was shown by Klump et al. (1991).

That eutrophication has an effect on levels of persistent pollutants in biota was suggested by Olsson and Jensen (1975). The hypothesis was based on the "dilution concept"; that a similar dose of persistent pollutants in an eutrophic and an oligotrophic lake, would result in lower pollutant levels in the larger biomass of the eutrophic lake, and higher contaminant levels in the smaller biomass of the oligotrophic lake. The concept has later been used to explain levels of organochlorine pollutants in zooplankton (Taylor et al 1991). Lower levels of pollutants were found in zooplankton in lakes with high total-phosphorus content compared to lakes with lower phosphorus concentrations. Similar conclusions were drawn by Richer and Peters (1993) in a laboratory study with the phytoplankton species *Selenastrum capricornutum* , where lower biomass of plankton resulted in higher levels of 2,2',4,4',5,5'-hexachlorobiphenyl.

Also the water quality of the lake may influence on pollutant levels in gill-breathing biota. Humic substances, originating from degradation of wood, leaf, etc. in the catchment area of the lake, adsorb the contaminants thereby increasing the apparent solubility in water (Chiou et al. 1986). However, the pollutants are so firmly adsorbed that they are less available for uptake in fish or their prey (McCarthy 1983, Carlberg et al. 1986).

Eutrophication and persistent pollutants

Atmospheric transport has long been regarded as one of the major transport routes of persistent pollutants to organisms in aquatic and terrestrial ecosystems (e.g. Oehme and Manö 1984, Tanabe et al. 1983) . In the atmosphere these pollutants are mainly present in the gasphase, which faciliate their transport over long distances. During rain events the compounds may partition to airborne water or adsorb to particles resulting from human activity or released by natural sources.They subsequently reach the ground as airborne fallout . Presumably they are additionally transported to terrestrial and aquatic ecosystems in their volatilized stage.

In Holarctic ecosystems, as in Sweden, the atmospheric fallout is different from south to north as regards specific pollutants and quantity (Larsson and Okla 1989), depending on climatic factors and closeness to the sources. In many areas airborne inflow is the only source for persistent pollutants, as shown by the relation between atmospheric fallout and uptake in organisms (Larsson et al. 1990) and the similarity in pollution pattern in atmospheric fallout and aquatic and terrestrial animals

Consequently, in a space limited, homogenous geographical area the atmospheric inflow of persistent pollutants may be constant when atmospheric inflow dominates (Eisenreich et al. 1981). Lakes within this area will thus obtain the same "dose" of pollutants, qualitatively and quantitatively. Other factors that govern uptake of persistent pollutants may then be recognized, mechanisms of principal importance for the partitioning of the substances between biota and it's environment.

Population dependent variables for uptake of persistent pollutants

Northern pike (*Esox lucius*) is a common aquatic predator present in most parts of Europe and North America. Predators have higher organochlorine contaminant levels than their prey, because they live longer (longer exposure time) and obtain an integrated effect of pollutant uptake via water and food, which make them suitable for pollutant studies (Larsson et al. 1993). To study age- and sexdependent differences in pollutant uptake in one pike population inhabiting a lake in southern Sweden, a larger material was taken.

Before spawning 10-15% of the body weight of females consists of gonadal tissue, while male testes are considerably less of the total body weight. This is the only fat deposits in pike, the fish "invest" a large part of it's energy reserves in gonads. Possibly also fat is accumulated in the liver, but this phenomenon has not been closely examined. The spawning has consequences for levels of persistent pollutants. Since the roe contain much fat, the females are detoxified during spawning. About half of the females' total fat- and pollutant amounts are eliminated by roe. The proportion roe/total body weight is constant during the major part of a female's life. However, the roe production ceases after ca 10 years of age (at a weight of about 10 kg) in an eutrophic lake (as Lake Ringsjön in southern Sweden), as a result of physiological aging. Levels of pollutants then increase in these old fishes. The milt of male pike does not contain the same high amounts of fat, and the proportion of pollutants released by spawning is thereby small.

The female pike's yearly elimination of pollutants by spawning, results in a pollutant decrease by age. This is also the case for aquatic mammals, primarily because females eliminate the pollutants by foetus and the fatcontaining milk. For male pike gonadal elimination is considerably less; pollutant levels increase by age.

If pollutant levels in fish from different lakes are examined , only fish of the same sex and in a narrow age interval can be compared. The levels of pollutants are also skewed in the population, both for females and males, with a higher frequency of lower levels. The phenomenon has been observed in the literature, e.g. for birds (Newton 1988). Thus, the data has to be transformed if parametric statistics is to be used. An adequate transformation of data for pollutants in the pike population was the logaritmic, that resulted in normal distibution.

Consequently, male and females should be separated when pollutant data of pike from different lakes are analysed, only pike in a narrow age interval should be utilized and, finally, pollutant data should be log-transformed before the statistical evaluation.

Persistent pollutants in a trophic gradient

Pikes (*Esox lucius*) were caugth in about 60 lakes in the province of Skåne (southern Sweden, Larsson et al. 1992). The lakes are characterized by a wide range of salinities and nutrient contents. Lenght, weight, age and sex were determined for the fishes. From each pike a segment of the muscle behind the back fin was taken, and analysed for PCB and DDT compounds. Results from the earlier study from one pike population revealed that the sexes were to be separated, only fish aged 3-7 years old were used, and pollutant data were log-tranformed before statistical analysis (Larsson et al. 1993). For the lakes, transparency (Secci depth), and concentrations of phosphorus and chlorophyll a were used as trophic indicators. Water color was used as a measure of the humic content.

Levels of PCB and p,p-DDE in pike decreased as the trophic status of the lake increased (Larsson et al. 1992, Fig 1). There was an inverse relationship between the concentrations of total-phosphorus or chlorophyll a in the lake water and levels of persistent pollutants in pike. Since the Secci-depth is lower in eutrophic lakes than in oligotrophic, the levels of pollutants were directly proportional to the transparency. The difference in pollutant levels between fishes in oligotrophic and eutrophic lakes was large, up to 200 times higher in lakes with low phosphorus and chlorophyll values.

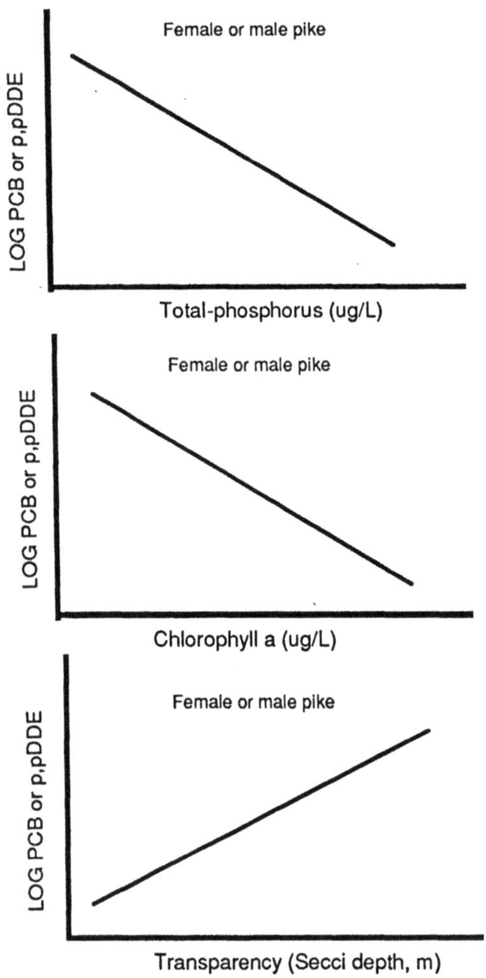

Figure 1. Principal relation between concentrations of PCB or p,p-DDE in fish in lakes of various trophic status. The trophic indicators are total-phosphorus, chlorophyll a and water transparency (redrawn from Larsson et al. (1992)).

Turnover of persistent pollutants in eutrophic and oligotrophic ecosystems

The turnover of persistent, organic pollutants may differ in lakes with various trophic status (Larsson et al. 1992), and could be one of the mechanisms why pollutant levels decrease in fish in a trophic gradient. The amount of organic particles (phytoplankton) differ between an eutrophic and an oligotrophic lake. In eutrophic lakes there are a large amount of phytoplankton. These phytoplankton adsorb (take up) the pollutants very effectively. When the phytoplankton die and sediment the pollutants are transported to the bottom. A large amount of the phytoplankton biomass is not degraded before it reaches the sediment surface (Fig 2, Wetzel 1983), which means that it is incorporated within the sediment together with the pollutants. A major part of the pollutants is, therefore, withdrawn from the water mass and is not available for fish. Since lakes in Holarctic ecosystems are frozen during winter, the large inflow of pollutants will occur during the vegetation period. The phytoplankton succesion , with diatom-bloom in the spring, green-algae bloom during summer and blue-green algae during autumn will maintain this transport of pollutants towards the sediment. On the other hand, in oligotrophic lakes the biomass of phytoplankton is lower, and adsorption of pollutants is less. The major part of the phytoplankton biomass is degraded during sedimentation (Wetzel 1983) and the pollutants, that mainly are sorbed to organic carbon, are released. Consequently, in oligotrophic lakes the atmospherically derived pollutants are more available for uptake in fish and prey.

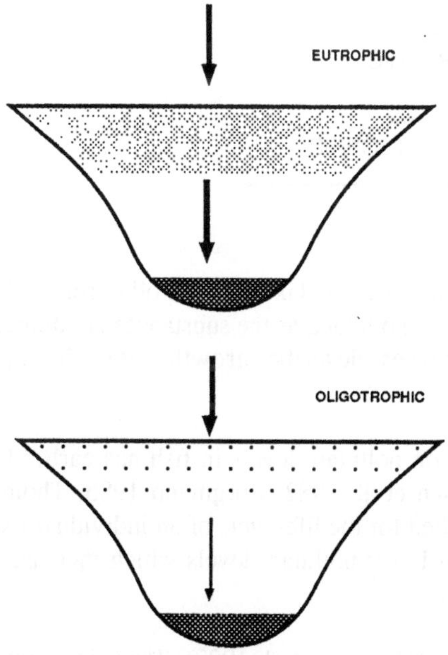

Figure 2. The sedimentation of organic matter in eutrophic lakes is higher compared to oligotrophic, which may cause a higher sedimentation of hydrophobic, persistent pollutants. This result in lower bioavailability of the pollutants in eutrophic lakes (from Wetzel 1983).

Growth rate

Increasing growth rate of piscivorous predators can be another of the mechanistic factors that results in lower pollutant levels in fish as the trophic status of the lake increase (Larsson et al. 1992). The growth rate of pike is higher in eutrophic lakes than in oligotrophic, and levels of pollutants were significantly lower for fast growing fish. In lakes with a total-P >85 ug/L the growth rate of male pike was 190 g/y and the growth rate of females 222 g/y. In lakes with total-P <15 ug/L was the growth rate 162 g/y for males and 197 g/y for females. When the fish grow fast, the pollutants are "diluted" in the expanding tissue, under the assumption that assimilation of the pollutants is similar as for a slow growing fish (Fig. 3). A slow growing fish will not have the same dilution effect, and pollutant levels increase faster.

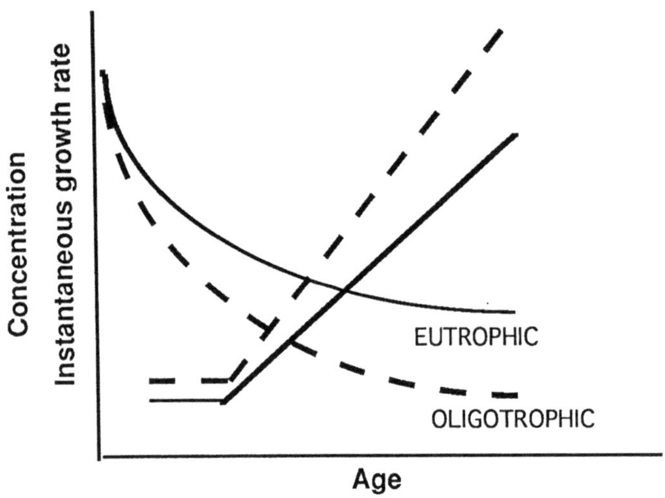

Figure 3. The growth rate of pike is higher in eutrophic lakes than in oligotrophic. This results in lower concentrations of persistent pollutant, as the substances are diluted in the growing biomass. The lower curves describe growth rate, the upper concentrations of pollutants

That the growth rate should have an effect on pollutant levels in fish has earlier been discussed and theoretically modelled (Jensen et al. 1982, Borgmann 1985, Thomann 1989). The models have generally been applied for the life-cycle of an individual fish; a faster growth rate for juvenile fish result in lower pollutant levels which increase fast when the growth rate decrease at higher ages.

The results were confirmed in another study (Hammar et al. 1993). There is an unique opportunity to study the effect of growth rate on pollutant levels in fish. Char (*Salvelinus alpinus*) in isolated, alpine lakes may develop two different life strategies; one normal growing form that delay spawning several years (to grow faster) and later in life become cannibalistic and one dwarf form that spawn much earlier in life and therefore grow much slower. Up to about five years of age the diet for the two forms is similar, mainly macroinvertebrates and zooplankton are consumed. The habitat of the two forms differ; the dwarf is commonly found in the pelagial while the normal form mainly feed in the litoral zone.

We caught these two forms in lake Blåsjön (northern Sweden). The lake receives pollutants only from the atmosphere, and the two forms of char are exposed in the same way, since they inhabit the same lake. The dwarf had significantly higher levels of persistent pollutants than the normal growing form. The total amounts of pollutants per fish (ng/fish) did not differ between the forms. The relation indicated that the concept "dilution of pollutants in growing biomass" for individual fish is true, the normal growing form has lower levels of persistent than the slow growing dwarf. Uptake of pollutants for the two forms does not differ, since total amounts per fish are the same. The results from the study should be principal, and valid for fast and slow growing individuals of any fish population.

Influence of humic substances

Levels of PCB and p,p-DDE in fish decreased as the humic content of the lake water increased (Larsson et al. 1992). The phenomenon can be explained by two theories. Firstly, the humic compounds adsorb persistent pollutants and function as carriers, which results in increased solubility (Chiou et al. 1986). However, the pollutants are so firmly adsorbed that they are no longer available to fish (McCarthy 1983, Carlberg et al. 1986). The second explanation is microbial degradation. Even if these pollutants are very persistent degradation occur by microorganisms and fungi. The microbial degradation of aromatic pollutants are higher in humic lakes compared to clear-water lakes Larsson et al. 1988, Tranvik et al. 1991). In oligotrophic, humic lakes a microorganism flora has evolved that can utilize naturally persistent, aromatic substances (humic compounds) as a carbon source. At ring-cleavange also aromatic pollutants are degraded, presumably as a result of co-metabolism (when the natural substrate is broken down, the enzymes working are so un-specific that also other aromates are attacked). Consequently, aromatic pollutants like PCB and DDT are present in lower concentrations in brown-water lakes and uptake in fish decrease.

Implications for the Baltic sea

The Baltic Sea is an aquatic ecosystem highly contaminated with organochlorine compounds. Levels of pollutants in biota can be compared with the Great Lakes in North America. PCB levels in large predatory, pelagic fish, for the Baltic Sea represented by salmon (Salmo salar) was 0.6 (s.d 0.9) ug/g fresh weight (Thuren et al. 1993) and for Lake Ontario represented by lake trout was 2.5 ug/g (Borgmann and Whittle 1991).

The Baltic Sea has during the recent decades moved towards a more eutrophic stage, as shown by that the total nutrient input has doubled, the primary production has increased, increasing oxygen deficiency in the bottom waters and changed bentic communities (e.g.

Wulff et al. 1986, Elmgren 1989, Jonsson et al. 1992). There are several processes coupled to eutrophication that would lead to decreased exposure of persistent pollutants (i.e. PCBs) to biota: 1. Increased sedimentation of organic matter rich in organic carbon withdraw the pollutants from water. 2. Increasing amounts of non-degraded organic matter in the sediment enhance the adsorption capacity of the contaminants. Thus the partitioning (desorption) towards the water will decrease (Steen et al. 1978, Larsson 1983). 3. As oxygen deficency in the bottom water expand, the number of macroinvertebrates in the sediment decrease. This will lead to lowered bioturbation, a process that has been shown to be important for pollutant release (Södergren and Larsson 1982). Possibly this process will be counter-balanced by release of pollutants by gas-convection from anaerobic sediment that release particles to the water phase, particles that contain the substances (Södergren and Larsson 1982, Larsson 1983). 4. Increased growth rate of organisms favoured by eutrophication will lead to lower levels of pollutants, already explained by the "growth dilution concept". The "biomass dilution" concept (Olsson and Jensen 1975, Taylor et al. 1991) that contaminants like PCB and DDT will dilute in the large biomass of a eutrophic system and concentrate in the smaller biomass of an oligotrophic system are valid only under one of two assumptions: a) the system exposed is static and no inflow or outflow of pollutants occur. In this case pollutant concentration in a biomass is determined by the biomass itself and the amount of pollutant within the system. To model an aquatic system as static seems to be too simple, especially in thought of studies on atmospheric deposition of the pollutants (Larsson and Okla 1989), sedimentation (Broman 1990), volatilization and other processes that have been shown to affect mass-balance of persistent pollutants to an aquatic ecosystem. b) The uptake processes of pollutants in the biomass is so fast that concentration in the water is lowered and limiting. This may occur under a restricted time such as a phytoplanton bloom, but will be balanced by inflow of pollutants in a longer time period.

There is indications of increased trapping of PCBs and DDT in layers of sediment cores dated from 1960 and onwards (Jonsson 1992). This could be an effect of increased sedimentation, although the number of samples to verify this are limited. Additionally, there is a clear trend of decreasing DDT-levels in biota in the Baltic Sea which also seems to be the case for levels of PCBs (Olsson and Reutergård 1986). This could be an effect of eutrophication, but the loads of these persistent pollutants to the Baltic Sea have probably decreased as DDT was banned in 1972 and restictions for PCB use was introduced the same year in Sweden. However, restrictions against these contaminants have not been valid in the Baltic countries or in Poland.

There is a decreasing trophic gradient from the Baltic Sea towards the Bothnian Bay (Wulff, pers.comm.), which would suggest that the eutrophication effect would be greatest in the southern part of the Baltic. However, such an effect should be hard to discover, as the atmospheric and river loads of PCB and DDT are considerably higher on this part of the Baltic and the atmospheric deposition decreases in a gradient towards the north (Larsson unpublished). The sources of pollutants are also important for distribution of the pollutants. Deposition of pollutants from the atmosphere will lead to distribution over the whole Baltic, also out in open sea. The pollutants hit the sea surface and are adsorbed to phytoplankton or dissolved depending on their chemical properties.

The phytoplankton constitutes an effective first link to mediate pollutants to the rest of the food web. Pollutants deposited from the atmosphere will, consequently, be directly available for biota. On the other hand, pollutants that are transported to the sea by rivers, do not have the same bioavailability. The river water contain particles to which the pollutants will adsorb. The contaminant containing particles will sediment in a gradient in the river mouth, depending on size. Resuspension and other processes will transport the particles further out into the sea, processes that take time. The pollutants are to a large extent inactivated in the sediment.

The gradient of humic substances increases from the Baltic Sea to the Bothnian Bay. As humic substances adsorb persistent pollutants, bioavailability should decrease towards the north. Other physical/chemical factors of importance for pollutant distribution and exposure to biota are i) the long water exchange time for the Baltic, which results in a long residence time for pollutants ii) the development of thermo- and halocline may increase residence time of the pollutants in the productive, upper zone of the sea which increase exposure time to phyto- and zooplankton as well as pelagic fish.

The interaction between eutrophication and persistent pollutants in biota leads to a paradox. One of the environmental goals in the Baltic is to reduce nutrient load and stop the eutrophication process. This may result in increasing levels of persistent pollutants in biota, levels that already can be considered high. Consequently, loads of chlorinated contaminant must be reduced simultanuously as nutrients, if conditions in the Baltic shall improve.

References

Borgmann U (1985). Predicting the effects of toxic substanses on pelagic ecosystens. Sci Total Environ 44:111-121.

Borgmann U, Whittle DM (1991) Contaminant concentration trends in lake Ontario lake trout (*Salvelinus namaycush*): 1977 to 1988. J Great Lakes Res 17:368-381.

Borgmann U, Whittle DM (1992) Bioenergetics and PCB, DDE and mercury dynamics in Lake Ontario lake trout (*Salvelinus namaycush*): A model based on surveillance data. Can J Fish Aquatic Sci 49:1086-1096.

Broman D (1990) Transport and fate of hydrophobic organic compounds in the Baltic aquatic environment. PhD-thesis, University of Stockholm, Sweden.

Carlberg GE, Martinsen K, Kringstad A, Gjessing E, Grande M, Källqvist T, Skåre JU (1986) Influence of aquatic humus on the bioavailability of chlorinated micropollutants in atlantic salmon. Arch Environ Contam Toxicol 15:543-548.

Chiou CT, Malcolm RL, Brinton TI, Kile DE (1986) Water solubility enhancement of some organic pollutants and pesticides by dissolved humic and fulvic acids. Environ Sci Technol 20:502-508.

Eisenreich SJ, Looney BB, Thornton JD (1981) Airborne organic contaminants in the Great Lakes ecosystem. Environ Sci Technol 15:30-38.

Elmgren R (1989) Man's impact on the ecosystem of the Baltic Sea: energy flows today and at the turn of the century. Ambio 18:326-332.

Hamelink JK, Waybrant RC, Ball RC (1971) A proposal: Exchange equilibria control the degee chlorinated hydrocarbons are biologically magnified in lentic environments. Transactions of the American Fisheries Society.100:207-214.

Gobas FAPC, Muir DCG, Mackay D (1988) Dynamics of dietary and faecal elimination of hydrophobic organic chemicals in fish. Chemosphere 17: 943-962.

Hammar J, Larsson P, Klavins M (1993) Persistent pollutants in normal and dwarf forms of an Arctic Charr population (*Salvelinus alpinus* L.). Under revision in Can J Fish Aquatic Sci.

Jensen AL, Spigarelli SA, Thommes MM (1982) PCB uptake by five species of fish in Lake Michigan, Green Bay of Lake Michigan, and Cayuga Lake, New York. Can J Fish Aquatic Sci 39:700-709.

Jonsson P (1992) Large scale changes of contaminants in the Baltic Sea. Ph.D.-thesis. University of Uppsala, Sweden.

Klump JV, Kaster JL, Sierszen ME (1991) *Mysis relicta* assimilation of hexachlorobiphenyl from sediments. Can J Fish Aquatic Sci 48:284-289.

Larsson P (1983) Transport of ^{14}C-labelled PCB compounds from sediment to water and air in laboratory model systems. Wat Res 17:1317-1326.

Larsson P, Okla L, Tranvik L (1988) Microbial degradation of xenobiotic, aromatic pollutants in humic water. Appl Environ Microbiol 54:1864-1867.

Larsson P, Okla L (1989) Atmospheric transport of chlorinated hydrocarbons to Sweden 1985 compared to 1973. Atmos Environ 23:1699-1711.

Larsson P, Okla L, Woin P (1990) Atmospheric transport of persistent pollutants governs uptake by Holarctic biota. Environ Sci Technol 24:1599-1601.

Larsson P, Okla L, Collvin, L, Meyer G (1992) Lake productivity and water chemistry as governors of the uptake of persistent pollutants in fish. Environ Sci Technol 26:346-352.

Larsson P, Collvin L, Okla L (1993) Reproductive status and lipid content as factors in PCB, DDT and HCH contamination of a pike population (*Esox lucius* L.). Environ Toxicol Chem 12:855-861.

McCarthy JF (1983) Role of particulate organic matter in decreasing accumulation of polynuclear aromatic hydrocarbons by *Daphnia magna*. Archives of Environ ContamToxicol 12:559-568.

Newton I (1988) Determination of critical pollutant levels in wild populations, with examples from organochlorine insecticides in birds of prey. Environ Pollut 55: 29-40.

Oehme M, Manö S (1984) The long-range transport of organic pollutants to the Arctic. Fresenius Zeitschrift für Analytische Chemie 319:141-146.

Olsson M, Jensen S (1975) Pike as a test organism for mercury, DDT and PCB pollution. A study of the contamination in the Stockholm Archipelago. Institute of Freshwater Research, Drottningholm 54:83-106.

Richer G, Peters RH (1993) Determinants of the short-term dynamics of PCB uptake by phytoplankton. EnvironToxicol Chem 12:207-218.

Rasmussen JB, Rowan DJ, Lean DRS, Carey JH (1990) Food chain structure in Ontario lakes determines PCB levels in lake trout (*Salvelinus namaycush*) and other pelagic fish. Can J Fish Aquatic Sci 47:2030-2038.

Södergren A, Larsson P (1982) Transport of PCBs in aquatic laboratory model ecosystems from sediment to the atmosphere via the surface microlayer. Ambio 11:41-45.

Tanabe S, Hidaka H, Tatsukawa R (1983) PCBs and chlorinated hydrocarbon pesticides in antarctic atmosphere and hydrosphere. Chemosphere 12:277-288.

Taylor WD, Carey JH, Lean, DRS, McQueen DJ (1991) Organochlorine concentration in the plankton of lakes in southern Ontario and their relationship to plankton biomass. C J Fish Aquatic Sci 48:1960-1966.

Thomann, RV (1989) Bioaccumulation model of organic chemical distribution in aquatic food chains. Environ Sci Technol 23:699-707.

Thuren A, Asplund L, Björklund I, Jansson B, Järnberg U, Litzen K, Olsson A. (1993) Chlorinated organic compounds in Baltic salmon showing reproductive disturbances. Manuscript.

Tranvik L, Larsson P, Okla L, Regnell O (1991) In situ mineralization of chlorinated phenols by pelagic bacteria in lakes of differing humic content. Environ Toxicol Chem 10:195-200.

van der Oost R, Heida H, Opperhuizen A (1988) Polychlorinated biphenyl congeners in sediments, plankton, molluscs, crustaceans, and eel in freshwater lake: implications of using reference chemicals and indicator organisms in bioaccumulation studies. Arch Environ Contam Toxicol 17:721-729.

Wetzel RG (1983) Limnology. Sanders College Publishing, New York, 767 pp.

Wulff F, Ärtebjerg G, Nicolaus G, Neimi Å, Ciszewski P, Schulz S, Kaiser W (1986) The changing pelagic ecosystem of the Baltic Sea. Ophelia 4, 299-319.

Effects of Toxic Metal Pollutants on the Ecology of the Baltic Sea

L. Brügmann
Schützenkamp 10, D-22880 Wedel, Germany

In marine environmental studies, the term *pollution* is used if any *contamination* causes detrimental *effects*. In the last decade, the development of suitable methods and searching for such effects has deserved increasing attention in monitoring and in related scientific investigations (IOC 1988, 1990). In relation to inputs of nutrients, plastic litter and persistent organic chemicals, the contamination of the ocean and its marginal seas by heavy metals is either classified now as 'less important' (GESAMP 1990) or not considered at all (FAO 1989). The high annual input of potentially toxic metals is often considered an ecological problem for the Baltic Sea. Are effects already visible or to be suspected?

Peculiar hydrographical conditions, including extreme gradients (salinity, temperature, oxygen content, turbidity...), exclude many marine species entirely from the Baltic Sea. Other species can penetrate only to certain zones. Since the ecosystems in the Baltic Sea are simple with few species and few links between them, even a relatively small environmental change may cause severe imbalance in the entire ecosystem. Additional stress by contaminants is of great importance in such an environment.

The freshwater input (mean 472 km^3/year) drains through an area of more than 1.67×10^6 km^2, inhabited by about 80 million people. Intensive agriculture and industries, including metal mining and processing contribute to the load of contaminants. The residence time of the Baltic Sea water (25-40 years) favours the accumulation of contaminants therein. Only about 2-3 % of the input of a particular contaminant would leave the Baltic Sea annually. The present paper aims (1) to provide background information on the occurrence of metals, (2) to have a closer look at compartments and *hot spots* where the metals could cause effects following excessive enrichment by natural and anthropogenic processes, and (3) to discuss scenarios under which critical conditions for the well-being of organisms and/or significant changes of the metal inventories could occur.

Metals in the Baltic Sea

The metal distribution in Baltic waters is governed by external (atmospheric and land-based inputs) and internal factors (bio-geochemical cycling within the water body and exchange processes at its interfaces). This results in broad concentration ranges. Metal levels in Baltic waters are not generally higher than in the Atlantic Ocean or in the North Sea. For instance, Mo, U and V show lower concentrations in brackish or freshwater than in real sea water. Metals like Cu, Cd and Pb, are strongly diminished in stagnant, i.e. permanently or frequently anoxic, deep waters. Otherwise, in the surface water layer, the high anthropogenic load of Cd, Cu, Ni and Zn results in up to ten times higher values than for the ocean.

Ideally, the gradients of metal concentrations in mixed surface waters from zero salinity to sea water should exhibit about five major types of curves, i.e.
(A) conservative mixing behaviour from the freshwater source to the sea water , (B) conservative mixing behaviour from the sea water source to the freshwater end member, (C) as type *A* with strong depletion at the contact zone river/sea due to (co-) precipitation, adsorption..., (D) as type *A* with some gradual depletion in productive seasons due to uptake by phytoplankton, and (E) as type *A* but with a slower decrease due to significant atmospheric inputs.

In the 'real world' of a European sea, which is semi-enclosed by nine industrialized countries, types *A-E* are seldom met according to theory. This is mainly due to high loads with suspensions and 'dissolved' organic matter of greatly unknown composition. Moreover, any causal relationships are overshadowed by the patchiness in organic production and in anthropogenic inputs.

Table 1. Typical concentrations, budgets and residence times for 'dissolved' metals (<0.4 μm), in the Baltic Sea

	Concentration range ng/l	Budget tons	Residence time years
Fe	300 - 1,500[a,b]	>50,000	0.4
Zn	700 - 1,400[b]	22,000	2
As	600 - 1,300	22,000	
Ni	700 - 850	16,000	6
Cu	500 - 800[c]	13,000	5-6
Mn	150 - 1,500[a]	>15,000	
Pb	25 - 80[c]	1,100	0.3
Cd	15 - 60[b,c]	650	5-6
Co	6 - 12[a]	>500	0.6
Hg	1 - 10	70	3

[a] >10 times higher in anoxic basins
[b] lower in productive surface layers
[c] lower in anoxic basins

The residence time of metals in the water body of the Baltic Sea was estimated to be in the range of less than one year up to six years (Brügmann 1986, Table 1). This enables a fast reaction of the system on reduced inputs and excludes the accumulation of high metal pools in the water column. It favours, however, high metal contents in the sediments. Outside the coastal regions, the particulate metal fraction is partly very low (Ni, Cd, Cu...). Suspended Fe and Pb forms may represent a very significant part of the total concentrations.

The input of metals is not well known. This is mainly due to (a) a weak data base, (b) uncertainties about how much of the gross input finally arrives in the open sea,

and (c) the very limited understanding about the nature and extent of the exchange at the interfaces with the sea bottom and with the atmosphere. In relation to natural background contents in dated sediments the present metal input into the Baltic Sea seems to be 7 times (Cd, Pb), 5 times (Hg) and 2 to 3 times (Cu, Zn) higher. The input of elements like Co, Ni, Cr and As has increased only in the order of 10 to 40 %. A comiplation was published on the metal input from land (HELCOM 1987) which is still the first 'official' data set (in t/yr):
Cr >0.2; Hg >5; Cd 60; Ni 110; As 180, Pb 300; Cu 4200; Zn 9000

For the Baltic Sea, the majority of the atmospheric input is provided by non-Baltic countries. For 1986, the input of metals with atmospheric deposition was estimated as follows (HELCOM 1989; in t/yr):

	wet	dry	total
Cd	31	4	35
Cu	445	20	465
Pb	1305	255	1560
Zn	3090	310	3400

Metals from wet deposition clearly dominate.

Trends. Data from nine expeditions (1980-1991) on dissolved trace metals in the Baltic Sea (Brügmann 1992) reflected the high temporal and spatial variability. A few preliminary conclusions on tendencies over the observational period can be drawn:

a) Ni, Cu and Cd showed only minor fluctuations in the water column and low enrichment factors for the surface microlayer.

b) In the surface microlayer, the enrichment of Pb decreased during recent years whereas Zn was continuously strongly enriched.

c) Zinc increased below 10 m under 'winter conditions'.

d) In the Gdansk Bight, the concentrations of cadmium were 1980 and 1981 several times higher than nowadays.

Regarding the trace metal contents in biota from the Baltic Sea first trends were identified (HELCOM 1990). The lead content in fish from the Kattegat and Belt Sea decreased. This probably indicates the reduction in use of leaded gasoline in Europe. Swedish data from 1981-86 showed a doubled increase of Cd contents in liver of herring from the Bothnian Bay, i.e. from about 1 to 2 $\mu g/g$ dry tissue. The reason for these significantly higher levels ,in relation to other parts of the Baltic Sea , has been attributed to differences in bioavailability. The Hg content in herring muscle from the Baltic Sea is close to values from the open North Sea. However, for the Sound and the southern Bothnian Sea between 4 and 10 times higher contents were reported.

Risks for Effects caused by Metals

Toxicity of Metals in Organisms. The toxicity of metals in natural waters is influenced by many variables. These variables may be ranked in different ways. However, the metal *speciation* is always the major factor which provides the necessary conditions for the functioning of the other factors. Metals occur in water:

* dissolved (charged or uncharged ions and molecules; complexes {aquo, chelate, covalent...}; metal-organic compounds {$(CH_3)_4$-Pb, CH_3-Hg-X...}),
* colloidal dispersed (metal oxides/hydroxides, sulphides...; adsorbed or absorbed by colloids; complexes with high molecular organic molecules), or
* suspended ({co-}precipitates; particulate organic, organically coated and uncoated inorganic matter).

One of the 'essential' metals, copper, is already toxic for marine organisms at very low activities (Table 2). In oceanic upwelling areas this toxicity may prevent the rapid phytoplankton development even if the other background conditions would favour blooming. The upwelling water seems to contain a high percentage of ionic Cu forms which then become increasingly organically complexed. Finally, in productive areas of the ocean over 90 % of the dissolved copper may be present in such non-toxic forms.

To allow the development of phytoplankton in the Baltic Sea, the copper activity should represent less than 2 % of the total dissolved *concentration* in the euphotic zone. Whether upwelling water creates a temporary problem will be unknown until proper Cu speciation studies have been performed. Possibly, the permanently high pool of dissolved and colloidal humic and other organic substances in Baltic waters will buffer a low Cu activity in all depths and seasons. At hot spots contaminated with waters of low organic and high copper loads, significant reductions in productivity could occur, too.

Besides toxicity of inorganic Cu species for bacteria and phytoplankton, excessive ('non-buffered') metal forms have been discussed as one of the possible causes for exceptional or even toxic algal blooms (e.g., *Chrysochromulina polylepis* bloom in May/June 1988).

Different from Cu or Ni, the successful competition of Ca and Mg ions for organic ligands results in relatively high Cd activities in sea water. This could increase the toxicity and create a *cadmium problem.* For marine organisms, the US-NAS proposed a *minimum risk concentration* of 0.2 μg Cd/l. This takes into consideration the long residence time of cadmium in humans and many animals.

Table 2. Effects of low Cu activities on marine organisms (References in Brügmann 1984)

Organisms	pCu(a)[a]	Effects
Bacteria (gram-neg.euryhal.)	9.1 8.3	50 % inhibition of glucose upt. 100 % " "
Bacteria (5 nat.ocean.comm.)	8.7 9.3	50 % inh.of amino acid upt. 100 % " "
Diatoms (*Thalass.pseudon.*)	14.6 10.6 8.3	Cu deficit Onset of growth inhibition Complete growth inhibition
Dinoflagellates (*Gonyaulax tamar.*)	10.4	50 % loss of motility
Phytoplankton (11 spec./4 class.)	<9-11.5 8.5	Onset of growth inhibition Complete growth inhibition
Phytoplankton (nat.ocean.comm.)	8.4	20 % inhibition of ^{14}C uptake
Phytoplankton (nat.estuar.comm.)	ca.7[b]	50 % inhibition of assimil.rate
Green algae (*Nannochlor.atomus*)	10.4 8.7	Onset of growth inhibition Complete growth inhibition
Kelp cultures (*Macrocystis pyrif.*)	11.1	Inhibit.of gametophytic dev.
Crustaceans	ca.7	$^{48}LD_{50}$ value

[a] pCu(a)=-log a_{Cu}
[b] pCu(c)=-log c_{Cu} (total concentration in coastal water)
Remarks: pCu(c) for oceanic deep water: ca. 8.8
pCu(c) for Baltic water: ca. 7.9

The TBT Story. Tributyl tin (TBT) the most effective anti-fouling agent, is a textbook example on the surprising impacts of contaminants on non-target organisms. Shell malformations in the Japanese oyster (*Crassostrea gigas*) and reproduction failures in the European flat oyster (*Ostrea edulis*) were first detected at about 200 ng TBT/l. Consequently, the further use of TBT based paints for pleasure crafts was restricted and the environmental non-effect level defined to 20 ng/l. However, further studies showed that already at 2.5 ng TBT/l a marine snail, the common dogwhelk (*Nucella lapillus*) developed *imposex* (GESAMP 1990a).

This introduced TBT into the list of the most toxic chemicals for aquatic organisms. To develop adequate methods with extremely low detection limits is now a challenge for marine analytical chemists. Unfortunately, restrictions for the application of TBT based paints may not function well because soon similar toxic substitutes (e.g., triphenyltin formulations) will appear on the market.

For the Baltic Sea, information on TBT concentrations in areas affected by marinas is limited. From available data (Björklund 1989; Cheng 1987) significant biological effects must be assumed at least locally. The observed TBT concentrations are comparable to values measured at other localities (Table 3). The effects may be amplified if the animals themselves or their target organisms have close contacts to the surface microlayer. Under certain circumstances, TBT may accumulate therein more than ten times, exceeding thereby possible effect levels. Studies on Bermuda (Hamilton Harbour) exhibited enrichment factors between 2 and 20 from 0.5 m depth to the surface microlayer. The concentration levels in the microlayer (45-450 ng TBT/l) reflected both the distance from the source and the wind and wave strengths (IOC 1990b).

Table 3. Concentration of TBT in coastal waters (ng/l) (References in Brügmann 1993)

San Diego Bay (1986)	<1 - 788
Chesapeake Bay (1986)	<1 - 1342 (1171)[a]
15 US harbours (1986)	2 - 97
Crouch estuary (1987)	17 - 430
Essex coastal waters (1983)	430 - 2000
Arcachon Bight (1986)	200 - 300[b]
Belt Sea (1987)	<40[b]
Kattegat (1987)	<40 - 80[b]
Swedish west coast (1986-88)	<50 - 430[c]

TBT $(C_4H_9)_3SnX$, with X as O, Cl, etc.
a concentration in the surface microlayer
b given as 'organotin'
c in ng Sn/l

Metal Accumulation in the Surface Microlayer. Under calm conditions, the sea is covered with a stable 0.1-0.5 mm *skin*, the *surface microlayer*. Microlayers may be composed by hydrophobic lipids or hydrophilic proteins and glycoproteins. The uppermost part of this mainly naturally formed interface, the *surface films*, are important for the exchange of matter and energy between the atmosphere and the hydrosphere. In the surface microlayer, organic contaminants may be enriched up to 10^4 times in relation to concentrations in subsurface water. The enrichment of metals in the microlayer is lower and exceeds seldom a factor of 10.

The inhabitants of the microlayer (*neuston*) include some of the most sensitive organisms in the aquatic environment, i.e. larvae and other juvenile forms (Table 4). Whereas other factors (UV radiation, organic toxicants...) may be already harmful for these organisms, additional stress, e.g. higher metal levels, could cause their extinction. Zaitsev (1986) found for the Mediterranean and Black Sea a significant correlation between increasing contaminant levels in the microlayer and the decrease of the diversity and abundances of neuston species. Certainly, except for the bacterio-neuston, the organisms will not spend their entire life in the microlayer.

Table 4. Toxicicty of metals on crustracean embryos ($^{48}EC_{50}$ in μg/l; Martin et al. 1981)

	Crassostrea gigas	*Mytilus edulis*
Cu	5.3	5.8
Hg	6.7	5.8
Zn	119	175
Ni	349	891
Pb	758	476

Risks Caused by Metal Accumulation in Sediments. Sediments are the final sink for metals introduced into the sea. Due to their potential bioavailability for benthic organisms and remobilization into the water column, the geo-accumulated metals represent a potential threat for aquatic ecosystems. Estimates on oceanic inventories show that more than 99 % of most metals are bound in the sediments. For the hydrosphere, the metals in the pore water contribute more than 90 % of the dissolved pools. Mass balances for the Baltic Sea resulted in similar findings (Brügmann 1986).

In about 30 % of the sea bottom of the Baltic Sea, a net-sedimentation of fine-grained material takes place. These muddy deposits have built up already to layers with maximum thicknesses between 10 and 20 m. Only the upper part of the sediment, i.e. 0.1-1 m, is relevant regarding possible biological effects. From the pore water, the organisms may take up the dissolved metals actively or by diffusion via the cell membranes of their body surface. The bioavailability of particulate metals in the sediment depend on both the solid matter speciation and the ability of the organisms to release the metals from those phases after ingestion. The present state of knowledge on solid matter speciation is still somewhat unsatisfactory with respect to the assessment of the quantitative extent of bioavailable element contents (Salomons 1987). Despite a first indication of possible problems for benthic organisms, which may be obtained from selective leaching studies, the potential metal uptake cannot be derived from purely chemical data. Additional biological accumulation studies are needed (Luoma and Bryan 1979).

Potentially toxic metals exceeding natural background levels are not strongly fixed in the lattice, but can be easily mobilized by applying relatively weak agents. This has been clearly demonstrated by numerous leaching studies. Therefore, with chances in the redox conditions of the sediments or after ingestion of sediments by organisms a significant part of these *excess metals* may be released.

One example shall demonstrate the existence of partly excessive high metal contents in sediments of the Baltic Sea which could create problems for the ecosystem. In 1985, from the eastern part of the Lübeck/Mecklenburg Bight, 121 sediment cores were taken. Their uppermost parts (0-3 cm and 3-6 cm) were analyzed. In Table 5 some results on the total and 0.5 N HCl-extractable metal

Table 5. Metal contents in surface sediment samples (0-3 cm) from the Lübeck/Mecklenburg Bight, 1985 (in μg/g)

Zn *total*	170	30 - 540
HCl	130	30 - 370
Pb *total*	110	37 - 330
HCl	85	18 - 260
Ni *total*	39	7 - 88
HCl	15	3 - 19
Cu *total*	31	4 - 50
HCl	20	3.3 - 48
Cd *total*	2.0	0.4 - 43
HCl	1.0	0.2 - 3.2
Hg *total*	0.20	0.04 - 0.70

For the southwestern part of the sampling area, indication of heavy contamination of the sediments was found. Several samples taken in the inner Lübeck Bight showed up to 43 μg Cd_{total}/g. This exceeded natural background values (0.2-0.4 μg/g) in pre-historical sediment layers of the same area (Brügmann and Lange 1990) more than 100 times. Leaching with 0.5 N HCl did not release major parts of the (probably sulphidic) Cd. The Cd-enriched sediments showed higher contents of Cu, Pb and Zn, too. Between 70 % and 80 % of these metals were extracted by the diluted HCl.

Extraordinary high metal contents in sediments of the inner Lübeck Bay have been attributed (Müller et al. 1980) to severe local contamination. Up to 1150 μg Zn/g and 1110 μgPb/g were found in the 5-6 cm layer of a core representing the years around 1950. In 1983, surface samples (0-1 cm) were taken on a 5 n.m. grid from the same area (Irion 1984). Again, extremely high contents of Zn (mean 558, max. 6250 μg/g), Pb (mean 377, max. 5100 μg/g), Cu (mean 61, max. 344 μg/g) and Cd (max. 33 μg/g) were found in the grain size fraction $<$20 μm. The heavy contamination was attributed to point sources, e.g. wastes from the production of military equipment during the 2nd World War. In the Trave river sediments, similar burdens (4000 μg Zn/g, 1300 μg Pb/g, 270 μg Cu/g, 30 μg Cd/g) were observed. In December 1984 (11 stations) and April 1985 (15 stations), sediments from the western Lübeck Bight were surveyed by the DHI/Hamburg for 10 metals (DHI 1986, 1987). Up to 940 μg As/g, 95 μg Cd/g, 92 μg Cr/g, 27 μg Hg/g, 870 μg Ni/g, 3.8 % Pb and 5.1 % Zn were found in one of the layers (6-8 cm) of a box core. This was explained now with the dumping of wastes from a metal ore processing plant in earlier decades. The waste containing layers are now covered with less contaminated material, but may have been spread before over a larger area.

In the Baltic Proper, high Cd contents in cores from anoxic basins were observed. This is probably due to (a) Cd 'stripping' from the euphotic layer by phytoplankton and (b) the fixation of sulphidic Cd forms in the sediment (Brügmann and Lange, 1990). For the Framvaren (Skei 1986) and even for the Atlantic Ocean off Namibia (Brongersma-Sanders et al. 1980; maximum 80 μg Cd/g!) this mainly natural phenomenon has been described, too.

The previous stagnation period, at least in the eastern Gotland Basin, must be regarded as one of the largest and most serious stagnation intervals ever recorded during this century. This has caused such extreme changes in the deep layers that never have been observed since the start of oceanographical observations in the Baltic Sea. Now, with the inflow of more saline water into the deep basins, the biogeochemical 'climate' will change again. Mobilization and (co-)precipitation of metals may occur in vast amounts. The rapid change in the concentration and pool of dissolved metals may have consequences for the ecosystem. A rough estimation shall reflect possible threats:
Allow the change from anoxic to oxic conditions to occur for 50 % of the deeper parts (> 150 m) of the Gotland Basin. This would affect about 3.300 km^2 (Fonselius 1969). Let further assume (a) a mean linear sedimentation rate of 1 mm/year, (b) deposits with 0.1 g dry material/ cm^3 (Boström et al. 1983) containing 2-10 μg Cd/g, and (c) a 1 cm thick layer of surface sediments to be affected by the re-oxidation of sulphidic Cd forms. By flushing processes 6.6-33 t Cd could be released in a relatively short time span. The uniform distribution of this Cd in the Gotland Basin below 150 m (110 km^3; Ehlin and Mattisson 1976) would add about 60-300 ng/l. If the Cd would be distributed solely in a one meter thick bottom layer, the concentrations would increase to a horrible 2-10 μg Cd/l!

Conclusions

At the end of the 80s, epidemic diseases diminished the seal population in the Baltic Sea much more than reproduction failures caused probably by the accumulation of certain chlorinated hydrocarbons in the animal's blubber. Similarly, benthic organisms may be influenced much more by changes in their food supply and in the redox state of the bottom water and of surface sediments than by the metal contents in their surroundings. Nevertheless, the metal contamination of the Baltic Sea environment may be *one* of the critical stress factors for some organisms. At hot spots, under extraordinary background conditions and/or assuming critical pathways, for the early stages in the development of the organisms, some metals may even become the controlling factor. Research is needed (1) as a long-term task to derive trends and to develop reliable mass balances, (2) to fill gaps in knowledge on speciation and bioavailability of metals for organisms in solid and aqueous phases, (3) to derive early warning signals from properly designed effect studies, (4) to separate the natural from the anthropogenic causes for observed changes and their effects, and (5) as a short-term task to follow the consequences of the present inflow of more saline and oxygen containing water into the central deep basins of the Baltic Sea.

References

Björklund I (1989) Organotin in the Swedish aquatic environment KEMI Report 8/89, Stockholm 50 pp

Boström K, Burman J-O, Ingri J (1983) A geochemical mass balance for the Baltic. In: Hallberg R (ed.) Environmental Biogeochemistry, Ecol Bull, Stockholm, 35:39-58

Brongersma-Sanders M, Stephan KM, Kwee TG, DeBruin M (1980) Distribution of minor elements in cores from the southwest African shelf with notes on plankton and fish mortality. Mar Geol 37:91-132

Brügmann L (1984) Electrochemical speciation of trace metals in sea water. The Science of the Total Environment 37:41-60

Brügmann L (1986) The influence of coastal zone processes on mass balances for trace metals in the Baltic Sea. Rapp P-v Reun Cons int Explor Mer 186:329-342

Brügmann L (1988) Some pecularities of the trace-metal distribution in Baltic waters and sediments. Mar Chem 23:425-440

Brügmann L (1992) Research & monitoring in the Baltic Sea - two sides of one coin? ICES C M 1992/E:44/Sess U

Brügmann L (1993) Meeresverunreinigung: Ursachen, Zustand, Trends und Effekte. Akademie Verlag GmbH, Berlin, 294 pp

Brügmann L, Lange D (1990) Metal distribution in sediments of the Baltic Sea. Limnologica (Berlin) 20:15-28

Bryan GW, Gibbs PE, Hummerstone LG, Bunt GR (1986) The decline of the gastropod *Nucella lapillus* around southwest England: evidence for the effect of tributyltin from anti-fouling paints J Mar Biol Assoc UK 66:611-640

Cheng Z (1987) Monitoring and assessment of trace metals in seawater and marine organisms. Miljöstyrelsens havforureningslab, Charlottenlund

DHI (1986) Überwachung des Meeres. Bericht für das Jahr 1984, Teil II: Daten, Hamburg, 93 pp

DHI (1987) Überwachung des Meeres. Bericht für das Jahr 1985, Teil II: Daten, Hamburg, 144 pp

Ehlin U, Mattisson I (1976) Volymer och areor i Östersjöomradet. Vannet i Norden 9:16-20

FAO (1989) Long-term consequences of low-level marine contamination An analytical approach. GESAMP Reports and Studies No 40, 14 pp

Fonselius S (1969) Hydrography of the Baltic Deep Basins III. Fish Board Swed Ser Hydrogr Rep 23

GESAMP (1990a) Technical Annexes to the Report on the State of the Marine Environment. UNEP Regional Seas Reports and Studies No 114/1, 319 pp

GESAMP (1990b) The state of the marine environment. UNEP Regional Seas Reports and Studies No 115, 111 pp

HELCOM (1987) First Periodic Assessment of the State of the Marine Environment of the Baltic Sea Area, 1980-1985; General Conclusions. Baltic Sea Environment Proceedings No 17 A, Helsinki

HELCOM (1987) First Baltic Sea Pollution Load Compilation. Baltic Sea Environment Proceedings No 20, Hamburg, 56 pp

HELCOM (1989) Deposition of Airborne Pollutants to the Baltic Sea Area, 1983-1985 and 1986. Baltic Sea Environment Proceedings No 32, Helsinki, 62 pp

HELCOM (1990) Second Periodic Assessment of the State of the Marine Environment of the Baltic Sea Area, 1984-1988; General Conclusions. Baltic Sea Environment Proceedings No 35 A, Helsinki, 432 pp

IOC (1988) IOC Workshop on the Biological Effects of Pollutants, Oslo 11-29 August 1986, Workshop Report No 53, UNESCO, Paris

IOC (1990) Second IOC Workshop on the Biological Effects of Pollutants, Bermuda 10 September-2 October 1988, Workshop Report No 61, UNESCO, Paris

Irion G (1984) Schwermetallbelastung in Oberflächensedimenten der westlichen Ostsee. Naturwissenschaften 71:536-538

Luoma SN, Bryan GW (1979) Trace metal bioavailability: modelling chemical and biological interactions of sediment bound zinc. In: Jenne EA (ed.) Chemical modelling in aqueous systems. Am Chem Soc Symp Ser 93:577-609

Müller G, Dominik J, Reuther R, Malisch R, Schulte E, Acker L, Irion G (1980) Sedimentary record of environmental pollution in the western Baltic Sea. Naturwissenschaften 67:595-600

Salomons W (1987) Sediment pollution in the EEC. Report of Delft Hydraulics Laboratory/Institute for Soil Fertility No T 244, 136 pp

Skei J (1986) The biogeochemistry of Framvaren, a permanent anoxic fjord near Farsund, South Norway. Data Report 1981-1985 NIVA-Report No F-80400, Oslo, 256 pp

SNV (1988) Monitor 1988. Sweden's marine environment - ecosystems under pressure. National Swedish Environmental Protection Board, Solna, 207 pp

SNV (1990) Marine Pollution '90. Action Programme. Swedish Environmental Protection Agency, Solna, 165 pp

Zaitsev YuP (1986) Contourbionts in ocean monitoring. Proc Symp Integrated Global Ocean Monitoring Vol 2, Leningrad, pp 74-81

EFFECTS OF PERSISTENT ORGANIC POLLUTANTS ON BIOTA IN THE BALTIC SEA

by

MATS OLSSON

INTRODUCTION

In the end of the 1960s the detrimental effects of metals and organochlorines on Baltic biota were observed and the increased concentrations of anthropogenic substances in biota were correlated to decreasing populations of many fish feeding species (Hook and Johnels 1972, Sandegren et al. 1980, Olsson 1963, Helander 1975, Lindberg 1975, Odsjö and Sondell 1976).

Among species which have declined since the end of World War II are all the marine mammals inhabiting the Baltic: grey seal (*Halichoerus grypus*), harbor seal (*Phoca vitulina*), ringed seal (*Phoca hispida*) and common porpoise (*Phocaena phocaena*). Also some other populations of fish feeding mammals like the mink (*Mustela vison*) and the European otter (*Lutra lutra*) were decreasing. In fact, there were no fish feeding mammalian species in the Baltic area which did not show a decreasing population. Also some species of birds of prey such as white tailed sea eagle (*Haliaetus albicilla)*, peregrine falcon (*Falco peregrinus)*, eagle owl (*Bubo bubo*) and osprey *(Pandion haliaetus*) have decreased in numbers. Among the bird species, the eggshell thickness of several Baltic species were found to decrease after World War II, strongly indicating an effect of DDT compounds. Species like guillemot (*Uria aalge*), razorbill (*Alca torda*), white tailed sea eagle and osprey, all got thinner eggshells (Odsjö and Edelstam 1972). As a very rough description of the general opinion on the Baltic condition, DDT can be said to have caused impaired bird reproduction whereas PCB mostly have been coupled to decreasing mammalian populations.

The awareness of the ecotoxicological implications of organochlorines and some heavy metals implied efforts from the various Baltic societies to ban or reduce the use of pesticides like DDT, dieldrin and aldrin, to reduce or ban the use of mercury both in industrial processes and as a pesticide and to ban the use of PCB.

POPULATION DEVELOPMENT AFTER THE 1970s

During the end of the 1980s there have been several reports on slightly increasing populations for many of the species mentioned above. There are obvious signs of population increase for Baltic white tailed sea eagle (Helander 1991a), osprey (Odsjö 1993), eagle owl and peregrine (Lindberg 1990). The Baltic populations of both grey and harbor seals are now slightly recovering (Helander 1991b, Helander and Bignert 1992). However, the development of the ringed seal population is still uncertain although the negative trend seems broken (Härkönen and Lunneryd 1992). The mink population is recovering (Anonymus 1991) but the otter is still a very rare species in the Baltic archipelagos and there is no indication of a recovery (Olsson and Sandegren 1991b).

Several different measures taken in Baltic countries have contributed to the positive development indicated above. There have been hunting restrictions and bans, supporting feeding, implantation of juvenils and protection of breeding areas and residence areas. Among the most important measures are the restrictions and bans against the use of organochlorines which has been taken both nationally and internationally. These measures have implied a sometimes dramatic decrease of the environmental concentrations of many organochlorines such as DDT, PCB, toxaphene, polychlorinated naphthalenes and dioxins in Baltic biota (Olsson and Reutergårdh 1986, Bignert et al. 1993, Järnberg et al. 1993, Wideqvist et al. 1993, de Wit et al. 1993). The decreases have been significant but it shall be stressed that there are still unacceptable high concentrations of various anthropogenic substances found in Baltic biota and the efforts to reduce the contaminant and pollutant burden of the environment have to continue. There is also an urgent need for a better understanding on the mechanisms which cause the impaired reproduction and diseases in wildlife. The story on the effects of contaminants on Baltic seals might illustrate the complications to deduce the impact of anthropogenic substances on wildlife.

CONTAMINANTS AND BALTIC SEALS

In the end of the 1960s DDT compounds and PCBs were suspected to cause declining seal populations (Hook and Johnels 1972). The suspicion was based on the fact that the concentrations were high and both compounds were known to have toxic effects. In the middle of the 1970s this suspicion was further strengthened by the findings of aborted grey seal pups (Olsson et al. 1975) and uterine lesions causing sterility at high prevalence among both ringed and grey seals (Helle et al. 1976a). From literature a comparison of concentrations of DDT and PCB in various international seal populations (California and Holland) suffering from reduced pup survival or impaired reproduction indicated that PCB and not DDT was responsible (Helle et al. 1976b). Laboratory experiments on mink supported the suspicion that PCB and not DDT was the responsible agent (Kihlström et al. 1976).

In the middle of the 1980s a disease complex was described on basis of autopsies on Baltic grey and ringed seals (Bergman and Olsson 1986, Bergman and Olsson 1989). The complex was occurring in high prevalence in adult seals and was described as a primary lesion in the adrenals (adrenocortical hyperplasia) causing secondary reactions and lesions in other organs like skin, claws, skeleton, blood vessel, intestine, kidney and, among females, uterus. The character of the complex, the high incidence and the knowledge on contaminant concentrations in Baltic biota supported the earlier suspicion that PCB was the responsible agent. The findings of skeleton lesions made it possible to study the time perspective of the disease complex since Baltic seal skulls saved in museum collections could be used for retrospective studies (Bergman et al. 1986). It was found that the prevalence of skull bone lesions increased from about 10% before 1950 to more than 50% after 1970. Significantly increased prevalence was found during the 1960s (Bergman et al. 1992). The timing of the disease complex fitted well with the increased use of PCB during the 1950s.

In the end of 1987 the Swedish seal project, named -SEAL and SEAL PROTECTION, started with the aim, in a broad sense, to increase the understanding of the mechanisms behind the decreased seal populations. The project was based on both experimental studies and studies on seals from the Baltic. The experimental studies aimed at studying predominantly the toxic effects of PCB on the reproduction of the mink. Part of the experimental studies were, however, also directed to comparative studies on adrenocorticolytic DDT-metabolites in order to achieve an understanding of the toxic mechanisms of methyl sulphonated metabolites of organochlorines. The seal studies included population studies, studies on contaminant concentrations in seals from Swedish waters and studies on pathology.

WHICH PCB CONGENERS ARE CAUSING REPRODUCTION IMPAIRMENT?

Earlier studies have shown that commercial PCB have a strong negative effect on the mink reproduction even at fairly low PCB concentrations in the feed (Ringer et al. 1972, Kihlström et al. 1976). In the end of the 1980s there were two major explanations to the toxicity of PCBs. One explanation was that the contaminants in the commercial PCB, namely the polychlorinated dibenzofurans, were responsible for the toxicity. Another explanation was that the coplanar congeners in the commercial PCB mixture, having a toxicity similar to the dioxins, were responsible. Thus, an important question for the project was to evaluate whether or not the toxicity in commercial PCB and in PCBs found in nature was a toxicity only based on the similarities to dioxins. Commercial PCB was fractionated into different PCB fractions according to the number of chlorines substituted in the *ortho* position to the biphenyl bond (Athanasiadou et al. 1992). Fractions containing 2-4, 1 and no chlorines respectivly in *ortho* position to the biphenyl bond were produced as well as a fraction containing the bi- and tricyclic contaminants in PCB, the dibenzofurans and the polychlorinated naphtalenes. The various fractions were fed to groups of mink bitches at a rate corresponding to the amount found in a daily dose of 2 mg of the commercial product. One group was fed 2 mg of the commercial product, and one group was not exposed and used as a control. The group exposed to commercial PCB had a very poor reproduction whereas the groups exposed to the single PCB fraction reproduced at a normal rate. The group exposed to the bi- and tricyclic contaminants had a significantly higher reproduction than the control (Kihlström et al. 1992). In a second experiment groups of mink bitches were exposed to two or more fractions of the commercial PCB. This experiment showed that the effect on reproduction could be found when two or more fractions were combined. Thus the toxic effect was not found only in the fractions containing the coplanar chlorinated biphenyls, (CBs), or the *mono ortho* CBs but also in the fraction containing 2-4 *ortho* CBs. There were also indications that the fraction containing the bi- and tricyclic contaminants, the dibenzofurans, worked as an antagonist to the CB fractions increasing the reproductive rate (Kihlström et al. 1992). Thus one conclusion from the experiments was that also nonplanar CBs have a toxicological impact on the reproduction in the mink.

WHICH CONTAMINANTS EXPLAIN THE DISEASE COMPLEX AMONG SEALS?

Up to 1975 all three seal populations in the Baltic decreased since the end of the 1950s whereas the harbor seal population along the Swedish west coast have increased from 1950 up to 1988. If environmental contaminants are responsible for the disease complex among Baltic seals then the concentrations of contaminants ought to be higher in Baltic seals than in seals from the Swedish west coast. An important question for the project was to find out which bioaccumulated contaminants are present at higher concentrations in Baltic seals compared to seals from the west coast and which contaminants are higher in diseased animals compared to apparently healthy animals.

Juvenile specimens from the Swedish coasts (1 year old) were analysed and used for comparison between areas and species and autopsied adult grey seal females were divided into three groups according to health condition. Following groups were formed: **a,** apparently healthy females in good nutritional status (in the following called "normal, unstarved"); **b,** females with adrenocortical hyperplasia and with occluded uterines but in a good nutritional status, ("diseased, unstarved"); **c,** females with adrenocortical hyperplasia, with occluded uterines and in a poor nutritional status, ("diseased and starved").

Fourteen elements were analysed in kidney cortex and liver: Al, Ca, Cd, Co, Cr, Cu, Fe, Mg, Mn, Ni, Pb, V, W and Zn. Hg and Se were analysed in liver and As in blubber. No significant difference between areas in concentrations of the various elements were found with the exception for Cr (Frank et al. 1992). The concentrations of Cr were slightly higher in Baltic harbor seals compared to harbor seals from the west coast. Studying species variations in the Baltic, grey seals contained the highest concentrations of Pb, Hg and Se. Concentrations of Ca, Fe, Al differed in diseased females compared to normal, unstarved females but the differences were interpreted as a consequence of the disease, not as a cause of the disease. As a conclusion from this study there were no indications of severe metal contaminations of Baltic seals.

Several different halogenated organic compounds were analysed within the project. The same groups as mentioned above were analysed. Following compounds were studied: DDT compounds and PCBs (Blomkvist et al. 1992), methyl sulphonated metabolites of DDT and PCBs (Haraguchi et al. 1992), polychlorinated camphenes (toxaphen), chlordane compounds and polybrominated diphenyl ethers (Andersson and Wartanian 1992), as well as polychlorinated dibenzo-*para*-dioxin and dibenzofurans (Bergek et al. 1992), the two latter are in the following called dioxins. All haloorganic compounds studied, except the dioxins, were found in higher concentrations in the Baltic seals compared to seals from the Swedish westcoast (Olsson et al. 1992). The concentrations of dioxins were similar in the two areas compared. There was an increase of halogenated compounds in the group of diseased, starved females except for the concentrations of dioxins which in all samples remained remarkably low. Studies on seven individual CB congeners showed that several of them were easily metabolised in seals. Among those, it was interesting to see that 2,3',4,4',5-penta CB (I-118), a mono-*ortho* CB, could not be found in the most heavily contaminated diseased and starved females whereas it was found in less contaminated groups of seals. This results show that dioxins are effectivly metabolised by seals. The studies on I-118 indicate that high concentrations of contaminants increase the metabolic degradation of coplanar CBs.

In a recent report comparing concentrations of coplanar CBs and dioxins in various marine mammals from Swedish waters as well as in fish consuming birds with the concentrations in their feed it was shown that the mammals did not bioaccumulate most of the coplanar CB congeners and the dioxins whereas there was a substancial bioaccumulation in the birds (de Wit 1993).

As a conclusion from these studies it is obvious that the substances which are so far regarded as the most potent toxicants, the dioxins and coplanar CBs, are not bioaccumulated in the Baltic seals; the concentrations are fairly low. However, the concentrations of sDDT and the PCBs (mainly 2-4 *ortho* CBs) were very high in Baltic seals. Also, the concentrations of methylsulfonated metabolites of DDT and especially PCB were high in Baltic seals in general and especially in the diseased groups of female grey seals. The concentrations of toxaphene, chlordanes and polybrominated diphenylethers were, although higher in Baltic seals, regarded as fairly low.

The studies on Baltic seals disclosed the presence of *tris*(chlorophenyl)methane in both Baltic ringed seal (Zook et al. 1992) and grey seal (Haraguchi et al. 1992). This compound has earlier been found in Arctic, north American and Australian wildlife (Walker et al. 1989, Jarman et al. 1990). The toxicological implication of the substance is unknown as well as its sources.

THE DISEASE COMPLEX AMONG BALTIC SEALS AND CONTAMINANTS

As has been mentioned above the disease complex found in Baltic seals seems to be caused by a primary lesion in the adrenal cortex. One of the most common lesions among various animal species at exposure to dioxins and coplanar CBs is a thymus atrophy. During the autopsies of Baltic seals no obvious signs of thymus atrophy have been found but many other severe lesions at high prevalence, building up the disease complex. The chemical analysis does not give support for a hypothesis that high concentrations of the dioxins or the coplanar CBs are responsible for the development of a disease complex among Baltic seals.

In the comparative *in vivo* and *in vitro* studies on adrenocorticolytic DDT-metabolites using several different animal species, 3-methylsulfonyl-DDE was shown to be covalently bound in the adrenal cells and highly potent, causing adrenocortical necrosis (*Zona fasciculata*) in some animal species (Brandt et al. 1992). Studies on PCB metabolites still remain to be done, but we know that both methylsulfonated DDT and PCB have been found in considerable amounts in seal adrenals (Haraguchi et al. 1990). The interesting result is that we now can postulate a possible mechanism behind the disease complex, primary caused by adrenocortical lesions, although much remains to be done before we have conclusive evidence on the mechanisms. What is interesting is that the PCB story among Baltic seals have included rapid changes in the tentative explanatory concepts. What seemed to be an obvious concept in the end of the 1980s, the toxicity of coplanar molecules, now seems less probable. A new concept to be investigated has arose based on the presence of the methylsulfonated metabolites of DDT and PCB, chemically discovered in Baltic seals by Jensen and Jansson already 1976. The findings by Brandt and coworkers imply that DDT, as an environmental pollutant, has to be brought into the Baltic seal discussion again.

BALTIC ECOSYSTEMS IN THE FUTURE

Even if there are many indications on population recovery in the Baltic still problems remain. The pup survival among Baltic grey and harbor seals is low and the reasons for this are unknown. It might be due to climate variations, but it may also be due to immune suppression or other physiological alterations caused by environmetal contaminants. The eggshell thickness in various bird populations, thinner during the 1960s and 1970s, most probably because of the high DDT concentrations, have now increased, but the thickness is not back to the conditions before 1940 (Bignert et al. 1993). Whether this is due to remaining DDT in the feed or to other compounds is unknown. It shows, however, that problems remain. The positive knowledge that marine mammals do not bioaccumulate dioxins and most coplanar CBs, has no relevance for the birds living in the Baltic environment since these quite efficiently bioaccumulate these components. We do not know whether the otters have a similar metabolic capability as the seals and the common porpoise.

The cause of the increasing incidence of reproductive failure among Baltic salmons *(Salmo salar)* reported during the last two decades, named M 74, is still unknown. Since most of the known organochlorine contaminants (DDT, PCB, coplanar PCB, toxaphene, polychlorinated naphthalenes, dioxins) seem to decrease in the Baltic proper since the 1970s it is probably not caused by these contaminants. But is it a new contaminant or are there other causes for the reproduction failure?

We do not know which contaminants will be important in future, and we do not know which effects poor economies and political instabilities in European countries will have on the present efforts to reduce the use and output of DDT, PCB etc. The ecotoxicological implications of *tris*(chlorophenyl)methane and the flame retardants on the Baltic sea remain to be seen.

Since the 1970s we know that eutrophication reduce the impact of bioaccumulative persistent pollutants (Olsson and Jensen 1975) simply by the fact that the biomass increase and disolve the output of bioaccumulative pollutants. We know that the otter population in southern Sweden have survived only in waters where measures against eutrophication have been less successful and where the biomasses have increased in such a way that PCB concentrations in fish are very low (Olsson and Sandegren 1991a+b). In surrounding oligotrophic 'natural' waters the concentrations of PCB in fish were found to be ten times higher and the otter population have disappeared. The high concentrations are caused by airborne fallout. What will happen with the fishconsumers if the efforts to reduce nutrients to the water are more effective than the efforts to reduce the output of harmful persistent chemicals?

So far, we have only a vague knowledge on the synergistic, additive or antagonistic effects of the many contaminants in nature. Furthermore, we can be sure that we also in the future will make mistakes, and unwanted and unexpected harmful substances will be released to the environment. Thus, the many signs of improvements, which can be seen in the Baltic fish feeding animal populations during recent years, cannot be an argument to reduce our efforts to study ecotoxicology, especially the toxicological aspects of this scientific discipline. This has to be stressed since it is not unusual that scientists from other disciplines argue that ecotoxicology today is less important in view of the positive developments during the last decades.

LITTERATURE

Andersson, Ö. and Wartanian, A. 1992. Levels of Polychlorinated Camphenes (Toxaphene), Chlordane Compounds and Polybrominated Diphenyl Ethers in Seals from Swedish Waters. Ambio 21(8):550-552.

Anonymus, 1991. Hunting statistics from the Swedish Sportsmen's Association.

Athanasiadou, M., Jensen, S. and Klasson-Wehler, E. 1992. Preparative fractionation of a commercial PCB product. Chemosphere 23:957-970.

Bergek, S., Bergqvist, P.-A., Hjelt, M., Olsson, M., Rappe, C., Roos, A. and Zook, D. 1992. Concentrations of PCDDs and PCDFs in Seals from Swedish Waters. Ambio 21(8):553-556.

Bergman, A. and Olsson, M. 1986. Pathology of Baltic Grey Seal and Ringed Seal females with special reference to adrenocortical hyperplasia: Is environmental pollution the cause of a widely distributed disease syndrome? Proc. from the Symposium on the Seals in the Baltic and Eurasian Lakes. Savonlinna, 1984-06-05--08. Finnish Game Research No 44: 47-62.

Bergman, A., Olsson, M. and Reiland, S. 1986. High frequency of skeletal deformaties in skulls of the Baltic grey seal. ICES, Marine Mammals Committee. C.M. 1986/N:15. Even published In: 1989. Yablokov, A. and Olsson, M. (Eds.) Influence of Human Activities on the Baltic Ecosystem. Proceedings of the Soviet-Swedish Symposium, Moscow, USSR, April 14-18, 1986, pp 87-95.

Bergman, A. and Olsson, M. 1989. Pathology of Baltic Grey and Ringed Seal Males. Report Regarding Animals Sampled 1977-1985. In: A. Yablokov and M. Olsson (Eds.). Influence of Human Activities on the Baltic Ecosystem. Proceedings of the Soviet-Swedish Symposium, Moscow, USSR, April 14-18, 1986. pp 74-86.

Bergman, A., Olsson, M. and Reiland S. 1992. Skull-Bone Lesions in the Baltic Grey Seal (*Halichoerus grypus)*. Ambio 21(8):517-519.

Bignert, A., Göthberg, A., Jensen, S., Litzén, K., Odsjö, T., Olsson, M. and Reutergårdh, L. 1993. The need for adequate biological sampling in ecotoxicological investigations: a retrospective study of twenty years pollution monitoring. The Science of the Total Environment 128:121-139.

Bignert, A., Litzén, K., Odsjö, T., Olsson, M., Persson, W. and Reutergårdh, L. 1993. Time-related factors influence the concentrations of sDDT, PCBs and shell parameters in eggs of Baltic guillemot (*Uria aalge*) 1861-1989. Manuscript in preparation.

Blomkvist, G., Roos, A., Jensen, S., Bignert, A. and Olsson, M. 1992. Concentrations of sDDT and PCB in Seals from Swedish and Scottish Waters. Ambio 21, (8):539-545.

Brandt, I., Jönsson, C.-J. and Lund, B.-O. 1992. Comparative Studies on Adrenocorticolytic DDT-Metabolites. Ambio 21(8):602-605.

Frank, A., Galgan, V., Roos, A., Olsson, M., Petersson, L. R. and Bignert, A. 1992. Metal Concentrations in Seals from Swedish Waters. Ambio 21(8):529-538.

Haraguchi, K., Bergman, Å., Athanasiadou, M., Jakobsson, E., Olsson, M. and Masuda, Y. 1990. PCB Methyl Sulphones in Grey Seal and Otter from Swedish Environment. In: Organohalogen Compounds, Vol. 1:415-418. Dioxin'90 - EPRI-SEMINAR. Toxicology, Environment, Food, Exposure-Risk. Ed. by O. Hutzinger and H. Fiedler. Bayreuth: Ecoinforma Press, 1990.

Haraguchi, K., Athanasiadou, M., Bergman, Å., Hovander, L. and Jensen, S. 1992. PCB and PCB Methyl Sulfones in Selected Groups of Seals from the Swedish Waters. Ambio 21(8):546-549.

Helander, B. 1975. Havsörnen i Sverige. With English summary: The White-tailed Sea Eagle (*Haliaeetus albicilla L.*) in Sweden. Svenska naturskyddsföreningen. 79 pp.

Helander, B. 1991a. Havsörn i Sverige. Populationsundersökningar. Rapport från verksamheten 1990. Statens naturvårdsverk Rapport nr 3932.

Helander, B. 1991b. Inventering av gråsäl och knubbsäl vid svenska Östersjökusten. Rapport från verksamheten 1990. Statens naturvårdsverk. Rapport 3938.

Helander, B. and Bignert, A. 1992. Harbor Seal (Phoca vitulina) on the Swedish Baltic Coast: Population Trends and Reproduction. Ambio 21(8):504-510.

Helle, E., Olsson, M. and Jensen, S. 1976a. DDT and PCB Levels and Reproduction in Ringed Seal from the Bothnian Bay. Ambio 5(4):188-189.

Helle, E., Olsson, M. and Jensen, S. 1976b. PCB Levels Correlated with Pathological Changes in Seal Uteri. Ambio 5(5-6):261-263.

Hook, O. and Johnels, A.G. 1972. The breeding and distribution of the grey seal (*Halichoerus grypus Fab.*) in the Baltic Sea, with observations on other seals of the area. Proc. R. Soc. Lond. B182:37-58.

Härkönen, T. and Lunneryd, S.G. 1992. Estimating Abundance of Ringed Seals in the Bothinian Bay. Ambio 21(8):497-503.

Jarman, W. M., Burns, S. A., Bacon, C. A., Risebrough, R. W. and Norstom, R. J. 1990. Levels of Tris(4-chlorophenyl)methanol in birds and mammals from North America, Antarctica, and Australia. Soc. Environ. Toxic. Chem. 11th Conf., Paper 034, Arlington, Virginia, USA.

Jensen, S. and Jansson, B. 1976. Anthropogenic substances in seal from the Baltic: Methyl sulfone metabolites of PCB and DDT. Ambio 5:257-260.

Järnberg, U., Asplund, L., de Wit, C., Grafström, A.-K., Haglund, P., Jansson, B., Lexén, K., Strandell, M., Olsson, M. and Jonsson, B. 1993. Polychlorinated Biphenyls and Polychlorinated Naphthalenes in Swedish Sediment and Biota: Levels, Patterns, and Time Trends. Environmental Science & Technology 27(7):1364-1374.

Kihlström, J.E., Olsson, M. och Jensen, S. 1976. Effekter på högre djur av organiska miljögifter. I: Organiska miljögifter i vatten. Tolfte nordiska symposiet om vattenforskning, Visby 11-13 maj 1976. Nordforsk, miljövårdssekretariatet. Publikation 1976:2. pp. 567-576.

Kihlström, J. E., Olsson, M., Jensen, S., Johansson, Å., Ahlbom, J. and Bergman, Å. 1992. Effects of PCB and Different Fractions of PCB on the Reproduction of the Mink (Mustela vison). Ambio 21(8):563-569.

Lindberg, P. 1975. Pilgrimsfalken i Sverige. With English summary: The peregrine falcon (*Falco peregrinus*) in Sweden. Svenska naturskyddsföreningen. 94 pp.

Lindberg, P. 1990. Lyckad falksäsong. Sveriges natur 2:62. In Swedish.

Odsjö, T. and Edelstam, C. 1972. Effect of Toxic chemicals on Bird Life. Swedish Research on Chemical Hazard to Birds. In: XII Bulletin of The International Council for Bird Preservation 1975. Presented at the ICBP European Continental Section Conference. Mamaia, Rumania, 15-19 May 1972.

Odsjö, T. and Sondell, J. 1976. Reproductive success in Ospreys *Pandion haliaetus* in southern and central Sweden, 1971-1973. Ornis Scand. 7:71-84.

Odsjö, T. 1993. Personal communication

Olsson, M. and Jensen, S. 1975. Pike as the Test Organism for Mercury, DDT and PCB Pollution. A Study of the Contamination in the Stockholm Archipelago. Reprinted from Institute of Freshwater Research. Drottningholm, No 54:83-106.

Olsson, M., Johnels, A.G. and Vaz, R. 1975. DDT and PCB levels in seals from Swedish waters. The occurrence of aborted seal pups. Proceedings from the Symposium on the Seal in the Baltic. June 4-6, 1974, Lidingö, Sweden. SNV PM 591:43-65.

Olsson, M. and Reutergårdh, L. 1986. DDT and PCB Pollution Trends in the Swedish Aquatic Environment. Ambio 15(2):103-109.

Olsson, M. and Sandegren, F. 1991a. Otter survival and toxic chemicals - implication for otter conservation programmes. In: Reuther, C. and Röchert, R. (Eds.): Proceedings of the V. International otter colloquium.- Habitat 6:191-200, Hankensbüttel.

Olsson, M. and Sandegren, F. 1991b. Is PCB partly responsible for the decline of the otter in Europe? In: Reuther, C. and Röchert, R. (Eds.): Proceedings of the V. International otter colloquium.- Habitat 6:223-227, Hankensbüttel.

Olsson, M., Andersson, Ö., Bergman, Å., Blomkvist, G., Frank, A. and Rappe, C. 1992. Contaminants and Diseases in Seals from Swedish Waters. Ambio 21(8):561-562.

Olsson, V. 1963. Berguv och havsörn i fara. Sveriges Natur 54:177-184.

Ringer, R. K., Aulerich, R. J. and Zabik, M. 1972. Effects of dietary polychlorinated biphenyls on growth and reproduction in mink. Proc. 164th Natl. Meeting, American Chemical Society 12:149-154.

Sandegren, F., Olsson, M. und Reutergårdh, L. 1980. Der Rückgang der Fischotterpopulation in Schweden. In: Reuther, C. & Festetics, A. (Eds.) 1980: Der Fischotter in Europa - Verbreitung, Bedrohung, Erhaltung. Selbstverlag, Oderhaus & Göttingen. pp 107-113.

Walker, W., Risebrough, R. W., Jarman, W. M., deLappe, B. W., Tefft, T. A. and DeLong, R. L. 1989. Identification of Tris(4-chlorophenyl)methanol in blubber of harbor seals from Puget Sound. Chemosphere 18:1799-1804.

Wideqvist, U., Jansson, B., Olsson, M., Odsjö, T., Reutergårdh, L. and Uvemo, U. 1993. Temporal Trends of PCC in Guillemot eggs from the Baltic. Submitted to Chemosphere.

de Wit, C. A. 1993. Personal communication.

de Wit, C. A., Järnberg, U. G., Asplund, L. T., Jansson, B., Olsson, M., Lindstedt, I. L., Bergek, S.. Hjelt, M., Rappe, C. and Andersson, Ö. 1993. Biomagnification of PCDD/PCDF, non-ortho and mono-ortho PCB in predators from aquatic and terrestrial ecosystems. Submitted to the International symposium on chlorinated Dioxins and related compounds. Dioxin'93.

Zook, D., Buser, H.R., Bergqvist, P.-A.,Rappe, C. and Olsson, M. 1992. Detection of Tris(chlorophenyl) methane and Tris(4-chlorophenyl) methanol in Ringed Seal (Phoca vitulina) from the Baltic Sea. Ambio 21(8):557-560.

Human Health Risk Assessment and Risk Perception Related to the Baltic Sea

U. G. Ahlborg
Institute of Environmental Medicine, Karolinska Institute, PO Box 60208, 104 01 Stockholm, Sweden

The Baltic Sea is contaminated with a variety of halogentaed hydrocarbons. An assessment of the risk implicated by human exposure to eg. contaminated fish should comprise all these compounds. However, toxicological information on many of the polyhalogenated compounds are very limited. This overview will thus cover only polychlorinated dioxins (PCDDs), dibenzofurans (PCDFs) as well as the polychlorinated biphenyls (PCBs). There are 75 PCDDs, 135 PCDFs and 209 PCBs. Only PCDD/Fs with chlorine in at least the 4 lateral positions (2,3,7,8) are stable in the environment and thus found in the Baltic biota.

Risk assessments using cancer as the enpoint

2,3,7,8-Tetrachlorodibenzo-p-dioxin (TCDD) has been the model compound for all our concerns and discussions about the human health risk of these different groups of compounds.

TCDD has been extensively tested. In all, 17 long-term animal bioassays have been done on that compound (for references see Ahlborg et al. 1988). That is enormous when we compare that to other chemicals in our environment. All of the studies were positive; that means that there were cancers in various organs at certain dosages in all studies. Several species have been studied. The hamster is especially interesting since that species is very resistant to the ordinary toxicity of TCDD. From all these studies it is apparent that TCDD is a multi-site carcinogen. It is not only a liver carcinogen; tumours are found in various other organs in the tested animals, and there are also clear sex differences. This is most pronounced when considering liver cancer, where the female rats seem to be much more sensitive than the male rat.

There have been several studies showing very clearly that TCDD does not form DNA adducts. A tremendous amount of tests have been done on genotoxicity at various levels, most of them were negative. It is thus not a genotoxic carcinogen. Furthermore, there are several studies in various types of two-stage models in the rat and mouse that very clearly have shown TCDD to be a tumour promoter.

Except for TCDD, animal bioassays have only been done for two other dioxins and that was a mixture of two hexachlorodioxins (HxCDDs), which also came out positive, both in rats and mice (NTP 1980). There was a negative study on dermal application in mice. It seems that HxCDDs have a carcinogenic action similar to TCDD, but the potency is much lower, roughly about 20 times less potent than TCDD. In addition, other PCDD/Fs have been demonstrated to be potent tumour promotors (Waern et al 1991).

Studies on TCDD have also indicated low dose effects on reproduction and the immune system. However, the Kociba cancer bioassay in rats (Kociba et al 1978) has been the basis for all the risk assessment that have been done up to now. In this study there is a clear dose-response relationship, but the response falls off very much when it comes down to the lower doses. Furthermore, the incidence of preneoplastic and neoplastic lesions in the liver goes down at the lowest dose and the incidence is lower than in the controls. This change is not very significant in statistical terms, but is comes up in almost all cancer bioassays that have been done with dioxins. It thus seems likely that at very low doses there is some kind of inhibitory effect of TCDD. This is not only true for liver tumours, it also apparent for some of the spontaneous tumours that occur in the animals, like tumours of the uterus and tumours of the mammary gland. Some of these changes are probably related to interference with estrogens (Lucier et al. 1991). Another dose-response curve from an animal bioassays have been done with TCDD in Osborne-Mendel rats (NTP 1982). The dose-response curve for lesions in the liver of female rats is very similar to the one in the Kociba study. However, this study also revealed thyroid lesions in the male rats that occurred at very low doses.
There is no NOEL for the thyroid effect in the male rats and the LOEL for the carcinogenic lesion in the male rats is 1.4 ng/kg/bw and day. The thyroid cancer did not come up in any other studies and thus seems to be specific to these Osborne-Mendel rats. The shape of these dose-response curves has made several toxicologists in Europe reluctant to apply some kind of linear multi-stage straight line extrapolation.

The distribution of TCDD is depending on the dose. At the lowest dose tested in the Kociba study, the concentration in liver and fat is at the same level. With an increase in dose, much more is going into the liver as compared to the fat. This is obviously something that has to be taken into consideration when extrapolating these data to low dose exposure in human.

Using the Kociba study (Kociba et al. 1978), several attempts to estimate a tolerable daily intake (TDI) for TCDD have been made. Assuming a no adverse effect level of 1 nanogram/kg bodyweight and day and applying various safety factors, TDIs in the range 1-10 picogram/kg bodyweight have been derived in Canada, Germany, Great Britain, the Nordic countries and the Netherlands. In the United States, a virtually safe dose of 0.006 picogram/kg bodyweight and day has been determined by applying linear multi-stage extrapolation (US/EPA 1985).

The PCBs are somewhat different from the dioxins. As far as we know, the dioxins and the benzofurans work in the same way and lead to the same type of toxicity. They only really differ in potency and in kinetics. But it when it comes to the PCBs, they are different. There are some PCBs that act very much like the dioxins. These are the PCBs that lack chlorine atoms in the ortho positions, if they have one chlorine in the ortho position, they still can bind to the AH receptor and have a dioxin-like activity. With further substitution the dioxin-like activity drops off.

There have been some risk assessments of PCBs, but they have not been very fruitful. US/EPA did an exercise of risk assessment on a commercial mixture two years ago (US/EPA 1990). The FAO/WHO JECFA Committee could not establish an ADI for PCBs (FAO/WHO 1990). IARC has classified the PCBs in general as probably carcinogenic to humans (IARC 1978, 1987).

There have been several bioassays for cancer with different PCB mixtures (for overview, see Ahlborg et al 1992). There seem to be clear indications that the higher chlorinated mixtures (e.g. Arochlor 1260, Clophen A60) are carcinogenic to rats. The evidence for carcinogenicity is much less clear for lower chlorinated mixtures. However, most of these studies indicate that there are lesions of the liver that might be looked upon as preneoplastic lesions. Furthermore, there has been some two-stage rat liver testing for tumour promotion on mixtures that have come out positive (for references, see Ahlborg et al. 1992).

Regarding individual PCB congeners, there has not been one single bioassay for cancer done on any of the congeners. However, there have been several tumour-promoting tests of individual congeners, some non-ortho-substituted or mono-ortho-substituted, but also of some PCB congeners that lack dioxin-like activity (Table 1) .

Table 1: PCB congeners that have been shown to be tumour promoters in two-stage liver models (for references, see Ahlborg et al 1992)

Non-*ortho*-substituted	3,3',4,4' (IUPAC 77)
	3,3',4,4',5 (IUPAC 126)
Mono-*ortho*-substituted	2,3,4,4',5 (IUPAC 114)
	2,3,3',4,4' (IUPAC 105)
Di-*ortho*-substituted	2,2',4,4' (IUPAC 47)
	2,2',5,5' (IUPAC 52)
	2,2',4,5' (IUPAC 49)
	2,2',4,4',5,5' (IUPAC 153)

A multitude of evidence thus speaks in favour of a non-genotoxic mode of action for the carcinogenic process caused by PCDDs, PCDFs and PCBs.

The relative tumour promotion activity of individual congeners can be studied in the two-stage model for liver carcinogenesis in the rat (Pitot et al 1987) and if TCDD is used as a positive control. This has been done for some PCDDs and PCDFs (Wærn et al 1991), and for several individual PCBs (Hemming et al 1993). This approach can also be applied to the non-dioxinlike PCBs (e.g. PCB 153). Several studies have indicated that non-dioxinlike PCBs interact with TCDD or dioxinlike PCBs in at least an additive manner (Bager et al 1993, Sargent et al 1991). Similar findings have been reported for closely related polybrominated biphenyls (Jensen et al 1986).

The outcome of the studies of Wærn et al 1991 and Hemming et al 1993 have been transformed to toxic equivalency factors (TEFs) and are given in table 2 and compared to the international TEFs for PCDD/Fs and the TEFs for PCBs suggested by Safe (1990).

Table 2: Toxicity equivalency factors for certain PCBs, based on tumour promotion index (TCDD = 1.0)

Congener	I-TEF	Safe 1990	Promotion
1,2,3,7,8-PeCDD	0.5		1.0
2,3,4,7,8-PeCDF	0.5		0.2
3,4,5,3`,4'-PCB (IUPAC 126)		0.1	0.06
2,3,4,3',4'-PCB (IUPAC 105)		0.001	0.0003
2,,4,5,2',4',5'-PCB (IUPAC 153)		0.00002	0.00015

These data have been applied to a study comparing the normal population with a group of professional fisherman from the Baltic Sea who eat a lot of fatty and contaminated fish (Svensson et al. 1991, Asplund et al. 1992). Blood samples were taken and analysed for both dioxins, PCDDs/PCDFs and PCBs. When comparing the relative contribution to total dioxin-like activity from PCDD/Fs and PCBs it is quite apparent that the contribution from non-ortho-substituted PCBs is higher than that of the combined PCDDs and PCDFs, 3,3',4,4',5-PCB (IUPAC 126) being the most important single congener. Of the mono-ortho-substituted congeners, 2,3,3',4,4',5-PCB (IUPAC 156) is the most important. When comparing the tumour promoting potency, the non-dioxin-like di-ortho-substituted 2,2',4,4',5,5'-PCB (IUPAC 153) will be as important as the combined PCDD/Fs or non-ortho-substituted PCBs. As long as cancer being the enpoint used for the risk assessment of these compounds, it is thus imperative that risk assessments take all these compounds into account when assessing the potential carcinogenicity of environmental pollution.

Risk assessment using neurotoxicological enpoints

Behavioural effects, mainly noticed as slight decreases in cognitive functions has been observed in children exposed to PCBs through mothers with a high intake of contaminated fish from lake Michigan (Jacobson et al. 1985, 1990 a, b). The findings are supported by other studies in children and in animals) for review see Tilson et al. 1990). Analytical data on PCB concentrations in cord blood demonstrates that women in the Nordic countries have levels (Skåre et al 1988) that are comparable to those where behavioural effects could be expected in children to those mothers. The available data are at the moment difficult to utilize for a risk assessment but certainly deserves a lot of attention. Further studies are ongoing in USA, the Netherlands, Germany and in the Nordic countries.

References

Ahlborg UG, Holmstedt B, Håkansson H, Rappe C (1989) WHO/IPCS Environmental Health Criteria No. 88 (1–409), on Polychlorinated dibenzo-para-dioxins and dibenzofurans.

Ahlborg UG, Hanberg A, Kenne K (1992) Risk assessment of polychlorinated biphenyls (PCBs). Nordic Council of Ministers. Nord 1992:26.

Asplund L, Ericksson U, Jansson B, Wideqvist U, Jensen S, Svensson B-G, Nilsson A (1992) Levels of PCB, PCN and DDE in human plasma in relation to fish consumption. Presented at Dioxin -92, Tampere, Finland.

Bager Y, Hemming H, Flodström S, Wicklund-Glynn A, Wärngård L, Ahlborg UG (1993) Liver tumour promoting activity of 3,4,5,3',4'-pentachlorobiphenyl and its interaction with 2,3,7,8-tetrachlorodibenzo-*p*-dioxin. (In preparation)

FAO/WHO (1990) Evaluation of certain food additives and contaminants. Polychlorinated biphenyls (PCBs), pp 30–33. World Health Organization, Geneva.

Hemming H, Flodström S, Wärngård L, Bergman Å, Kronevi T, Nordgren I, Ahlborg UG (1993) Relative tumour promoting activity of polychlorinated biphenyls (3,4,5,3',4'-, 2,3,4,3',4'- and 2,4,5,2',4',5'-CB) in rat liver. (In manuscript)

IARC (1978) Polychlorinated biphenyls and polybrominated biphenyls. IARC monographs on the evaluation of the carcinogen risk of chemicals to humans, vol. 18.

IARC (1987) Overall evaluations of carcinogenicity: An updating of IARC monographs volumes 1 to 42. IARC monographs on the evaluation of the carcinogenic risk of chemicals to humans, Suppl. 7, PCBs, pp 322–326.

Jacobson SW, Fein CG, Jacobson JL, Schwatz PM and Dowler JK (1985) The effect of intrauterine PCB exposure on visual recognition memory. Child Development 56:853–860.

Jacobson JL, Jacobson SW and Humphrey HEB (1990a) Effects of in utero exposure to polychlorinated biphenyls and related contaminants on cognitive functioning in young children. J Pediatr 116:38–45.

Jacobson JL, Jacobson SW and Humphrey HEB (1990b) Effects of exposure to PCBs and related compounds on growth and activity in children. Neurotoxicol Teratol 12:319–326.

Jensen KJ, Sleight SD (1986) Sequential study on the synergistic effects of 2,2',4,4',5,5'-hexabromobiphenyl and 3,3',4,4',5,5'-hexabromobiphenyl on hepatic tumour promotion. Carcinogenesis 7, 1771.

Kociba RJ, Keyes DG, Beyer JE et al (1978) Results of a two-year chronic toxicity and oncogenicity study of 2,3,7,8-tetrachlorodibenzo-p-dioxin in rats. Toxicol Appl Pharmacol 46, 279-303.

Lucier GW, Tritscher AM, Goldsworthy T, Foley JJJJ, Clark G, Goldstein J and Maronpot R (1991) Ovarian hormones enhance TCDD-mediated increases in cell proliferation and preneoplastic foci in a two stage model for hepatocarcinogenesis. Cancer Res. 51, 1391-1397.

NTP (National Toxicology Program) (1980) Bioassay of a mixture of 1,2,3,6,7,8-hexachlorodibenzo-p-dioxin and 1,2,3,7,8,9-hexachlorodibenzo-p-dioxin for possible carcinogenicity (gavage study). US DHHS, PHS, Research Triangle Park, NC Technical Report, Series No 198.

NTP (National Toxicology Program) (1982) Bioassay of 2,3,7,8-tetrachlorodibenzo-p-dioxin for possible carcinogenicity (gavage study). US DHHS, PHS, Research Triangle Park, NC, Technical Report, Series No 201.

Pitot HC, Goldsworthy TL, Moran S, Kennan W, Glauert HP, Maronpot R, Campbell HA (1987) A method to quantitate the relative initiating and promoting potencies of hepatocarcinogenic agents in their dose-response relationships to altered hepatic foci. Carcinogenesis 8, 1491.

Safe S (1990) Polychlorinated biphenyls (PCBs), dibenzo-p-dioxins (PCDDs), dibenzofurans (PCDFs), and related compounds: Environmental and mechanistic considerations which support the development of toxic equivalency factors. CRC Crit Rev Toxicol 21, 51–88.

Sargent L, Dragan YP, Erickson C, Laufer CJ, Pitot HC (1991) Study of the separate and combined effects of the non-planar 2,5,2',5'- and the planar 3,4,3',4'-tetrachlorobiphenyl in liver and lymphocytes in vivo. Carcinogenesis 12, 793–800.

Skåre JU, Tuveng JM and Sande HA (1988) Organochlorine pesticides and polychlorinated biphenyls in maternal adipose tisssue, blood, milk and cord blood from mothers and their infants living in Norway. Arch Environ Contam Toxicol 17:55–63.

Svensson B-G, Nilsson A, Hansson M, Rappe C, Åkesson B, Skerfving S (1991) Exposure to dioxins and dibenzofurans through the consumption of fish. New Engl J Med 324, 8–12.

Tilson HA, Jacobson JL and Rogan WJ (1990) Polychlorinated biphenyls and the developing nervous system: Cross-species comparisons. Neurotoxicol Teratol 12:239–248.

US/EPA (1985) Health Assessment document forr polychlorinated dipenzo-p-dioxins (final report) EPA 600/8-84-014F. US Environmental Protection Agency, Office of Environmental Assessment, Cincinatti, OH, USA.

US/EPA (1990) Drinking water criteria document for polychlorinated biphenyls (PCBs). Environmental Protection Agency, Cincinnatti, OH, USA.

Wærn F, Flodström S, Busk L, Kronevi T, Nordgren I, Ahlborg UG (1991) Relative liver tumour promoting activity and toxicity of some polychlorinated dibenzo-p-dioxin- and dibenzofuran congeners in female Sprague-Dawley rats. Pharmacol Toxicol, 69, 450.

Contact and Respiratory Sensitisation

(Chair: G. Bylin, Sweden and I. Kimber, United Kingdom)

Immunoregulation of T cell-mediated skin hypersensitivity

Rik J. Scheper and B. Mary E. von Blomberg

Dept of Pathology, Free University Hospital
De Boelelaan 1117, 1081 HV, Amsterdam, the Netherlands

Development of contact allergic skin reactions

Allergic contact dermatitis is based on the increased frequency of allergen-specific effector T cells throughout the body of a sensitized individual, as well as on the enhanced capacity of these effector cells to enter peripheral tissues. The basic mechanism of allergic contact dermatitis has been recently outlined (Kapsenberg et al., 1992; Scheper et al., 1992).

In brief, at sites of primary allergenic skin contact, too few specific T cells are available locally to allow for macroscopically detectable skin reactivity. Allergen molecules bind, however, to antigen-presenting cells (Langerhans' cells) in the skin, which then migrate to the regional lymph nodes and present antigen to lymphocytes, leading to clonal expansion. Upon renewed skin contact, allergen-presenting cells now will meet sufficient numbers of specific T cells locally in the skin, causing rapid release of cytokines, notably IFNγ. Moreover, contact allergens may be particularly active in triggering Langerhans' cells to release IL-1ß, and keratinocytes to produce a broad repertoire of proinflammatory cytokines, including IL-1α and TNFα, and chemotactic cytokines (Barker et al., 1991; Enk et al., 1992, 1993) **(Table 1)**. In particular the latter cytokines, such as RANTES and MIP-1ß, cause the arrival of more T cells, thus further amplifying local mediator release (Taub et al., 1993). Gradually an eczematous reaction develops which reaches a maximum after 18-48 h.

Evidence has been obtained that, with few exceptions, the effector T cells mediating allergic contact dermatitis recognize allergen-modified self peptides present in the grooves of MHC Class II molecules, abundantly present on Langerhans' cells (Sinigaglia et al., 1985; Romagnoli et al., 1991). These T cells, therefore, are essentially the same as those mediating bacterial antigen (protein/ peptide)-induced delayed type hypersensitivity reactions, that is CD4+ T cells. The clonal expansion of such allergen-specific T cells obviously depends on contacting the allergen. The site of first encounter plays, however, a critical role in determining the final outcome: immunological tolerance or skin hypersensitivity.

In the following paragraphs we will focus on recent insights in the role of immunological tolerance in regulating the development of these T cell-mediated immune effector functions. Many of these insights have come from studies in the fields of infectious and autoimmune disease, but are of great relevance to allergic contact dermatitis.

TABLE 1

PROPERTIES OF CONTACT ALLERGENS DETERMINING THEIR SENSITIZING POWER

Property	Reference
Skin penetration (cf $K_2Cr_2O_7$ vs $CrCl_3$)	Polak et al., 1983
Binding to proteins/ peptides (through covalent binding, coordination bonds, etc)	Benezra et al., 1992
Induction of IL-1ß synthesis by LC (causing secondary cytokine release in KC, upregulation MHC Class II, etc)	Enk et al., 1993
Induction of expression of endothelial CAMs (ELAM-1, ICAM-1, VCAM-1)	Goebeler et al., 1993

Immunological tolerance

Control of T effector cell functions is required to prevent undesirable immune reactivities potentially leading to disease. In healthy individuals such control is obtained by both non-specific and specific mechanisms. Non-specific mechanisms typically do not depend on antigen-specific signals for their development nor for their action, such as exemplified by the endogenous, retrovirally related, p15E-like peptides (Oostendorp et al., 1992). In contrast, specific mechanisms do not affect the T cell repertoire as a whole, but rather control antigen-specific T cells. Specific T cell tolerance can be induced both in the thymus and periphery by clonal deletion and anergy (Jones et al., 1990; Schonrich et al., 1991; Rocha et al., 1991; Ohashi et al., 1991; Melamed et al., 1993). These latter mechanisms can, however, not account for our and others' findings that adoptive transfer of T cells from animals tolerant to a given antigen can downregulate immune responses to that antigen in syngeneic (naive) recipients (Gershon et al., 1970, Parker et al., 1976; Polak et al., 1978; Limpens et al., 1991; van Hoogstraten et al., 1992, 1993a; Gieni et al., 1993; Qin et al., 1993). Although 'professional' suppressor T cells may not exist (Bloom et al., 1992; Arnon et al., 1993) the above mentioned data supports the possible development of 'regulatory' T cells that suppress distinct immune functions. Depending on the experimental models such regulatory T cells can belong to either or both the CD4+ or CD8+ subsets (Bloom et al., 1992).

Regulatory T cells may exert their suppressive actions through different pathways, including the shedding of TCR-α chains (Kuchroo et al., 1991) or hapten-binding TCR (Fairchild et al., 1993), or through anti-idiotype reactivities (Bloom et al., 1992). Evidence is accumulating now that regulatory T cells often exert their role, after antigen-specific activation, by releasing distinct cytokines which (non-specifically) down-regulate effector T cell functions **(Table 2)**

TABLE 2

POSSIBLE NATURE AND FUNCTION OF ANTIGEN-SPECIFIC REGULATORY CELLS IN DELAYED CUTANEOUS HYPERSENSITIVITY

1. (Idiotypic) CD4+ T cells: shedding T Cell Receptors/ TCR-α chains: high affinity for hapten/ antigen: capable of non-MHC antigen-binding.
2. (Idiotypic) CD4+ T cells (Th2): releasing non-specific suppressor factors/ cytokines, like IL-10
3. (Idiotypic) CD8+ cells: releasing suppressive cytokines, or killing hapten/ antigen presenting cells
4. (Anti-idiotypic) CD4+/ CD8+ T cells: eliminating effector T cells

T cell cytokine patterns

Recently, two major cytokine patterns have surfaced with pronounced, often mutually exclusive, effects on different features of the immune response (Romagnani et al., 1992; Bloom et al., 1992, Scott et al., 1993). In brief, CD4+ T cell responses in mice that are characterized by substantial IL-2, IFNγ and TNFß, but little IL-4, IL-5 and Il-10 production (socalled type 1 Th, or Th1 responses), result in delayed type hypersensitivity (DTH) reactions, activation of macrophages to kill intracellular parasites, and in IgG2a, but not IgG1 and IgE synthesis. CD4+ T cell responses in which the production of IL-4, IL-5 and IL-10 predominate, generate IgG1 and IgE-secreting cells, and eosinophilia (type 2 Th, or Th2 responses). Th1 and Th2 cells probably arise from a common lineage since they were found to use the same T cell receptor repertoire (Reiner et al., 1993). Despite distinct differences, essentially similar patterns occur in humans (Kapsenberg et al., 1991; van Reijsen et al., 1992; de Waal Malefijt et al., 1992; Bloom et al., 1992; DelPrete et al., 1993). CD4+ T cell clones with an intermediate cytokine profile, known as Th0, were also found in mice and humans. The phenomena of distinct cytokine profiles for T cells may result from an intrinsic, mutually exclusive regulation of Th1 and Th2 cytokine genes. Type 2 cytokines such as IL-4 shift T cell differentiation away from the production of type 1 cytokines, whereas the type 1 cytokine IFNγ is very potent in preventing the development of Th2 cells (Romagnani et al., 1992; Bloom et al., 1992; Hsieh et al., 1993). These effects can be regarded as 'suppressor', or better: 'regulatory', activities. Also CD8+ T cell clones have been found to exhibit effector and suppressor functions depending on their predominant secretion of IFNγ or IL-4 respectively (Salgame et al., 1991; Bloom et al., 1992). The concept of autocrine stimulation of T cell subsets implies that, once a direction has been selected, further antigenic contacts corroborate this selection. This could explain why, once established, T cell tolerance can last life-long, and is not broken by repeated antigenic contacts (van Hoogstraten et al., 1992; Qin et al., 1993). In addition to type 2 cytokines, TGFß has also been identified as a cytokine causing suppression of T effector cell functions (Karpus et al., 1991; Oswald et al., 1992; Khoury et al., 1992; Meade et al., 1992). TGFß production does not unambiguously associate with type 1 or 2 patterns. Likewise, IL-13, showing 20-25% sequence homology with IL-4, behaves as a Th2 cytokine in blocking inflammatory monokine synthesis, but synergizes with IL-2 in stimulating IFNγ synthesis in large granular lymphocytes (Minty et al., 1993).

Antigen-presentation to naive T cells stimulates differentiation and clonal expansion of antigen-specific T cells. When differentiating, these T cells develop major cytokine profiles which depend on the site of primary contact. Along mucosal surfaces local IL-4 release, possibly by mast cells, basophils or locally residing T cells, causes predominant development of Th2 cells. In some individuals this response may be excessive, leading to mucosal allergies, such as respiratory hypersensitivity. Along the skin route local release of IL-12, a macrophage derived cytokine stimulating the production of IFN-γ from T cells, leads to predominant development of Th1 cells Here, allergic contact dermatitis may ensue **(Table 3)**.

TABLE 3.

SITE-DEPENDENT CYTOKINE-DRIVEN DIFFERENTIATION OF NAIVE T HELPER CELLS.

Site	Primary cytokine (source)	Mature Th-subset	Effector-cytokines	Primary function
Skin	IL-12 (Mph/NK)	Th1	IL-2, IFN-γ	Cell-mediated immunity
Mucosa	IL-4 (mast cells?)	Th2	IL-3,4,5,10	Humoral immunity

Mucosal tolerance

Current data suggest that intracellular bacteria and viruses, as well as most contact allergens, preferentially induce specific type 1 T cell responses. The profile of the natural response they initially evoke may play a crucial role in this development, notably through the strong release of IFNγ by IFNα/IL-12 activated NK cells and macrophages. The role of Langerhans' cell derived IL-1ß in this context is still unclear. On the other hand, allergenic contacts and helminthic infections along the mucosal surfaces, rather induce the differentiation of type 2 T cells (Romagnani et al., 1992; Lehner et al., 1992, Scott et al., 1993). Here, IL-4 production by cells of the mast cell/ basophil lineages, in the absence of local IFNγ release, is thought to represent a primary step determining further functional T cell differentiation. The frequent induction of type 2 T cell responses after antigen introduction along mucosal surfaces is probably further promoted by high local densities of B cells as compared to the skin compartment: B cells are excellent IL-10 producers, and antigen-presentation by B cells is known to favour Th2 responses (de Waal Malefijt et al., 1992; Eynon et al., 1992). Development of antigen-specific type 2 T cells along mucosal surfaces bears the advantage that responsiveness focuses on rapid antibody (IgE/IgA)- mediated effector mechanisms, preventing microorganisms to enter the body, in the absence of potentially damaging tissue inflammatory reactions. The latter reactions (mainly related to type 1 T cell function) are actively suppressed instead, and may preferentially develop to antigens not encountered along the gastro-intestinal tract, in particular entering from the skin. Indeed, the prevention of experimental encephalomyelitis in rats by oral administration of encephalitogenic protein was found to be associated with the production of type 2 cytokines (Khoury et al., 1992). Notably, similar mucosal tolerance could also be induced by oral administration of synthetic peptides (Sayegh et al., 1992).

These findings suggest that antigen-specific memory Th2 cells generated within mucosal lymphoid tissues may, despite their preferential recirculation through mucosal tissues, still enter skin-draining lymphoid tissues in sufficient numbers to prevent an eventual local Th1 response. It is tempting to speculate that the persistent unresponsiveness to contact allergens, readily obtained in non-sensitized individuals by oral allergen administration, also depends on this mechanism (van Hoogstraten et al., 1992, 1993a). Notably, after primary sensitization along the skin route no lasting state of tolerance can be achieved. Thus, feeding contact allergens to previously sensitized individuals could only induce transient 'desensitization' (van Hoogstraten et al., 1993b). In the light of the foregoing this would mean that mature Th1 cells, in contrast to their precursors, are highly refractory to further Th2 control **(Table 4)**.

Summary and conclusions

The recent extensive research on the different functions of T cells differing in cytokine production profiles has opened promising venues for further research on mechanisms and therapeutic options. Clearly, the routing hypothesis as described above still leaves many questions unanswered, such as the question why some chemicals may elicit strong Th2 responses and IgE antibody production even when applied to the skin, without apparent reduction of delayed allergic reactivity (Dearman et al., 1991). The preliminary understanding of regulatory mechanisms in allergic contact dermatitis has not yet led to further therapeutic progress. So

TABLE 4.

EFFECTS OF VARIOUS ROUTES OF ALLERGEN ADMINISTRATION ON THE CLINICAL OUTCOME

Allergen routes	Non-sensitized individuals		Hypersensitive individuals	
	Sensitization	Tolerance	Eczema	Flare ups/ Desens
* Skin	++ 1)	-	++	(+)
* Intravenous, Oral 2)	-	++	-	++

1) In this table, effects of different doses of allergen and frequencies of allergenic contacts, have been omitted. Generally, higher doses and frequencies favour involvement of cell-mediated tolerance and flare-up/desensitization phenomena, also with the skin route.

2) These tolerogenic routes may include skin deficient in, or depleted of, functionally active Langerhans' cells.

far, no methods of permanent desensitization have been devised. Nevertheless, major cell types and mediators involved in allergic contact dermatitis have been identified. How T cells specifically recognize distinct allergens, and how these and other inflammatory cells interact to generate inflammation has begun to be understood. Moreover, the recently defined cellular interaction molecules and mediators provide promising targets for anti-inflammatory drugs. Obviously, drugs found to be effective in preventing severe T cell-mediated conditions, e.g. rejection of a vital organ graft, should be very safe before their use in allergic contact dermatitis would seem appropriate. To date, prudence favours any measure to prevent allergic contact dermatitis, be it through legal actions to outlaw the use of certain materials, or through avoiding personal contacts with these materials. In the meantime, for difficult-to-avoid allergens, further studies on the potential value of tolerogenic treatments should be intensified.

REFERENCES

- Arnon R and Teitelbaum D. On the existence of suppressor cells. Int Arch Allergy Immunol. 100: 2-7, 1993
- Barker, JNWN, Mitra, RS, Griffiths, CEM, Dixit, VM, Nickoloff, BJ. Keratinocytes as initiators of inflammation. The Lancet 1991; 337: 211-214.
- Benezra C, Sigman CC, Bagheri D, Fraginals R, Maibach HI. Molecular aspects of allergic contact dermatitis. In: Rycroft RJG, Menné T, Frosch PJ and Benezra, C. (Eds.) Textbook of Contact Dermatitis, Springer-Verlag, Berlin, pp.105-117 (1992).
- Bloom BR, Salgame P, Diamons B. Revisiting and revising suppressor T cells. Immunol Today 13, 4: 131-136, 1992.

- Dearman RJ, Kimber I. Differential stimulation of immune function by respiratory and contact chemical allergens. Immunology 72: 563-570, 1991.
- Del Prete G, De Carli M, Almerigogna F, Guidizi MG, Biagiotti R, Romagnani S. Human IL-10 is produced by both type 1 helper (Th1) and type 2 helper (Th2) T cell clones and inhibits their antigen-specific proliferation and cytokine production. J Immunol 150, 2: 353-360, 1993.
- Enk AH, Katz SI. Early molecular events in the induction phase of contact sensitivity. Proc. Natl. Acad. Sci. USA. 1992: 1398-1402.
- Enk AH, Angeloni VL, Udey MC, Katz SI. An essential role for Langerhans cell-derived Il-1ß in the initiation of primary immune responses in skin. J Immunol 150: 3698-3704, 1993.
- Eynon BE, Parker DC. Small B cells as antigen-presenting cells in the induction of tolerance to soluble protein antigens. J. Exp. Med. 175: 131-138, 1992.
- Fairchild RL, Palmer E, Moorhead JW. Production of DNP-specific/ Class I MHC-restricted suppressor molecules is linked to the expression of T cell receptor α- and ß-chain genes. J Immunol 150, 1: 67-77, 1993.
- Gershon RK and Kondo K: Cell interactions in the induction of tolerance. The role of thymic lymphocytes. Immunology 18: 723-737, 1970.
- Gieni RS, Yang X, HayGlass KT. Allergen-specific modulation of cytokine synthesis patterns and IgE responses in vivo with chemically modified allergen. J Immunol 150, 1, 302-310, 1993.
- Goebeler M, Meinardu-Hager G, Roth J, Goerdt S, Sorg C. Nickel choride and cobals chloride, two common contact sensitizers, directly induce expression of intercellular adhesion molecule-1 (ICAM-1), vascular cell adhesion molecule-1 (VCAM-1) and endothelial leukocyte adhesion molecule (ELAM-1) by endothelial cells. J Invest Dematol 100: 759-765, 1993.
- Hoogstraten, IMW van, Boden, D. Blomberg, BME von, Kraal, G, Scheper, RJ. Persistent immune tolerance to nickel and chromium by oral administration prior to cutaneous sensitization. J. Invest. Derm. 608-617, 1992.
- Hoogstraten IMW van, Boos, C, Boden, D, Blomberg, BME von, Scheper RJ, and Kraal, G. Oral induction of tolerance to nickel sensitization in mice. J. Invest. Derm. in press, 1993a.
- Hoogstraten, IMW van, von Blomberg, BME, Boden, D. Kraal, G, Scheper, RJ. Non-sensitizing epicutaneous skin contacts prevent subsequent induction of immune tolerance. J. Invest. Derm., in press, 1993b.
- Hsieh CS, Macatonia SE, Tripp CS, Wolf S, O'Garra A, Murphy KM. Development of Th1 CD4+ T cells through IL-12 produced by Listeria-induced macrophages. Science 260: 547-549, 1993.
- Jones LA, Chin LT, Longo DL, Kruisbeek AM. Peripheral clonal elimination of functional T cells. Science 250: 1726-1729, 1990.
- Karpus WJ, Swanborg RH. CD4+ suppressor cells inhibit the function of effector cells of experimental autoimmune encephalomyelitis through a mechanism involving transforming growth factor-ß. J Immunol 146, 15: 1163-1168, 1991.
- Kapsenberg ML, Wierenga EA, Bos JD and Jansen HM. Functional subsets of allergen-reactive CD4+ T cells. Immunol. Today 1991.
- Kapsenberg ML, Bos JD, Wierenga EA. T cells in allergic responses to haptens and proteins. Springer Semin Immunopathol 13: 303-314, 1992.
- Khoury SJ, Hancock WW, Weiner HL. Oral tolerance to myelin basic protein and natural recovery from experimental autoimmune encephalomyelitis are associated with downregulation of inflammatory cytokines and differential upregulation of transforming growth factor ß, interleukin 4 and prostaglandin E expression in the brain. J Exp med 176: 1355-1364, 1992.
- Kuchroo VK, Byrne MC, Atsumi Y, Greenfield E, Connoly JB, Whitters MJ, O'Hara RM, Collins M, Dorf ME. T-cell receptor α chain plays a critical role in antigen-specific suppressor cell function. Proc Natl Acad Sci USA 88: 8700-8704, 1991.
- Limpens, J and Scheper, RJ. Inhibition of T suppressor cell function by local administration of an active CY-derivative at the sensitization site. Clin Exp Immunol 84: 383-388 (1991).
- Lehner T, Bergmeier LA, Panagiotidi C, Tao L, Brookes R, Klavinskis LS, Walker P, Ward RG, Hussain L, Gearing AJH, Adams SE. Induction of mucosal and systemic immunity to a recombinant simian immunodeficiency viral protein. Science 258: 1365-1369, 1992.
- Meade R, Askenase PW, Geba GP, Neddermann K, Jacoby RO, Pasternak RD. Transforming growth factor-ß1 inhibits murine immediate and delayed type hypersensitivity. J Immunol 149, 2: 521-528, 1992.

- Melamed D, Friedman A. Direct evidence for anergy in T lymphocytes tolerized by oral administration of ovalbumin. Eur. J. Immunol. 23: 935-942, 1993.
- Minty A, Chalon P, Derocq JM, Dumont X, Guillemot JC, Kaghad M, Labit C, Leplatois P, Liauzun P, Miloux B, Minty C, Casellas P, Loison G, Lupker J, Shire D, Ferrara P, Caput D. Interleukin 13 is a new human lymphokine regulating inflammatory and immune responses. Nature 362: 248-250, 1993.
- Ohashi PS, Oehnen S, Buerki K, Pircher H, Ohashi CT, Odermatt B, Malissen B, Zinkernagel RM, Hengartner H. Ablation of "tolerance" and induction of diabetes by virus infection in viral antigen transgenic mice. Cell 65: 305-317, 1991.
- Oostendorp RAJ, Schaaper WMM, Post J, Blomberg-van der Flier, BME, Meloen RH, Scheper RJ. Suppression of lymphocyte proliferation by a retroviral p15E-derived hexapeptide. Eur J Immunol 22: 1505-1511, 1992.
- Oswald IP, Gazzinelli RT, Sher A, James SL. IL-10 synergizes with IL-4 and transforming growth factor-ß to inhibit macrophage cytotoxic activity. J Immunol 148, 11: 3578-3582, 1992.
- Parker, D, Turk, JL and Scheper, RJ Central and peripheral action of suppressor cells in contact sensitivity in the guinea pig. Immunol 30: 593-597 (1976).
- Polak L and Turk JL. Reversal of immunological tolerance by cyclophosphamide through inhibition of suppressor cell activity. Nature 249: 654-656, 1974.
- Polak L. and Rinck C. Persisting activity of suppressor cells in T cell mediated immune response. J. Immunol. 121: 762-766, 1978.
- Polak L. Immunology of chromium. In: Chromium metabolism and toxicity. Ed. D. Burrows, CRC Press Inc. Boca Raton, FA USA, pp 51-136.
- Qin S, Cobbold SP, Pope H, Elliott J, Kioussis D, Davies J, Waldmann H. "Infectious" transplantation tolerance. Science 259: 974-977, 1993
- Reijsen FC van, Bruijnzeel-Koomen CAFM, Kalthoff FS, Maggi E, Romagnani S, Westland JKT, Mudde GC. Skin-derived aeroallergen-specific T cell clones of Th2 phenotype in patients with atopic dermatitis. J Allergy Clin Immunol 90: 184-192, 1992.
- Reiner SL, Wang Z, Hatam F, Scott P,. Locksley RM. Th1 and Th2 cell antigen receptors in experimental Leishmaniasis. Science 259: 1457-1460, 1993.
- Rocha B, Boehmer H von. Peripheral selection of the T cell repertoire. Science 251: 1225-1231, 1991.
- Romagnani S. Induction of Th1 and Th2 responses: a key role for the 'natural' immune response? Immun Today 13, 10: 379-381, 1992.
- Romagnoli P, Labhardt AM, Sinigaglia F. Selective interaction of Ni with MHC-bound peptide. EMBO J 10: 1303, 1991.
- Salgame P, Abrams JS, Clayberger C, Goldstein H, Convit J, Modlin RL, Bloom BR. Differing lymphokine profiles of functional subsets of human CD4 and CD8 T cell clones. Science 254: 279-282, 1991.
- Sayegh MH, Khoury SJ, Hancock WW, Weiner HL, Carpenter CB. Induction of immunity and oral tolerance with polymorphic calss II major histocompatibility complex allopeptides in the rat. Proc. Natl. Acad. Sci. USA 89: 7762-7766, 1992.
- Scheper, RJ, Blomberg, BME von. Cellular mechanisms in allergic contact dermatitis. In: Rycroft RJG, Menné T, Frosch PJ and Benezra, C. (Eds.) Textbook of Contact Dermatitis, Springer-Verlag, Berlin, pp.11-28 (1992).
- Schönrich G, Kalinke U, Momburg F, Malissen M, Schmitt-Verhulst AM, Malissen B, Hämmerling GJ, Arnold B. Down-regulation of T cell receptors on Self-reactive T cells as a novel mechanism for extrathymic tolerance induction. Cell 65: 293-304, 1991.
- Scott P. IL-12: Initiation cytokine for cell-mediated immunity. Science 260: 496-497, 1993.
- Sinigaglia, F, Scheidegger, D, Garotta, G, Scheper, RJ, Pletscher, M, Lanzavecchia, A; Isolation and characterization of Ni-specific T cell clones from patients with Ni-contact dermatitis. J Immunol 135: 3929-3933 (1985).
- Taub DD, Conlon K, Lloyd AR, Oppenheim JJ, Kelvin DJ. Preferential migration of activated CD4+ and CD8+ T cells in response to MIP-1α and MIP-1ß. Science 260: 355-358, 1993.
- Waal Malefyt R de, Yssel H, Roncarolog MG, Spits H, Vries JE de. Interleukin-10. Curr Opinion Immunol 4: 314-320, 1992.

Test Methods for Contact Sensitization

Th. Maurer, A. Arthur, Ph. Bentley
Ciba-Geigy Ltd., Pharmaceuticals Division, Preclinical Safety, CH-4002 Basel, Switzerland.

Introduction

Predictive tests for the evaluation of skin sensitizing effects of chemicals have existed for nearly 50 years. The Draize test (Draize et al. 1944) was the first test which was included in a guideline (Draize 1959). The test was designed to screen out potent sensitizers. Negative results with known human sensitizers, such as benzocaine, neomycin or nickel sulfate (Magnusson and Kligman 1969) were the main reasons to develop other protocols.

A considerable change in the development of tests was induced by the introduction of Freund's complete adjuvant (FCA) injections during the induction phase. FCA enhances the immunological responses of the animals and it functions as a depot for the distribution of the allergen (Waksman 1979). With the introduction of FCA, it was possible to detect not only potent, but also moderate and weak contact allergens.

Guidelines for the testing of sensitizing effects were published in 1981 by the OECD and in 1984 by the EEC. Both guidelines included until recently 7 methods, 4 protocols using adjuvant in the induction phase and 3 tests which do not using FCA.

The "old" list of OECD and EEC guinea pig tests

Author	Route of treatment for Induction	Challenge
Non-adjuvant test		
Draize 1959	i.d.	i.d.
Buehler 1965	epid. occl.	epid. occl.
Klecak et al. 1977	epid. open	epid. open
Adjuvant test		
Magnusson 1969	i.d. + FCA and epid. occl.	epid. occl.
Maguire 1973	i.d. + FCA and epid. occl.	epid. occl.
Klecak et al. 1977	i.d. + FCA	epid. open
Maurer et al. 1980	i.d. and i.d. +FCA	i.d. and epid. occl.

The OECD guideline of 1981 listed all 7 methods without giving preference to a certain method. The EEC favoured to use the maximization test of Magnusson and Kligman (1969). Both guidelines described the test in tabular form and many modifications of the original protocol were possible.

Working groups of the Federal Health Office in Germany, of the European Industrial Ecology Toxicology Centre (ECETOC 1990) and of the OECD started to discuss changes of the guidelines and to reduce the number of recommended tests. From surveys in Europe and the USA it was apparent, that the maximization test is the most frequently used test in Europe but the Bühler test the most common in the USA. As a consequence of these surveys, the two mentioned tests are now the primary accepted tests in the new guidelines of the OECD (1992) and EEC (1992). The EEC still favours the maximization test as the first recommended test.

Beside the reduction of the number of primary recommended tests, the following major changes were included in the new guidelines:

- It is recommended to perform the tests according to the original literature.
- It is recommended to use moderate and/or weak allergens for validation of the method used in the own laboratory.
- Other protocols can be used when special reasons are given.
- The OECD accepts the mouse methods MEST (Gad et al. 1986) and LLNA (Kimber et al. 1989) as screening tests and all positive results obtained in one of the mouse tests have not to be confirmed in a guinea pig test. Negative results in mouse tests, in contrast, have to be confirmed in a guinea pig test.

Influence of induction concentration on sensitization

Based on long lasting experience in man on the high sensitivity of already allergic persons, it was thought that the elicitation of allergic reactions are not dose dependent. In the mean time we know that the induction and the challenge phases of sensitization are dose dependent. Magnusson and Kligman (1969, 1970) recommended quite high concentrations for the intradermal and epidermal applications. The selected concentrations should even have a slight irritant effect during induction and the challenge concentration should be just under the threshold of irritation. In combination with the adjuvant injections, the selection of high doses may lead to strong irritant effects. Further research demonstrated that such combination effects may diminish the response rate and not maximize it. Schäfer (1978) showed that the optimal concentrations for the testing of para-substituted benzenes are in the range of 1-2% and that higher concentrations have an inhibitory effect. Comparable experience were published by Andersen (1987) for formaldehyde and by Rohold et al. (1991) for nickel sulfate.

Choice of vehicles

The choice of vehicle for intradermal, epidermal open and epidermal occlusive application of test compounds is important and several publications have shown that the vehicle influences the outcome of sensitization tests (Magnusson and Kligman 1970, Wahlberg 1973). It is clear that with the vehicle best adapted to the physico-chemical properties of the test compound, the penetration of the compound is influenced and the incidence of positive animals is higher than with a standard vehicle for all test compounds. Our conclusion based on studies with water soluble and water insoluble compounds was, however, that the choice of vehicle will not have such a strong influence on the outcome of a positive or negative test (Maurer 1983). It is clear that beside the choice of the vehicle, the type of patch test material and the volume of solution or cream applied influences the reactivity of the skin to allergens.

Testing strategies

The surveys in Europe and the USA on the most frequent used sensitization methods also demonstrated differences in testing strategies between Europe and USA. Bühlers intension was to screen out strong and moderate sensitizers in the animal and to test negative or weakly positive compounds in humans in a second step (Bühler 1965). Magnusson and Kligman (1969), in contrast, wanted to test compounds in the most sensitive test, to omitt testing in humans. The testing strategy should also be dependent on the type of chemicals. Industrial compounds may just be tested for labelling purposes or transportation classification. It is commonly accepted that certain classes of industrial chemicals are skin sensitizers. On the other hand, it cannot be accepted that cosmetic ingredients or ingredients of pharmaceutical preparations are common sensitizers. In the case of dermatics the difference of sensitivity of intact and diseased skin must be considered.

Dependent on the importance of the sensitizing potential of a chemical on the market, screening for sensitizing potentials can be included in different stages of development of compounds. Three stages are possible today:

1) Screening of compounds in an early stage of development in the mouse MEST or LLNA test.
2) Testing of single compounds in a guinea pig test for labelling purposes or handling safety reasons within production units.
3) Testing of final formulations to get additional informations on the sensitizing potential of the actual use concentration and on the influence of auxiliary compounds in the formulation.

In the first screen it is possible to detect strong and moderate sensitizers in tests with short duration, but the threshold of sensitivity of these tests is presently unknown. However, the inclusion of these tests in the OECD guideline, with the acceptance of positive results for labelling purposes, was a major step forward.

Guinea pig tests of the second step will still be important for further development of new chemicals dependent on their intended use. The maximization test is a very sensitive method and will detect sensitizers with strong or weak potentials. The Bühler test is only sensitive when correctly performed (Robinson et al. 1989).
Additional testing of the final formulation in a third step would have some advantages. It would be possible to detect influences of the actual concentration of the allergen compared to the concentration used in step two and on the influences of other ingredients. The best comparison between the tests of step 2 and 3 is only possible if the same test protocol has been used or the same sensitivity of the test is achieved. In cases where the maximization protocol is used in the second step it would be advisable for step 3 to use a modified version with only epidermal occlusive application of the test compound during the induction phase instead of intradermal injections. This modified protocol has been validated and showed the same sensitivity as the original version (Maurer 1989).

Validation of test methods

The sensitivity of the guinea pig strain used must be checked at regular intervals with known skin sensitizers. Strong contact allergens have been used in recent years. During the discussions in the working groups of ECETOC, OECD and EEC, it was generally agreed that the use of moderate or weak sensitizers would be more efficient and would better demonstrate the sensitivity of the strain and the quality of performing the tests. But the recommended compounds from the different groups are not identical and some strong contact allergens are still included in the EEC as shown below.

Reference compounds recommended by ECETOC (1990):
Benzocaine, neomycin sulphate, hydroxycitronellal.

Reference compounds recommended by the EEC (1992):
p-phenylenediamine, 2,4-dinitrochlorobenzol, potassium dichromate, neomycin sulphate, nickel sulphate.

Reference compounds recommended by the OECD (1992):
hexyl cinnamic aldehyde, mercaptobenzothiazole, benzocaine.

The reference compounds recommended by the OECD have been tested recently in two laboratories (Basketter et al. 1993). The compounds were tested in the maximization, the Bühler as well as in the LLNA test. The interlaboratory variation in the selection of vehicles and test concentrations is well documented. The results obtained demonstrate that the classification of the sensitizing potential may differ from one laboratory to another and shows how important it is to perform inhouse validation with reference compounds of different potentials. The studies demonstrate also how important the quality of the performance of sensitization test is and its implications on the classification and labelling of compounds. It is stated: "Thus, a laboratory which runs a more stringent version of a given assay may achieve a positive classification for a weakly sensitizing chemical which another laboratory may fail to detect, while both laboratories nevertheless meet the minimum standards".

Classification of sensitizing potentials

A positive result in a guinea pig test shows the skin sensitizing potential of the test compound. Magnusson and Kligman (1969) classified compounds to 5 classes of potency. Even in the case of a nagative results in a guinea pig maximization test, the compound is classified as weak sensitizer.

Good correlation was found between experimental tests evaluating the contact allergenic potency in animals and man, but it is very difficult to predict whether compounds showing positive results will act as weak or strong sensitizers in humans under use conditions, because many factors other than the potency of the compound alone are involved in humans. Nickel sulphate, for example, is a weak to moderate sensitizers in guinea pigs, but one of the most frequent allergens in man. The main reason is its wide distribution and the high frequency of exposure. Dinitrochlorobenzene, the old classical reference compound with a very strong sensitizing potential, but induced only a few cases of allergy in man, because the industrial use is very limited.
The real sensitizing risk of a chemical can only be evaluated based on good epidemiological data in man. In a recent publication (Knudsen et al. 1993) criteria for the validation of published guinea pig and human data are discussed and a classification of significant allergens is proposed. A group of the Federal Health Office in Germany is also working on the classification of allergens based on animal and human data. Good progress has been achieved in a realistic classification of allergens based on guinea pig and human data, but much work has still to be done.

References

Andersen KE (1987) Testing for contact allergy in experimental animals. Pharmacol. & Toxicol. 61:1-8

Basketter DA, Selbie E, Scholes EW, Lees D, Kimber I, Botham PA (1993) Results with OECD recommended positive control sensitizers in the maximization, Buehler and local lymph node assays. Fd. Chem. Toxicol. 31:63-67

Bühler EV (1965) Delayed contact hypersensitivity in the guinea pig. Arch Dermatol. 91:171-177

Draize JH, Woodward G, Calvery HO (1944) Methods for the study of irritation and toxicity of substances applied topically to the skin and mucous membranes. J. Pharms. Exp. Ther. 82:377-389

Draize JH (1959) Intracutaneous sensitization test on guinea pig. In: Appraisal of the Safety of Chemicals in Food and Cosmetics. Association of Food and Drug Officials of the United States, Austin, Texas.

ECETOC (1990) European Chemical Industry Ecology & Toxicology Centre, Monograph No. 14, Skin Sensitization Testing.

EEC (1984), Commission Directive of 25 April 1984 adapting to technical progress for the sixth time Council Directive 67/548/EEC on the laws, regulations and administrative provisions relating to the classification, packaging and labelling of dangerous substances. Off. J. Europ. Commun. L 251, Vol 27:113-117

EEC (1992), Commission Directive of 31 July 1992 adapting to technical prgress for the 17th time, Council Directive 67/548/EEC. Off. J. Europ. Commun. L 383 A, Vol 35:131-136

Gad SC, Dunn BJ, Dobbs DW, Reilly C, Walsh RD (1986) Development and validation of an alternative dermal sensitization test: the mouse ear swelling test (MEST). Toxicol. Appl. Pharmacol. 84:93-114

Kimber I, Weisenberger C (1989) A murine local lymph node assay for the identification of contact allergens. Assay development and results of an initial validation study. Arch. Toxicol. 63:274-282

Klecak G, Geleick H, Frey JR (1977) Screening of fragrance materials for allergenicity in the guinea pig, comparison of four testing methods. J. Soc. Cosmet. Chem. 28:53-64

Knudsen BB, Wahlberg JE, Andersen I, Menne T (1993) Classification of contact allergens. Proposal for criteria. Dermatosen 41:5-9

Magnusson B, Kligman AM (1969) The identification of contact allergens by animal assay. The guinea pig maximization test. J. invest. Dermatol. 52:268-276

Magnusson B, Kligman AM (1970) Allergic Contact Dermatitis in the Guinea Pig. Charles C Thomas Publisher, Springfield, Illinois

Maguire HC (1973) The bioassay of contact allergens in the guinea pig. J. Soc. Cosmet. Chem. 24:151-162

Maurer T, Weirich EG, Hess R (1980) The optimization test in the guinea pig related to other predictive sensitization methods. Toxicol. 15:163-171

Maurer T (1983) Contact and Photocontact Allergens. A Manual of Predictive Test Methods. Marcel Dekker INC., New York and Basel

Maurer T, Hess R (1989) The maximization test for skin sensitization potential - updating the standard protocol and validation of a modified protocol. Fd. Chem. Toxic. 27:807-811

OECD (1981), Organisation for Economic Co-operation and Development. OECD guideline for Testing of Chemicals No. 406, Skin Sensitization.

OECD (1992) Organisation for Economic Co-operation and Development. OECD guideline for Testing of Chemicals No. 406, Skin sensitization, adopted by the council on 17th July 1992

Robinson MK, Stotts J, Danneman PJ, Nusair TL, Bay PHS (1989) A risk assessment process for allergic contact sensitization. FD. Chem. Toxicol. 27:479-

Rohold AE, Nielsen GD, Andersen KE (1991) Nickel-sulfate-induced contact dermatitis in the guinea pig maximization test: a dose-response study. Contact Dermatitis 24:35-39

Schäfer U, Metz J, Pevny I, Röckl H (1978) Sensibilisierungsversuche an Meerschweinchen mit fünf parasubstituierten Benzolderivaten. Arch Derm. Res. 261:153-161

Wahlberg JE (1973) Thresholds of sensitivity in metal contact allergy. Berufsdermatosen 21:151-159

Waksman BH (1979) Adjuvants and immune regulation by lymphoid cells. Springer Semin. Immunopath. 2:5-

Test methods for respiratory sensitization

Jürgen Pauluhn
Institute of Toxicology, BAYER AG, 42096 Wuppertal, Germany.

The increasing awareness of the potential of chemicals capable of inducing allergic respiratory hypersensitivity has resulted in attempts to develop test models that can be used to evaluate this endpoint (Botham et al., 1988; Griffiths-Johnson and Karol, 1991; Thorne and Karol, 1988; Karol et al., 1985, 1989; Karol and Thorne, 1988; Karol, 1988). To date, practically all such models have relied upon the use of the guinea pig, a species known to be sensitive for agents inducing bronchoconstriction and in which respiratory function and respiratory hypersensitivity can be measured readily. The models have utilized various modes of hapten or antigen administration and methods for detecting sensitization. It has been shown that guinea pigs sensitized by inhalation exposure to either a free or a protein-bound chemical can be induced to exhibit changes in respiratory patterns following inhalation challenge with the same chemical in the free or in the form of its hapten-protein conjugate. Respiratory hypersensitivity is characterized by a specific allergic hyperresponsiveness as well as increased non-specific airway hyperresponsiveness. Both responses may be clinically indistinguishable. Experimental and clinical studies have intimated that there is reason to suspect that acute exposure to brief high-level concentrations of asthmagenic chemicals and the development of increased airway hyperresponsiveness are associated.

The complex relationship between non-specific airway hyperresponsiveness and allergic hypersensitivity has received considerable attention (Gordon et al., 1985; Karol and Thorne, 1988; Poulter et al., 1990; Sheffer, 1991). The mechanisms underlying the alterations in airway hyperresponsiveness are controversial and the specific aspects of airway inflammation responsible for this increase in airway hyperresponsiveness are unknown. Epithelial shedding, subepithelial collagen thickening, airway mucosal inflammation, activation of airway afferent nerves, and the release of low-molecular-weight neuropeptides as mediators of inflammation are known to affect the tonus of the airway smooth muscle and may play a crucial role in the acute increase in airway hyperresponsiveness occurring after exposure to irritant or inflammatory stimuli (Barnes et al., 1991; Saetta et al., 1992).

More recently it has been found that changes in sensitive respiratory parameters can also be provoked in dermally sensitized guinea pigs by inhalation challenge with the free chemical or the hapten-protein conjugate (Botham et al., 1988; Hayes et al., 1992; Pauluhn and Eben, 1991; Sarlo and Clark, 1992). In attempting to derive an animal model that permits the identification of asthmagenic low-molecular-weight chemicals without the presence of overriding effects caused by toxic (irritant) airway inflammation the utility of the intradermal induction - inhalation challenge protocol has been evaluated. The intradermal route of induction was used to minimize the influence of absorption or particle-size specific factors. This route of induction also minimizes the risk of potential confounding effects attributable to irritation-induced bronchoalveolitis and bronchial hyperreactivity as a result of the release of neurogenic inflammatory mediators. For the validation of this animal model known human chemical respiratory allergens such as trimellitic anhydride (TMA) and diphenylmethane 4,4'-diisocyanate (MDI) were tested. Ovalbumin (OA) sensitized guinea pigs served as positive controls.

THE ANIMAL MODEL

Test protocol

Male or female guinea pigs of the Dunkin-Hartley strain from different breeders were used. All studies consisted of two groups each comprising at least 8 control (vehicle) and 8 animals sensitized with the free chemical. As shown in TABLE 1 guinea pigs were sentitized with TMA by single or repeated inhalation exposures to concentrations bracketing non-irritant and irritant concentrations. Particles of TMA were in the respirable range {mass median aerodynamic diameter (MMAD): 1.6-4.2 µm/geometric standard deviation (GSD): 1.4-1.6}. Approximately two weeks after the last encounter with the test compound guinea pigs were again subjected to the hapten. One week thereafter the animals were re-challenged with the TMA-guinea pig serum albumin (GSA) conjugate. Sensitization of guinea pigs with MDI by inhalation was achieved by a single brief high-level exposure (1 x 15 min) to monomeric MDI (concentration: 135 mg/m³, MMAD: 1.8 µm/GSD: 1.8).

The intradermal inductions were performed by repeated injections on days 0, 2, 4 (TMA: 2 x 100 µl/day, ca. 5-10% in corn oil; MDI: 2 x 50-100 µl/day, ca. 5-10% in corn oil:acetone). In guinea pigs sensitized with MDI one day after the challenge with MDI an acetylcholine (ACh) bronchoprovocation challenge with increasing concentrations of ACh was performed (for details see Figs. 3 and 6).

Prior to each challenge exposure baseline parameters were recorded for ca. 10-15 min. Individual baseline data were used to calculate the mean ± 3 x standard deviation (STD) as shown in Fig. 2. Responses exceeding the mean ± 3 x STD were considered positive. All signals were averaged over a period of 10 to 30 sec.

Respiratory measurements

After acclimatisation to the plethysmograph the baseline parameters were measured as already described (exposure to air). On average, the duration of the challenge with the hapten or hapten-GSA conjugate was 30 min, followed by post-challenge measurements of ca. 45 min. For the evaluation of immediate-onset hypersensitivity reactions occurring during challenge with the hapten, the hapten-GSA conjugate or ACh the following respiratory parameters are evaluated: respiratory rate (RR) [breaths/min], tidal volume (TV) [ml], respiratory minute volume (MV) [ml/min], peak expiratory flow rate (PEF) [ml/sec], inspiratory (IT) and expiratory times (ET) [sec], PEF x (ET+IT)/TV [dimensionless]. The last takes into account the inhaled volume, the duration of one breath and the corresponding expiratory flow rate. As demonstrated in Fig. 3 this derivatization makes evaluation of respiratory responses easier and less subjective. Measurements were taken in a nose-only animal restainer with wire mesh pneumotachograph and a differential pressure transducer (MP 45 ± 2 cm H_2O, Validyne) fitted directly onto the exposure tube. The head and the body compartment were separated using a double-layer latex neck seal. Precautions were taken to avoid artefacts due to restraint and tight fitting seals around the neck. Volumes were calculated by integration of the flow signal. Fig. 1 shows the dependence on the calculated volume as a function of frequency.

From Fig. 1 it is evident that the calculated volume based on flow needs particular attention due to its frequency dependence on volume. The resistance to air flow was adjusted in such a way that arteficial volume changes between pump rates of 50-250 cycles/min did not exceed 10%. Increasing the sensitivity of the system by increasing the resistance of the pneumotachograph was associated with a significant frequency dependence.

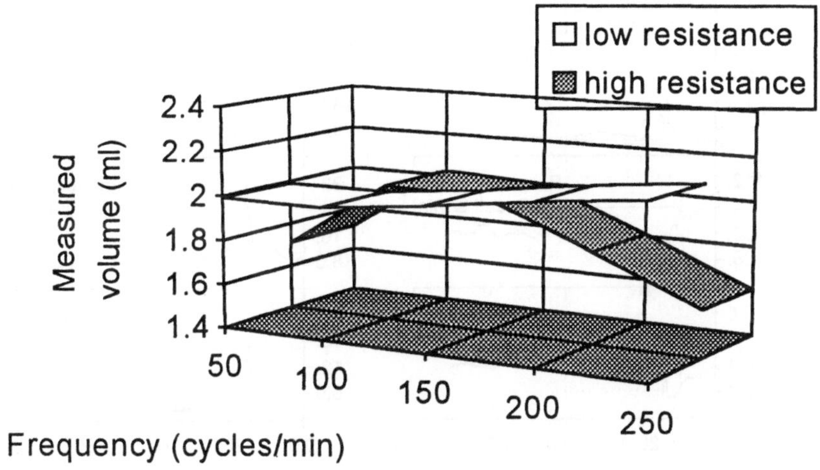

Fig. 1. Analysis of the frequency dependence of volume calibration of nose-only plethysmograph tubes (stroke volume: 2 ml). The initial calibration was performed at 100 strokes/min.

The analysis of changes in respiratory pattern necessitates the selection of sensitive, robust as well as of characteristic respiratory responses. Due to the non-specific nature of respiratory changes it also requires careful interpretation of effects observed. These considerations are of particular importance when animals are exposed to the free chemical since upper or lower respiratory tract irritation are known to induce irritation related non-specific respiratory responses. Previous inhalation exposures to irritant concentrations of the hapten may also lower the threshold for the elicitation of non-specific airway hyperresponsiveness. As demonstrated in guinea pigs sensitized and challenged with TMA changes in RR, MV, and in the dimensionless parameter [PEF x (IT+ET)/TV] are sensitive indicators for the evaluation of the pulmonary hypersensitivity responses. This was also true for challenge exposures to ACh (Fig. 3). As is evident from Figs. 2 and 3 the ratio of back-ground variability of the [PEF x (IT+ET)/TV] parameter is more favourable than RR. In many instances, changes in MV were more distinct and were easier to evaluate than the RR values. However, guinea pigs intradermally sensitized with toluene diisocyanate (TDI) and challenged with the TDI-protein conjugate by inhalation demonstrated a concomitant increase in RR and decrease of TV. Likewise only minute changes of the composite parameter MV were present.

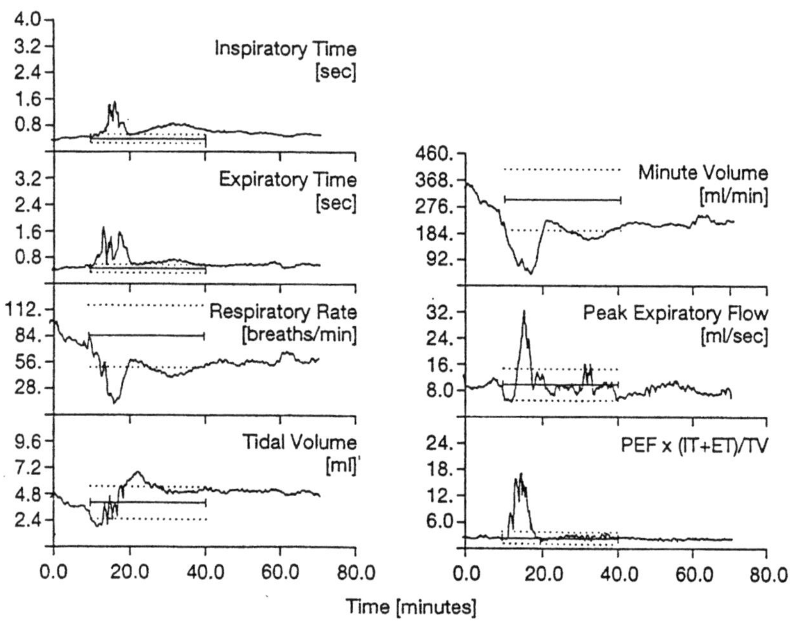

Fig. 2. The effect on respiratory parameters upon an inhalation challenge to TMA for 30 min (⊢⊣). Animals were sensitized with TMA by repeated intradermal injections.

However, a sudden decrease in RR and in particular in MV followed by a sudden increase produces a "*waveform pattern*" that is considered one of the hallmarks of an allergic hypersensitivity response (Fig. 4).

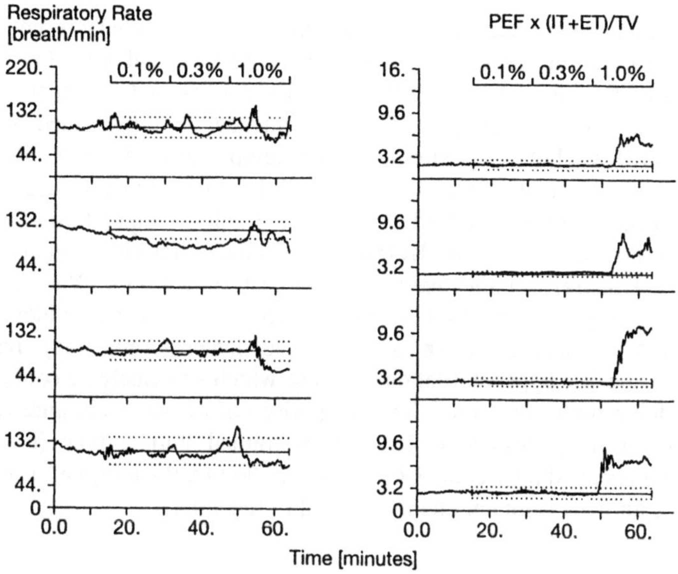

Fig. 3. The effect on the respiratory parameters RR and [PEF x (IT+ET)/TV] during the challenge to 0.1%, 0.3% and 1.% ACh (each 15 min). Animals were sensitized with OA by repeated intradermal injections. One day prior to the ACh-challenge the guinea pigs were challenged with OA by inhalation.

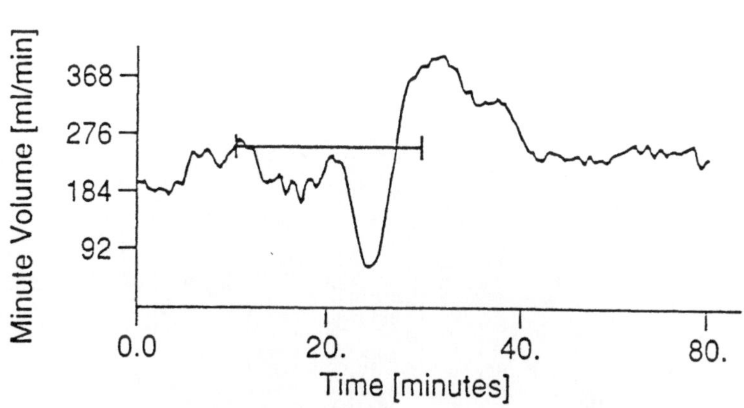

Fig. 4. Characteristic "*waveform*" breathing pattern indicating an allergic hypersensitivity. Intradermal induction with TMA and inhalation challenge to TMA-GSA (⊢⊣: duration of challenge).

This "*waveform*" breathing pattern was observed during challenge exposures to OA, TMA, and TMA-GSA. The onset of response upon challenge to OA or TMA was immediate that to TMA-GSA slightly delayed as depicted in Fig. 4. In animals challenged with OA or TMA some animals of the same group experienced an increased RR while others experienced a decreased in RR. Except for the "*waveform pattern*" it appeared that the respiratory hypersensitivity response is apparently not associated with an unequivocally unique respiratory effect (anaphylaxis not taken into account). Changes in respiratory patterns may also depend on other as yet unknown factors pertaining to the substance used for induction and challenge.

The delayed-onset of an increased RR was evaluated in unrestrained animals using water jacketed whole-body plethysmographs (temperature ca. 21 °C, duration of measurement: ca. 20 hours; bias air-flow rate 2 l min^{-1}). Early measurements performed in non-temperature controlled plethysmographs (temperature above 25 °C) revealed a significant increase in RR (onset ca. 6 hours after challenge, duration of increase ca. 2-3 hours) in either naive or sensitized guinea pigs. When a water jacketed plethysmograph was used no conclusive increase in RR was evident. This demonstrates that animals subjected to prolonged hyperthermia may possibly experience a non-specific hyperventilation response which can easily be confounded with a positive delayed-onset response. The comparison of RR values in unrestrained (whole-body bias flow plethysmographs) and restrained guinea pigs (nose-only plethysmographs) revealed a baseline RR value of ca. 90 breaths/min; i.e., it is quite apparent that the animals were not unduly stressed in the latter.

Determination of IgG Antibody

The association between the presence of homocytotropic total IgG specific antibody and the occurrence of immediate-onset hypersensitivity reactions was analyzed using published procedures (Pauluhn and Eben, 1991).

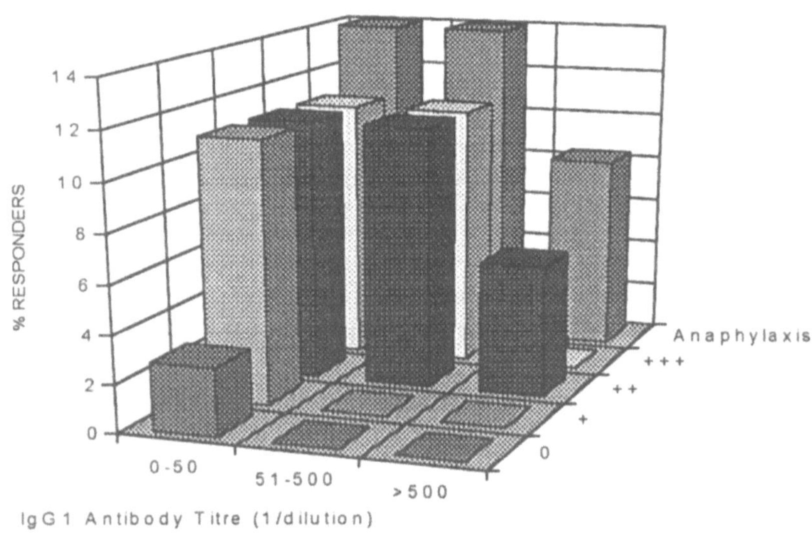

Fig. 5. Association of specific IgG antibody to OA and immediate-onset respiratory hypersensitivity reactions in guinea pigs sensitized and challenged with OA by inhalation. Classification of responses: + slight, ++ moderate, +++ severe, anaphylaxis (death).

Identification of antigen-specific IgG antibody in animals sensitized with OA was of interest for diagnostic reasons. The serologic response in the guinea pig was clearly dependent on the dose used for induction. However, as findings in Fig. 5 indicate some animals produced low titres of specific IgG antibody but did demonstrate respiratory hypersensitivity upon OA challenge. It appears that the antibody response can possibly be used as indicator of exposure and successful sensitization rather than an indicator of allergic respiratory hypersensitivity. The same conclusion is valid for studies performed with TMA and MDI.

Influence of Sensitization Regimen

Guinea pigs were sentitized with TMA by single or repeated inhalation exposures or by repeated intradermal injections. As summarized in TABLE 1 the respiratory responses occurring during challenge with ca. 40 mg TMA/m³ were most pronounced in animals sensitized intradermally.

Table 1. TMA - Comparison of different sensitization regimens.

Dose1)	Exposure regimen	Maximal concentration [mg/m³]	Positive response3)	Characteristic type of "*waveform*" response3)
0	5 x 3 hrs	-	0/8	0/8
0.7	5 x 3 hrs	2.6	2/8	0/8
3.6	5 x 3 hrs	13.5	3/8	0/8
16.82)	5 x 3 hrs	118	2/8	1/8
3.9	4 x 15 min	310	6/8	3/8
1.3	1 x 15 min	290	7/8	3/8
vehicle	intradermal	-	0/8	0/8
TMA	intradermal	-	8/8	8/8

1) Arbitrary dose per animal; assumption: MV 300 ml/guinea pig
2) 33% mortality during induction period (5 out of 15 animals)
3) #/#: number of responders/total number of animals challenged

Comparison of elicitation responses in guinea pigs sensitized by different inhalation induction regimens revealed that a 1 x 15 min exposure to a high-level concentration was more efficient than a 5 x 3 hrs exposure to nearly lethal concentrations. However, the highest incidence of characteristic changes in breathing patterns (see Fig. 4) was observed in animals sensitized intradermally. The challenge with the TMA-GSA conjugate did not increase the sensitivity of the test model. Conclusive delayed-onset respiratory responses were not observed.

Brief high-level exposures can, however, result in acute airway inflammation and bronchial hyperresponsiveness. To study the likely effects caused by irritation a monodisperse condensation aerosol of the pulmonary irritant MDI was investigated. The comparison of immediate-onset reactions in intradermally induced guinea pigs upon challenge with ca. 12 mg MDI/m³ (MMAD: 1.3 µm/GSD: 1.3) demonstrated no significant differences between animals of the vehicle or the MDI group (Fig. 6).

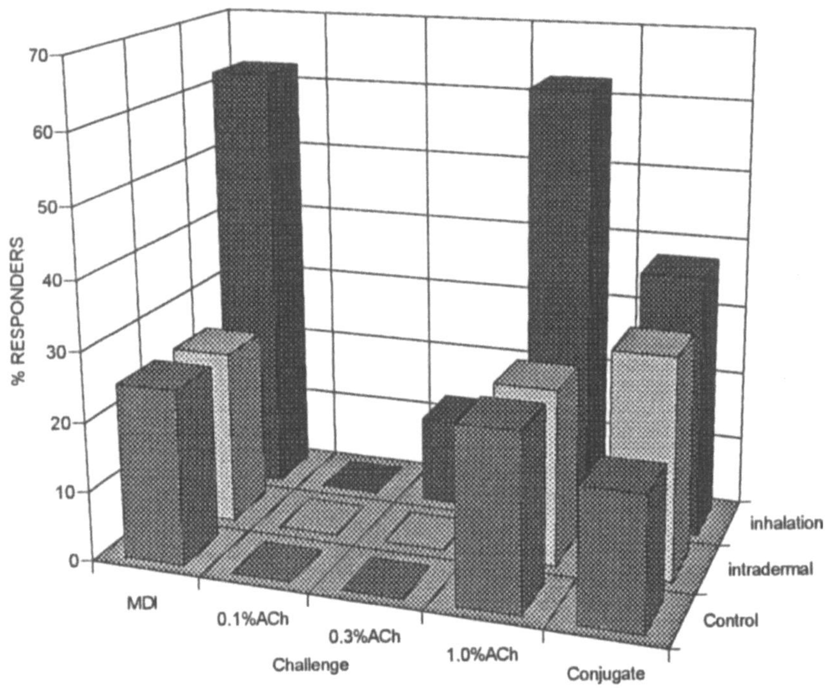

Fig. 6. Comparison of guinea pigs sensitized intradermally or by inhalation (1 x 15 min exposure to 135 mg MDI/m³).

Results obtained in animals induced by a single brief high-level exposure to MDI (1 x 15 min) confirm the conclusion that airway inflammation as a result of previous exposures may increase the susceptibility to specific (MDI) as well as non-specific (ACh) stimuli. The single brief high-level inhalation exposure to MDI also lowered the threshold of the ACh concentration necessary to elicit respiratory responses and hence appear to be related with non-specific hyperreactivity rather than specific hypersensitivity effects. However, a high incidence of respiratory responses was observed during the MDI-GSA conjugate challenge. Whether this is due to the specific epitopes of the conjugate or a non-specific effect of the carrier protein remains to be established. A "*waveform*" breathing pattern did not occur.

Conclusions

Experimental findings suggest that the pathogenesis of TMA- or MDI-induced airway disease relates more to brief high-level inhalation exposures occurring during accidental spills than to repeated low-level exposures. It would also appear that the pathogenesis of hapten induced airway hyperreactivity is not, in fact, immunologically mediated or linked to the development of hapten-specific IgG antibody. This possibility has been raised by other investigators before (Butcher et al. 1976; Cibulas et al., 1986; Mapp et al., 1985). The findings suggest that hyperresponsiveness after brief high-level exposure may be more closely related to epithelial damage in the airways and the inflammatory reactions that ensue. The precise mechanisms at play here are unclear at present. It is possible that tissue injury following inhalation sensitization results in airway hyperresponsiveness, and, thus, increased mucosal transit of aerosolized ACh to the airway nerve endings and smooth muscle targets. The same assumption may also be valid for the other agents used for challenge exposures. In addition, neuro(humoral) mediators may be released in conjuction with tissue injury that could directly affect target cells.

These findings suggest that the presence/absence of characteristic immediate-onset breathing responses, not occurring spontaneously in control guinea pigs or upon exposure to slightly irritant concentrations of the free chemical, apparently are of greater diagnostic relevance than the evaluation of changes in the respiratory rate alone. As demonstrated by the data obtained during the ACh-provocation test it would also appear that the composite dimensionless parameter PEF x (ET+IT)/TV represents a robust possibility to evaluate subtle immediate-onset responses. As a result of an increased non-specific hyperreactivity caused by previous inhalation exposures to the hapten it would also appear that the intradermal induction-inhalation challenge protocol is less susceptible to factors pertinent to irritation-induced airway hyperreactivity than the inhalation induction-inhalation challenge protocol. The advantages of the first approach are (1) that the sites of induction and of elicitation of response are different, (2) it is easier to standardize, and (3) that pulmonary irritation that may result from induction by inhalation is avoided (Briatico-Vangosa et al., 1993).

References

Barnes PJ, Baraniuk JN, Belvisi MG (1991) Neuropeptides in the respiratory tract (Part II). Am Rev Respir Dis 144:1391-1399

Botham PA, Hext PM, Rattray NJ, Walsh ST, Woodcock DR (1988) Sensitisation of guinea pigs by inhalation exposure to low-molecular-weight chemicals. Toxicology Letters 41: 159-173

Briatico-Vangosa G, Braun CJL, Cookman G, Hofmann T, Kimber I, Loveless SE, Morrow T, Pauluhn J, Sørensen T, Niessen HJ (1993) Respiratory Allergy. ECETOC Monograph No. 19.

Butcher BT, Salvagio JE, Weill H, Ziskind MM (1976) Toluene diisocyanate (TDI) pulmonary disease: immunologic and inhalation challenge studies. J Allergy Clin Immunol 58:89

Cibulas W, Murlas CG, Miller ML, Vinegar A, Schmidt DJ, McKay RT, Bernstein L, Brooks SM (1986) Toluene diisocyanate-induced airway hyperreactivity and pathology in the guinea pig. J. Allergy Clin Immunol 77:828-834.

Gordon T, Sheppard D, McDonald DM, Distefano S, Scypinski L (1985) Airway hyperresponsiveness and inflammation induced by toluene diisocyanate in guinea pigs. Am Rev Respir Dis 132:1106-1112

Griffiths-Johnson DA, Karol MH (1991) Validation of a non-invasive technique to assess development of airway hyperreactivity in an animal model of immunologic pulmonary hypersensitivity. Toxicology 65:283-294

Hayes JP, Daniel HR, Tee RD, Barnes PJ, Newman-Taylor AJ, Chung KF (1992) Bronchial Hyperreactivity after inhalation of trimellitic anhydride dust in guinea pigs after intradermal sensitization to the free hapten. Am Rev Respir Dis 146:1311-1314

Karol MH, Hillebrand JA, Thorne PS (1989) Characteristics of weekly pulmonary hypersensitivity responses elicited in the guinea-pig by inhalation of ovalbumin aerosols. Toxicol Appl Pharmacol 100:234-246

Karol MH, Thorne PS (1988) Pulmonary hypersensitivity and hyperreactivity: implications for assessing allergic responses in: Toxicology of the Lung, Gardner DE, Crapo JD, Massaro EJ (eds.). Raven Press, New York pp. 427-448

Karol MH, Stadler J, Magreni C (1985) Immunotoxicologic evaluation of the respiratory system: animal models for immediate- and delayed-onset pulmonary hypersensitivity. Fundam Appl Toxicol 5:459-472

Karol MH (1988) The development of an animal model for TDI asthma. Bull Eur Physiopathol Respir 23: 571-576

Mapp CE, Polato R, Maestrelli P, Hendrick DJ, Fabbri LM (1985) Time course of the increase in airway responsiveness associated with late asthmatic reactions to toluene diisocyanate in sensitized subjects. J Allergy Clin Immunol 75:568

Mullin LS, Wood CK, Krivanek ND (1983) Guinea pig respiratory response to isocyanates. Toxicol Appl Pharmacol 71:113-122

Pauluhn J, Eben A (1991) Validation of a Non-invasive technique to assess immediate or delayed onset of airway hypersensitivity in guinea pigs. J Appl Toxicol 11:423-431

Poulter LW, Power C, Burke C (1990) The relationship between bronchial immunopathology and hyperresponsiveness in asthma. Eur Respir J 3:792-799

Saetta M, Distefano A, Maestrelli P, DeMarzo N, Milani GF, Pivirotto F, Mapp CE, and Fabbri LM (1992) Airway mucosal inflammation in occupational asthma induced by toluene diisocyanate. Am Rev Respir Dis 145:160-168

Sarlo K, Clark ED (1992) A tier approach for evaluating the respiratory allergenicity of low-molecular-weight chemicals. Fundam Appl Toxicol 18:107-114

Sheffer AL (1991) Definition and diagnosis. In: Guidelines for the diagnosis and management of asthma. National Heart, Lung, and Blood Institute National Asthma Education Program Expert Panel Report. J Allergy Clin Immunol 88: (3 part 2), 427-438

Thorne PS, Karol MH (1988); Assessment of airway reactivity in guinea-pigs: comparison of methods employing whole body plethysmography. Toxicology 52:141-163

Newer Approaches to the Evaluation of Sensitization Phenomena

I Kimber

ZENECA Central Toxicology Laboratory, Alderley Park, Macclesfield, Cheshire, SK10 4TJ, UK.

Introduction

Sensitization is the immunological process whereby an individual is rendered hypersensitive to a particular allergen. Allergic reactions develop usually in two stages. Following first encounter with the inducing allergen the susceptible individual mounts a specific immune response which results in sensitization. Subsequent exposure to the same allergen provokes an accelerated and more aggressive response which in turn causes adverse health effects; the most common of which in the context of industrial toxicology and occupational medicine are allergic contact dermatitis and respiratory hypersensitivity.

Contact sensitization

Contact allergy (skin sensitization) is dependent upon the activity of T lymphocytes and their recognition of and response to the inducing allergen (Kimber, 1992). There is available a number of guinea pig predictive test methods. These assays seek to identify chemicals which have the potential to cause contact sensitization as a function of dermal hypersensitivity reactions induced by challenge of previously sensitized animals (Andersen and Maibach, 1985; Oliver et al, 1986). Although modifications to such methods (Maurer and Hess, 1989) and novel procedures (Kashima et al, 1993) have been proposed recently, newer approaches to the prospective identification of sensitizing chemicals have focused upon studies in the mouse. In mice also tests based upon the elicitation of contact reactions in previously sensitized animals have been described. The most thoroughly investigated of these being the mouse ear swelling test (MEST) described by Gad et al (1986). The sensitivity of the MEST has since been questioned (Cornacoff et al, 1988; Dunn et al, 1990) and recently a modified method, the noninvasive mouse ear swelling assay (MESA) has been proposed (Thorne et al 1991a;b). In this test mice are maintained on diets supplemented with vitamin A, a procedure shown previously to potentiate cell-mediated immune responses and contact sensitivity (Miller et al, 1984).

In experimental terms it is possible to define contact sensitization operationally as the ability of a chemical to cause in sensitized animals a dermal inflammatory reaction under conditions where a similar response is not seen in their naive (unsensitized) counterparts. It has proven possible, therefore, to measure contact hypersensitivity reactions as a function of systemic changes resulting from challenge-induced local inflammation. It has been shown that the elicitation of dermal reactions in previously sensitized mice is characterized by an increase in the plasma concentration of both acute phase proteins (APP; haptoglobin and serum amyloid A) and interleukin 6 (IL-6), a cytokine known to induce the synthesis of APP by hepatocytes (Kimber et al, 1989b;1990a). Moreover, contact hypersensitivity in mice is associated also with an elevated concentration of serum histamine (Kimber et al, 1991a). It is important to emphasize, however, that the serological analysis of systemic manifestations of inflammation has not yet been validated in the context of routine predictive testing.

An alternative approach, the murine local lymph node assay, is based upon evaluation of events during the induction rather than elicitation phase of contact sensitization (Kimber and Weisenberger, 1989; Kimber et al 1989a; Kimber and Basketter, 1992). The induction phase of skin sensitization initiated following first encounter with the chemical allergen is characterized by and dependent upon the activation and proliferation of T lymphocytes in lymph nodes draining the site of exposure (Kimber, 1992). Sensitizing activity in the standard local lymph node assay is measured as a function of replicative DNA synthesis induced in draining lymph node cells (LNC) by repeated topical application of the test material (Kimber and Basketter, 1992). On the basis of both internal and inter-laboratory studies this assay has proven reliable and found to correlate closely with guinea pig predictive test methods (Kimber et al, 1990b;1991b; Basketter et al 1991; Kimber and Basketter, 1992). In theory the local lymph node assay can be adapted to measure parameters of lymph node activation other than lymphocyte proliferation and recently the use of IL-6 production by LNC as an alternative readout for the assay has been described (Dearman et al, 1993).

Immune activation in draining lymph nodes is stimulated by the arrival of antigen-bearing dendritic cells (DC). These derive from epidermal Langerhans cells (LC) which, following sensitization, are induced to migrate from the skin to the lymph nodes via afferent lymphatics (Kimber and Cumberbatch, 1992). While in transit, LC are subject to a functional and phenotypic maturation such that on arrival in the draining lymph nodes they have acquired the characteristics of immunostimulatory DC and are able effectively to activate responsive T lymphocytes (Kimber and Cumberbatch, 1992). The migration and maturation of LC following skin sensitization is effected by epidermal cytokines synthesized and secreted by keratinocytes. Tumour necrosis factor α (TNF-α) provides an important signal for LC migration

(Cumberbatch and Kimber, 1992) and, by analogy with in vitro studies, it is likely that granulocyte/macrophage colony-stimulating factor (GM-CSF), possibly acting in concert with interleukin 1 (IL-1), causes the functional maturation of LC (Heufler et al, 1988). As effective skin sensitization is apparently dependent upon the local availability of epidermal cytokines the possibility is raised that contact allergens might be identified on the basis of their ability to provoke cytokine synthesis in the skin. Although it is clear that stimuli other than those provided by chemical allergens will provoke the synthesis by keratinocytes of TNF-α and GM-CSF, there is preliminary evidence that the production of some other epidermal cytokines may be upregulated only following exposure to sensitizing chemicals (Enk and Katz, 1992a;b). Furthermore, it has been shown recently that LC themselves are a source of cytokines and that their synthesis of interleukin 1β (IL-1β) and IL-6 is enhanced following activation (Heufler et al, 1992; Schreiber et al, 1992). The analysis of induced changes in the pattern of cytokine production by LC and keratinocytes may in the future provide a realistic approach to the identification of skin sensitizing activity in vitro.

Finally, the activation of T lymphocytes during the induction phase of skin sensitization and the resultant clonal expansion of allergen-reactive cells can be exploited for the investigation of skin sensitization in vitro. In the presence of the inducing allergen and an appropriate source of antigen presenting cells, cultured LNC prepared from sensitized mice will mount a secondary proliferative response. Based upon this phenomenon lymphocyte blastogenesis assays have been described which provide a method for confirming in vitro immunological sensitization (Robinson, 1989; Gerberick et al, 1990). The antigen presenting activity of DC and epidermal cells (containing LC) is such that they are able also to provoke measurable allergen-specific proliferative responses by lymphoid cells isolated from unsensitized mice (Hauser and Katz, 1988). In theory, therefore, it may prove possible to identify skin sensitizing chemicals on the basis of their ability, when presented by LC or DC, to stimulate primary immune activation in vitro.

Respiratory sensitization

A variety of chemicals, including some diisocyanates, acid anhydrides and reactive dyes, is known to cause occupational respiratory allergy in a proportion of the exposed workforce. Symptoms range from rhinitis to frank asthma. There are, as yet, no fully-validated or widely-applied methods available for the routine prospective analysis of respiratory sensitizing potential. A great deal of progress has been made, however, in the development of methods for the investigation of challenge-induced pulmonary hypersensitivity reactions in previously sensitized guinea pigs (Karol, 1992). It has been shown that inhalation exposure of guinea pigs to known respiratory allergens, in the form of either free chemical or as a conjugate with protein, results in respiratory hypersensitivity (an alteration in respiratory rate) when animals are challenged subsequently with atmospheres containing the relevant chemical-protein conjugate (Karol et al, 1985; Botham et al, 1988; Sarlo and Clark, 1992). More recently it has been reported also that respiratory hypersensitivity reactions can be provoked by inhalation challenge, with free or protein-bound chemical, of guinea pigs sensitized previously by intradermal exposure (Botham et al, 1989; Pauluhn and Eben, 1991; Hayes et al, 1992).

Immediate hypersensitivity reactions, including acute-onset respiratory hypersensitivity, are considered to be effected by IgE antibody. It has been demonstrated that chemical allergens vary with respect to the quality of immune response they induce in mice following topical exposure. Evidence available to date indicates that only respiratory chemical allergens provoke in mice IgE responses characterized by the appearance of specific IgE antibody and an increase in the serum concentration of this immunoglobulin. In contrast, those contact allergens which are known or suspected not to cause respiratory hypersensitivity fail to provoke similar changes (Dearman and Kimber, 1991;1992; Dearman et al, 1992b). It is suggested that the differential responses induced by contact and respiratory chemical allergens reflect the preferential stimulation of Th1- and Th2-type T helper cell responses, respectively (Kimber and Dearman, 1992). Interleukin 4 (IL-4), a cytokine product of Th2 cells, is known to be required for the initiation of IgE responses (Finkelman et al, 1988). These observations form the basis of a novel approach to the assessment of respiratory sensitizing potential in mice, the mouse IgE test, in which activity is evaluated by measurement of changes in the concentration of serum IgE (Dearman et al, 1992a).

Conclusions

Advances in our understanding of the immunobiological processes that result in the induction and elicitation of chemical allergy continue to provide new opportunities to design novel predictive test methods. The challenges for the future include the identification and classification of chemical allergens as a function of the molecular responses they induce and the development of methods for the prospective evaluation of sensitizing potential _in vitro_.

References

Anderson KE, Maibach HI (1985) Guinea pig sensitization assays. An overview. In: Anderson KE, Maibach HI (eds). Contact allergy predictive tests in guinea pigs. Current problems in dermatology Vol.14 Karger, Basel pp263-290.

Basketter DA, Scholes EW, Kimber I, Botham PA, Hilton J, Miller K, Robbins MC, Harrison PTC, Waite SJ (1991) Interlaboratory evaluation of the local lymph node assay with 25 chemicals and comparison with guinea pig test data. Toxicol Meth 1:30-43.

Botham PA, Hext PM, Rattray NJ, Walsh ST, Woodcock DR (1988) Sensitization of guinea pigs by inhalation exposure to low molecular weight chemicals. Toxicol Lett 41:159-173.

Botham PA, Rattray NJ, Woodcock DR, Walsh ST, Hext PM (1989) The induction of respiratory allergy in guinea-pigs following intradermal injection of trimellitic anhydride: a comparison with the response to 2,4-dinitrochlorobenzene. Toxicol Lett 47:25-39.

Cornacoff JB, House RV, Dean JH (1988) Comparison of a radioisotopic method and the mouse ear swelling test (MEST) for contact sensitivity to weak sensitizers. Fundam Appl Toxicol 10:40-44.

Cumberbatch M, Kimber I (1992) Dermal tumour necrosis factor-α induces dendritic cell migration to draining lymph nodes and possibly provides one stimulus for Langerhans cell migration. Immunology 75:257-263.

Dearman RJ, Basketter DA, Kimber I (1992a) Variable effects of chemical allergens on serum IgE concentration in mice. Preliminary evaluation of a novel approach to the identification of respiratory sensitizers. J Appl Toxicol 12:317-323.

Dearman RJ, Hope JC, Hopkins SJ, Debicki RJ, Kimber I (1993) Interleukin 6 (IL-6) production by lymph node cells: An alternative endpoint for the murine local lymph node assay Toxicol Meth, in press.

Dearman RJ, Kimber I (1991) Differential stimulation of immune function by respiratory and contact chemical allergens. Immunology 72: 563-570.

Dearman RJ, Kimber I (1992) Divergent immune responses to respiratory and contact chemical allergens: antibody elicited by phthalic anhydride and oxazolone. Clin Exp Allergy 22:241-250.

Dearman RJ, Spence LM, Kimber I (1992b) Characterization of murine immune responses to allergenic diisocyanates. Toxicol Appl Pharmacol 112 190-197.

Dunn BJ, Rusch GM, Siglin JC, Blaszcak DL (1990) Variability of a mouse ear swelling test (MEST) in predicting weak and moderate contact sensitization. Fundam Appl Toxicol 15:242-248.

Enk AH, Katz SI (1992a) Early molecular events in the induction phase of contact sensitivity. Proc Natl Acad Sci USA 89:1398-1402.

Enk AH, Katz SI (1992b) Identification and induction of keratinocyte-derived IL-10. J Immunol 149:92-95.

Finkelman FD, Katona IM, Urban JF Jr, Holmes J, Ohara J, Tung AS, Sample JG, Paul WE (1988) IL-4 is required to generate and sustain in vivo IgE responses. J Immunol 141:2335-2341.

Gad SC, Dunn BJ, Dobbs DW, Reilly C, Walsh RD (1986) Development and validation of an alternative dermal sensitization test: The mouse ear swelling test (MEST). Toxicol Appl Pharmacol 84:93-114.

Gerberick GF, Ryan CA, Fletcher ER, Sneller DL, Robinson MK (1990) An optimized lymphocyte blastogenesis assay for detecting the response of contact sensitized or photosensitized lymphocytes to hapten or prohapten modified antigen presenting cells. Toxic In Vitro 4:289-292.

Hauser C, Katz SI (1988) Activation and expansion of hapten- and protein-specific T helper cells from non-sensitized mice. Proc Natl Acad Sci USA 85:5625-5628.

Hayes JP, Daniel R, Tee RD, Barnes PJ, Chung KF, Newman Taylor AJ (1992) Specific immunological and bronchopulmonary responses following intradermal sensitization to free trimellitic anhydride in guinea pigs. Clin Exp Allergy 22:694-700.

Heufler C, Koch F, Schuler G (1988) Granulocyte/macrophage colony-stimulating factor and interleukin 1 mediate the maturation of murine epidermal Langerhans cells into potent immunostimulatory dendritic cells. J Exp Med 167:700-705.

Heufler C, Topar G, Koch F, Trockenbacher B, Kampgen E, Romani N, Schuler G (1992) Cytokine gene expression in murine epidermal cell suspensions. Interleukin 1β and macrophage inflammatory protein 1α are selectively expressed in Langerhans cells but are differentially regulated in culture. J Exp Med 176:1221-1226.

Kimber I, Hilton J, Weisenberger C (1989a) The murine local lymph node assay for identification of contact allergens: a preliminary evaluation of in situ measurement of lymphocyte proliferation. Contact Derm 21:215-220.

Kimber I, Ward RK, Shepherd CJ, Smith MN, McAdam KPWJ, Raynes JG (1989b) Acute-phase proteins and the serological evaluation of experimental contact sensitivity in the mouse. Int Arch Allergy Appl Immunol 89:149-155.

Kimber I, Weisenberger C (1989) A murine local lymph node assay for the identification of contact allergens. Assay development and results of an initial validation study. Arch Toxicol 63:274-282.

Maurer Th, Hess R (1989) The maximization test for skin sensitization potential - updating the standard protocol and validation of a modified protocol. Fd Chem Toxic 27:807-811.

Miller K, Maisey J, Malkovsky M (1984) Enhancement of contact sensitization in mice fed a diet enriched in vitamin A acetate. Int Arch Allergy Appl Immunol 75:120-125.

Oliver GJA, Botham PA, Kimber I (1986) Models for contact sensitization - novel approaches and future developments. Br J Dermatol 115:53-62.

Pauluhn J, Eben A (1991) Validation of a non-invasive technique to assess immediate or delayed onset of airway hypersensitivity in guinea-pigs. J Appl Toxicol 11:423-431.

Robinson MK (1989) Optimization of an in vitro lymphocyte blastogenesis assay for predictive assessment of immunologic responsiveness to contact sensitizers. J Invest Dermatol 92:860-867.

Sarlo K, Clark ED (1992) A tier approach for evaluating the respiratory allergenicity of low molecular weight chemicals. Fundam Appl Toxicol 18:107-114.

Schreiber S, Kilgus O, Payer E, Kutil R, Elbe A, Mueller C, Stingl G (1992) Cytokine pattern of Langerhans cells isolated from murine epidermal cell cultures. J Immunol 149:3525-3534.

Thorne PS, Hawk C, Kaliszewski SD, Guiney PD (1991a) The noninvasive mouse ear swelling assay. I. Refinements for detecting weak contact sensitizers. Fundam Appl Toxicol 17:790-806.

Thorne PS, Hawk C, Kaliszewski SD, Guiney PD (1991b) The noninvasive mouse ear swelling assay. II. Testing the contact sensitizing potency of fragrances. Fundam Appl Toxicol 17:807-820.

Karol MH (1992) Occupational asthma and allergic reactions to inhaled compounds. In: Miller K, Turk J, Nicklin S (eds:) Principles and practice of immunotoxicology. Blackwell Scientific Publications, Oxford pp228-241.

Karol MH, Stadler J, Magreni C (1985) Immunotoxicologic evaluation of the respiratory system: Animal models for immediate and delayed-onset pulmonary hypersensitivity. Fundam Appl Toxicol 5:459-472.

Kashima R, Oyake Y, Okada J, Ikeda Y (1993) Studies of new short-period method for delayed contact hypersensitivity assay in the guinea pig. 1. Development and comparison with other methods. Contact Derm 28:235-242.

Kimber I (1992) Contact sensitivity. In: Miller K, Turk J, Nicklin S (eds) Principles and practice of immunotoxicology. Blackwell Scientific Publications, Oxford pp104-124.

Kimber I, Basketter DA (1992) The murine local lymph node assay: A commentary on collaborative studies and new directions. Fd Chem Toxic 30:165-169.

Kimber I, Cumberbatch M (1992) Dendritic cells and cutaneous immune responses to chemical allergens. Toxicol Appl Pharmacol 117:137-146.

Kimber I, Cumberbatch M, Coleman JW (1991a) Serum histamine and the elicitation of murine contact sensitivity. J Appl Toxicol 11:339-342.

Kimber I, Cumberbatch M, Humphreys M, Hopkins SJ (1990a) Contact hypersensitivity induces plasma interleukin 6. Int Arch Allergy Appl Immunol 92:97-99.

Kimber I, Dearman RJ (1992) The mechanisms and evaluation of chemically induced allergy. Toxicol Lett 64/65:79-84.

Kimber I, Hilton J, Botham PA (1990b) Identification of contact allergens using the murine local lymph node assay: Comparisons with the Buehler occluded patch test in guinea pigs. J Appl Toxicol 10:173-180.

Kimber I, Hilton J, Botham PA, Basketter DA, Scholes EW, Miller K, Robbins MC, Harrison PTC, Gray TJB, Waite SJ (1991b) The murine local lymph node assay: results of an inter-laboratory trial. Toxicol Lett 55:203-213.

Risk Assessment of Sensitizing Agents

G. Frank Gerberick
Immunology Section, Human & Environmental Safety Division, The Procter & Gamble Company, Miami Valley Laboratories, Cincinnati, Ohio, USA

Introduction

Allergic contact sensitization is the term given to a cell-mediated immunological response to chemicals that contact and penetrate the skin. As a manufacturer of a variety of consumer and health care products that come in contact with the skin, one of our primary responsibilities to the consumer and our workers is to ensure that our products will not cause allergic contact sensitization. Product and ingredient safety testing should consider all types of human exposure situations. This includes the manufacture and distribution of the product as well as consumer use and reasonably foreseeable misuse. To evaluate the contact sensitization potential of a new ingredient prior to exposure of employees or consumers, we have developed a multi-step risk assessment process. A comprehensive review of this contact sensitization risk assessment process has been published previously (Robinson et al. 1989).

The basic process we employ for evaluating the skin sensitization risk of a new product ingredient is considered a comparative toxicological approach (Figure 1). It is the potential for an adverse effect to occur in humans exposed during manufacturing or product use that is being determined. This approach incorporates the assessment of both toxicity and exposure of the new ingredient followed by a comparison of these data with the toxicity and exposure data for known safe and potentially unsafe chemicals and product formulations. In assessing skin sensitization risk, the exposure level to the material as well as the material's inherent toxicity is determined. The key factors considered in exposure assessment for skin sensitization include chemical dose, chemical biovailability, concentration in dose/unit surface area (Upadhye, Maibach, 1992), duration of exposure, body location, presence of any skin penetration aid or vehicle, primary skin irritation potential, and the extent of occlusion of the exposed skin. Relevant to determining conditions of exposure for consumer use of products is the gathering of habits and practices data for each chemical or product type. These data include the frequency and duration of each product- use activity and the dose of product inherent to that particular habit or practice. All exposure information is incorporated into the planning of, and evaluation of the results from, predictive skin sensitization tests.

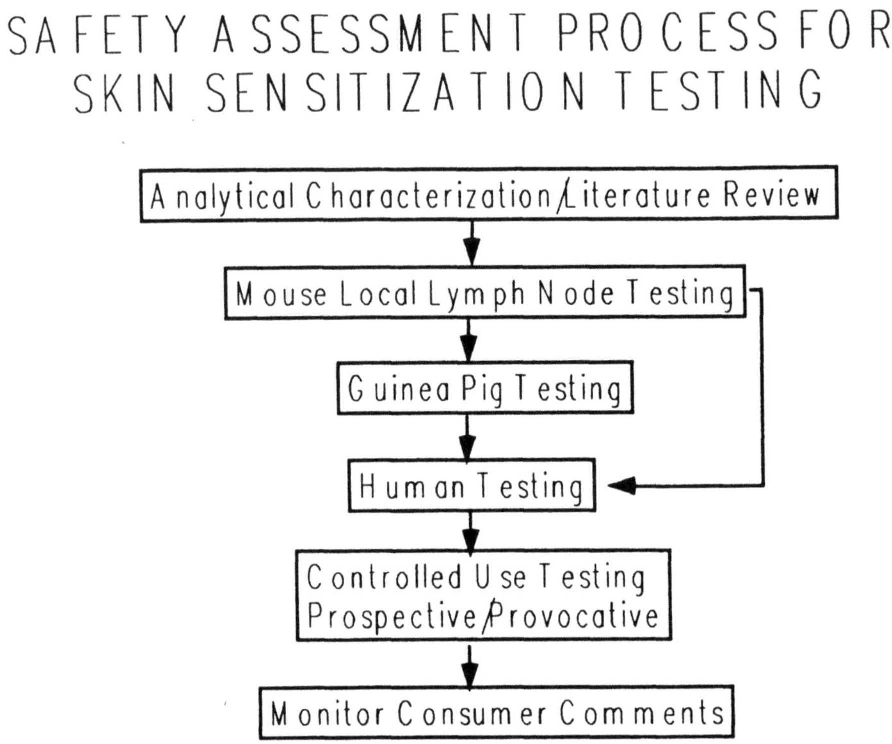

Figure 1. Risk assessment process for skin sensitizing agents

Analytical characterization and literature review

The first step of the risk assessment process involves review of the analytical data on the material and review of the published literature for any evidence suggestive of skin sensitization potential. Obviously, if adequate published literature on sensitization potential of the material of interest is available, preclinical testing is not conducted. Appropriate analytical information may come from the supplier or from the product development group. Important in the risk assessment process is the identification of materials that may raise issues regarding sensitization potential based on their physical and chemical properties (e.g., low molecular weight with functional groups that may interact with cell surface proteins) (Dupuis, Benezra, 1982). In addition, any expected contaminants or reaction by-products with known sensitization properties need to be evaluated prior to initiating pre-clinical testing. If any such materials are identified, analytical work can be conducted to have the allergic contaminants removed or controlled to safe, non-sensitizing levels.

Preclinical skin sensitization testing

Preclinical skin sensitization testing is recommended to provide safety data on new ingredients or formulations which lack adequate sensitization data. The initial predictive skin sensitization tests we recommended are the mouse local lymph node test (MLLNA) (Gerberick et al. 1992) and/or Buehler guinea pig skin sensitization test (Robinson et al. 1989).

The Buehler method has been used by us for more than 25 years for assessing inherent contact sensitization potential of materials prior to human patch test exposure. When properly conducted and interpreted, the Buehler test has demonstrated its ability to detect chemicals with moderate-to-strong sensitization potential as well as those with relatively weak sensitization potential (Robinson et al. 1990). The standard Buehler test protocol uses an occluded topical patch technique for the induction and elicitation of contact sensitization (Robinson et al. 1990; Ritz, Buehler, 1980). In addition to using the Buehler test in assessing inherent contact sensitization potential, the test has also been useful in addressing questions of sensitization potency or antigenic cross-reactivity of structurally similar chemicals, identification of chemical contaminants in raw materials (Ritz et al. 1975), or comparison of the sensitization potential of a raw material produced by different chemical processes. Very importantly, the Buehler test is recognized by regulatory agencies worldwide (Botham et al. 1991).

Recently, Kimber and others have developed and proposed the use of a mouse local lymph node assay for use in screening materials for their contact sensitization potential (Kimber, Weisenberger, 1989; Kimber et al. 1989; Kimber et al. 1986). The MLLNA measures the induction phase of contact sensitization by determining the ability of a chemical to stimulate the incorporation of radiolabeled thymidine into proliferating lymph-node cells in a small number of animals. The MLLNA offers the advantages of 1) requiring less time for completion, 2) an objective endpoint, 3) requiring approximately half the number of animals, and 4) being less costly than most currently employed guinea pig test methods. Results of our validation program (Gerberick et al. 1992), along with validation studies conducted by Kimber (Kimber, Weisenberger, 1989; Kimber et al. 1989), indicate that the assay is able to detect strong, moderate and some weak sensitizers. Moreover, interlaboratory comparisons have demonstrated that the MLLNA is a robust and reliable method for the identification of moderate and strong contact allergens and demonstrates good concordance with guinea pig test data (Kimber I, 1990; Kimber I, 1991; Basketter et al. 1991). Based on these results, we believe the assay is a useful test to incorporate into a scheme for contact sensitization risk assessment.

The Buehler guinea pig test and MLLNA are used as preliminary screens to determine inherent skin sensitization potential under exaggerated topical exposure conditions. The MLLNA results and/or guinea pig results, positive or negative, and the anticipated human exposure should be incorporated into an initial risk assessment. The MLLNA, as is the case with the Buehler test, can often define consumer skin sensitization risk. However, the final consumer skin safety assessment should incorporate data from a series of controlled human tests, the HRIP test, provocative or extended use tests, and postmarket monitoring of skin-related consumer comments.

Human repeat insult patch test (HRIPT)

Provided that the risk of inducing contact sensitization in volunteer panelists is judged to be minimal (based on assessment of preclinical sensitization test data), human skin sensitization testing is conducted. Generally, 80-120 test subjects are employed for HRIP testing (Stotts, 1980). The HRIPT protocol can provide an exaggeration of anticipated product use/misuse exposures through an extended duration of exposure, testing higher than use concentrations, minor skin irritation of the test material, and, for many product types, through the occluded patch. The procedure allows the detection of preexisting sensitization to the test material (chemical or final formulation) as confirmed by persistent skin reactions early in the induction period. Any positive or questionable response during the induction phase or only during the challenge phase is investigated further by a rechallenge patch study. It is important to note, however, that the HRIP test is not done to confirm sensitization potential but to confirm ingredient and/or product safety under exaggerated conditions relative to anticipated consumer exposure.

In addition to the important preclinical data generated, confirmation that humans will not adversely respond provides additional and essential data for safety programs designed to assess the skin sensitization risk of new chemicals and products. The data generated in HRIP testing are combined with data from the mouse and/or guinea pig, and worker and consumer exposure and compared with the data on relevant benchmark materials to give a revised assessment of risk. Depending on the sensitization data generated, the exposure calculations, and comparison with benchmarks, a decision may be reached to either stop development (Robinson et al. 1991) or to obtain additional clinical data in carefully controlled use tests under more realistic exposure conditions before proceeding with a new chemical ingredient for product development.

Provocative use testing of sensitized subjects

Individuals showing patch test sensitization reactions to a material may be asked to participate in extended provocative use testing of a product containing the putative sensitizing agent. The individual(s) may have been sensitized by HRIPT or identified by diagnostic patch test screening. Because of the relative rarity of sensitization in available study populations, the numbers of individuals in provocative use tests usually are small. However, test group size is not the critical factor, since the question being asked is whether elicitation of allergic contact dermatitis will occur when the test product is used by someone known to be sensitized to the formula or ingredient under patch-test conditions. A provocative use test is considered to be of minimal risk as long as there is sufficient exaggeration inherent in the HRIPT or exposure is highly localized. However, before conducting any use test, a formal risk assessment is done to verify there is minimal risk of eliciting allergic reactions. The risk assessment should include an evaluation of patch-test versus in-use exposure concentrations, frequency of exposure, a comparison with results obtained previously on benchmark formulations, and, if necessary, the results of an open application test. At the conclusion of the use test, even if no adverse skin effects were observed, a follow-up diagnostic patch test should be done to confirm that the test subject is still reactive under patch test conditions.

Extended prospective product use testing

The extended prospective use test is a method used to determine the potential for a product to induce sensitization or both induce sensitization and elicit allergic skin reactions under typical conditions of product use. As in other clinical studies, a formal risk assessment is done, written informed consent is obtained, and expert dermatologist monitoring and assessment of any skin reactions is included in the protocol. The number of test subjects and the length of a prospective use test is dependent on the product being evaluated. Generally, prospective use tests can include 100 to 500 subjects and extend three to six months. Diagnostic patch testing at the conclusion of the study should be done to verify lack of patch-test reactivity or to identify subclinical sensitization responses (i.e., subjects with induced sensitization but no clinical dermatitis). Any evidence of induced sensitization would necessitate consideration of withholding a developmental product (or ingredient) from the market.

Prior to marketing, all preclinical and clinical data are incorporated into a formal risk assessment for the chemical and product formulation in question. In the risk assessment, the data should again be compared with appropriate benchmarks. In some instances it is appropriate to obtain outside expert opinion and concurrence for any new ingredient that shows evidence of inducing skin sensitization in humans, even if only under exaggerated conditions of skin exposure.

Monitoring and follow-up of consumer comments

Contact sensitization risk assessment does not end with the decision to market the product. All comments related to skin safety that are received from consumers are evaluated and, if appropriate, investigated by medical personnel. In some instances, follow-up may involve diagnostic patch testing of the product and/or selected product ingredients to determine the relationship, if any, between the product in question and the skin reaction. Another method that can be used to obtain skin sensitization data among consumers is to conduct diagnostic patch test surveys and distribute questionnaires about skin effects among consumers who have used the product. Control subjects (non-users) of approximately equal age, gender, and other comparable factors, as well as control test substances and products, should be included in these studies. The data are useful not only in confirming the validity of the risk assessment but also in detecting potential problems. The post-market monitoring program provides both the ongoing assurance of product safety as well as additional benchmark data for comparison with other ingredient and product initiatives that follow.

Summary

This review describes an approach that has been used to assess the skin sensitization risk of new product ingredients prior to and after marketing. The risk assessment process utilizes a comparative toxicological approach in which data on the inherent toxicity of a material and the exposure to it through manufacturing or consumer use or foreseeable misuse are integrated and compared with data generated by 'benchmark' materials of similar chemistry or product application, or both. This approach has been valuable in providing an accurate assessment of skin sensitization potential and the basis for eventual safe marketing of a wide range of consumer household and personal care products and topical pharmaceuticals.

References

Basketter DA, Scholes EW, Kimber I, Botham PA, Hilton J, Miller K, Robbins MC, Harrison PTC, Waite SJ (1991) Interlaboratory evaluation of the local lymph node assay with 25 chemicals and comparison with guinea pig test data. Toxicol Methods 1: 30-43

Botham PA, Basketter DA, Maurer T, Mueller D, Potokar M, Bontinck WJ (1991) Skin sensitization--A critical review of predictive test methods in animals and man. Food Chem Toxicol 29: 275-286

Dupuis, G, Benezra, C (1982) Allergic Contact Dermatitis to Simple Chemicals. A Molecular Approach. Marcel Dekker, New York NY

Gerberick GF, House RV, Fletcher ER, Ryan CA (1992) Examination of the local lymph node assay for use in contact sensitization risk assessment. Fundam Appl Toxicol 19: 438-445

Kimber I, Hilton J, Botham PA (1990) Identification of contact allergens using the murine local lymph node assay: comparisons with the Buehler occluded patch test in guinea pigs. J Appl Toxicol 10: 173-180

Kimber I, Hilton J, Botham PA, Basketter DA, Scholes EW (1991) The murine local lymph node assay: results of an inter-laboratory trial. Toxicol Lett 55: 203-213

Kimber I, Mitchell JA, Griffin AC (1986) Development of a murine local lymph node assay for the determination of sensitizing potential. Food Chem Toxicol 24: 585-586

Kimber I, Hilton J, Weisenberger C (1989) The murine local lymph node assay for identification of contact allergens: a preliminary evaluation of in situ measurement of lymphocyte proliferation. Contact Dermatitis 21: 215-220

Kimber I, Weisenberger C (1989) A murine local lymph node assay for the identification of contact allergens. Arch Toxicol 63: 274-282

Ritz HL, Connor DS, Sauter ED (1975) Contact sensitization of guinea-pigs with unsaturated and halogenated sultones. Contact Dermatitis 1: 349-351

Ritz HL, Buehler EA (1980) Planning, conduct, and interpretation of guinea pig contact sensitization. In: Drill VA, Lazar P (eds) Current concepts in cutaneous toxicity. Academic Press, New York, pp 25-40

Robinson MK, Stotts J, Danneman PJ, Nusair TL, Bay PHS (1989) A risk assessment process for allergic contact sensitization. Food Chem Toxicol 27: 479-489

Robinson MK, Nusair TL, Fletcher ER, Ritz HL (1990) A review of the Buehler guinea pig skin sensitization test and its use in a risk assessment process for human sensitization. Toxicology 61: 91-107

Robinson MK, Parsell KW, Breneman DL, Cruze CA (1991) Evaluation of the primary skin irritation and allergic contact sensitization potential of transdermal triprolidone. Fundam Appl Toxicol 17: 103-119

Stotts J (1980) Planning, conduct, and interpretation of human predictive sensitization patch tests. In: Drill VA, Lazar P (eds) Current Concepts in Cutaneous Toxicity. Academic Press, New York, pp 41-53

Upadhye MR, Maibach HI (1992) Influence of area of application of allergen on sensitization in contact dermatitis. Contact Dermatitis 27: 281-286

Developmental Toxicity

(Chair: R. Wiger, Norway)

Nuclear retinoic acid receptors and regulation of gene expression

E. Ruberte
Department of Human Anatomy, University of Oxford. South Parks Road OX1 3QX Oxford, U.K.

Retinoids (vitamin A derivatives) play an important role in vertebrate morphogenesis and cellular differentiation. The mechanism underlying the wide spectrum of effects exerted by retinoids is still poorly understood. However, the discovery in recent years of multiple retinoid nuclear receptors that act as transcriptional trans-regulators, and the diversity of response elements to which they bind to regulate the transcriptional activity of specific sets of target genes, represents a great advance in our understanding of the mechanisms underlying the complexity of retinoid effects.

Two families of nuclear retinoic acid receptors: RARs and RXRs

Retinoic acid receptors (RARs)

Three types of retinoic acid receptors (RAR-α, RAR-β and RAR-γ) (Benbrook et al. 1988, Brand et al. 1988, Giguère et al. 1987, Krust et al. 1989, Petkovich et al. 1987, Zelent et al. 1989) which bind all-trans-retinoic acid and 9-cis-retinoic acid with high affinity, and which belong to the nuclear receptor superfamily, have been cloned. The amino acid sequence conservation of a particular RAR type is greater across species than that observed for the three RAR types in a given species (Fig. 1).

The comparison of the deduced amino acid sequence of the three RAR with each other, as well as with the other members of the nuclear receptor superfamily has lead to the division of their primary sequence into six regions (A-F) which show different degree of conservation (Green et al. 1988) (Fig.1).

The most highly conserved region C (95% homology between the three RARs) corresponds to the 66 amino acid sequence which is conserved throughout the superfamily of nuclear receptors, it contains two putative "zinc-fingers" and corresponds to the DNA binding domain, which is responsible for response element recognition.

Region E which contains the ligand binding domain is also highly conserved (85-90%), it contains in addition a ligand-dependent trans-activation function and a dimerization interface.

Region B is also highly conserved between the three RARs (75-86%), but not between the other members of the nuclear receptor superfamily, suggesting that it may perform a RAR-specific function.

Regions A and F differ markedly among the three receptors. Region A may also contain an activation function specific for some promoters. No function has yet been ascribed to the RAR C-terminal region which is lacking in some nuclear receptors (e. g. RXRs).

Retinoic acid receptor isoforms

Moreover, several isoforms have been identified for each RAR (Kastner et al. 1990, Leroy et al. 1991, Zelent et al. 1991). These isoforms differ in the sequence corresponding to their 5'-untranslated region (5'-UTR) and/or A region. The B to F sequences are common to all isoforms of a given RAR subtype. These isoforms are produced by transcription from two distinct promoters which operate in each RAR gene and by alternative splicing (Kastner et al. 1990, Leroy et al. 1991, Zelent et al. 1991). Two major isoforms exist for RAR-α and γ (Giguère et al. 1990, Kastner et al. 1990, Leroy et al. 1991). In the case of RAR-β there are 4 major isoforms (Zelent et al. 1991, Nagpal et al. 1992).

RAR isoforms transcribed from the upstream promoter P1 (RAR-α1, β1 and β3, and γ1), and isoforms transcribed from the downstream promoter P2 (RAR-α2, β2 and β4 and γ2) are designated as type 1 and 2 isoforms respectively. There are several similarities between RAR isoforms of a given class, such as A region relatedness and promoter structure (e.g. type 2 isoforms contain a RA response element) (Kastner et al. 1990, Leroy et al. 1991).

The overall similarity in the organization of the three RAR genes, taken together with the existence of amino acid homologies between the A regions of the isoforms transcribed from the upstream and downstream promoters, respectively, suggest that all the three RAR genes have evolved from a common ancestral RAR gene which already possessed two promoters controlling the transcription of two major isoforms (Kastner et al. 1991).

Retinoid X receptors (RXRs)

The second class of nuclear retinoic acid receptors, the retinoid X receptors (RXRs) (Mangelsdorf et al. 1990), specifically bind and are activated by 9-cis-RA. Three related genes (RXR-α, β and γ) have been identified. Sequence comparisons reveal that a given RXR type is also highly conserved across species (Fig. 1).

Similarly to RARs and other members of the nuclear receptor superfamily, RXR primary sequence can be divided into several regions designated A-E (the RXRs do not have a F region counterpart). Comparison of the amino acid sequence between the different RXRs shows a high degree of homology in regions C and E, whereas the regions A, B and D are more specific of a given RXR type (Fig. 1). There exist some preliminary evidence of existence of RXR-β isoforms that differ in the amino terminal A/B region (Leid et al. 1992, Fleischhauer et al. 1992).

Expression patterns of RARs and RXRs

The expression patterns of RARs and RXRs have been studied by Northern blot analysis (Kastner et al. 1990, Leroy et al. 1991, Zelent et al. 1989, Zelent et al. 1991, Mangelsdorf et al. 1990, Mangelsdorf et al. 1992) and in situ hybridization (Dollé et al. 1990, Ruberte et al. 1990, Ruberte et al. 1991, Ruberte et al. 1992, Ruberte et al. 1993).

The multiple RAR isoforms display differential expression patterns in mouse adult tissues. RAR-α1 is expressed in all tissues examined and has therefore the expression pattern of a housekeeping gene which is in agreement with the characteristic features of its promoter (Brand et al. 1990). The other RAR isoforms show a more restricted pattern of expression in adult mouse tissues (Kastner et al. 1990, Leroy et al. 1991, Zelent et al. 1989, Zelent et al. 1991).Most notably, RAR-γ1 expression is almost exclusively restricted to skin (Kastner et al. 1990, Zelent et al. 1989).

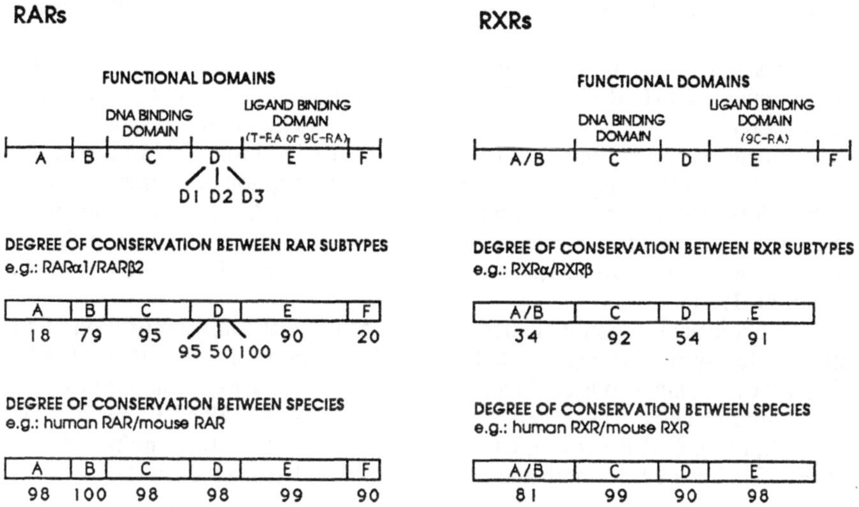

Fig. 1. Comparison of the amino acid sequences of RAR (a) and RXR (b) receptors among themselves, and between mouse and human. The amino acid sequence has been divided into A-F regions and the homologies are given in percentages.

The three RXRs display also different expression patterns. In the adult mouse, RXR-α is expressed at high levels in liver, skin and intestine, and at lower levels in a number of other tissues (Mangelsdorf et al. 1992). RXR-β appears to show a rather ubiquitous expression and RXR-γ is expressed at high levels in muscle and at lower levels in brain, liver and kidney (Mangelsdorf et al. 1992).

In the mouse embryo the transcripts of he RAR-α, β and γ genes display specific spatio-temporal patterns of distribution (Dollé et al. 1990, Ruberte et al. 1990, Ruberte et al. 1991, Ruberte et al. 1992, Ruberte et al. 1993). RAR-α is expressed almost ubiquitously whereas RAR-β and γ exhibit more restricted expression patterns. Among the most interesting observations are a general correlation between RAR-γ expression and appearance of cartilage and squamous keratinizing epithelia, the complementary expression patterns of RAR-β and RAR-γ in many embryonic structures, and expression of RAR-β in sites of programmed cell death.

RXR are also expressed in murine embryonic tissues. Northern blot analysis has revealed that all three RXRs are expressed at all stages of mouse gestation from day 10.5 to birth, however only RXR-γ, which is expressed in the myotomes, shows a clear pattern of transcript distribution by in situ hybridization, suggesting that these receptors may be expressed in a rather ubiquitous manner and at very low levels (unpublished results). Interestingly, RXR-γ in the chicken is expressed in spinal ganglia and nerves as well as neural crest cells (Rowe et al. 1991).

Regulation of transcription by the multiple RARs and RXRs.

The expression of a number of genes is known to be controlled by RA at the transcriptional level (LaRosa and Gudas 1988, Nicholson et al. 1990). Some of these genes contain specific cis-acting elements which mediate the RA response, some of these elements, however, do not bind RARs or RXRs in vitro (Nicholson et al. 1990, Okazawa et al. 1991) suggesting that control of transcription is indirect and may be mediated through other promoter-bound factors. Direct binding of RARs and /or RXRs to the response elements has been, however, demonstrated for a number of RA target genes (de Thé et al. 1990, Duester et al. 1991, Leid et al. 1992, Leroy et al. 1991, Smith et al. 1991).

The retinoic acid response element (RARE) consists of a direct repetition of a core motif (or closely related degenerated motifs) in the form of 5'-PuG(GT)TCA-(X)n-PuG(GT)TCA-3', and is present in a number of genes whose expression is modified by RA in cultured cells. These genes include the RAR-β (β2RARE, n=5), RAR-α (α2RARE, n=5), CRBP I (n=2), CRABP II (RARE1 and RARE2, n=2) and laminin B1 (n=4) (De Luca 1991, Lamb and MacKnight 1991, Leroy et al. 1991).

RAR/RXR heterodimers

Nuclear receptors are believed to bind to the response elements as dimers. The presence of repeated motifs in the RAREs suggested that RARs and RXRs could also bind to the response elements as dimers, and it has actually been shown that RARs and RXRs form heterodimers in vitro, and their stability is increased by binding to a RARE (Laudet and Stehelin 1992, Leid et al. 1992, Marks et al. 1992, Yu et al. 1991). The RAR/RXR heterodimers bind more efficiently in vitro than the respective homodimers to a number of response elements whose sequence, spacement and orientation may vary (Laudet and Stehelin 1992, Leid et al. 1992, Marks et al. 1992). Moreover, RXR is also able to form heterodimers with other receptors (TR, VDR, COUP-TF). The heterodimerization interfaces are localized mainly in the region E of each receptor and contain multiple repeated heptads of hydrophobic amino acids (Leid et al. 1992).The functional significance of RAR/RXR heterodimers could be the generation of diversity in the control of transcription with a reduced number of regulatory proteins.

Conclusions

Studies in the last five syears have revealed a great complexity of the molecular mechanisms involved in the retinoic acid signalling pathway. The high diversity in the effects of retinoids results from the combinatorial effects of the many elements exerting their effects on the control of gene transcription at different levels (the existence of multiple RARs and RXRs, the formation of RAR/RXR heterodimers which results in a large combinatorial number of binding species, the multiplicity and polymorphism of retinoic acid response elements). These combinatorial interactions may account for the multiple effects of retinoids on many biological systems, however, post-transcriptional and post-translational modifications may also be involved.

References

Benbrook D, Lernhardt E, Pfahl M (1988) A new retinoic acid receptor from a hepatocellular carcinoma. Nature 333:669-672

Brand N, Petkovich M, Chambon P (1990) Characterization of a functional promoter for the human retinoic acid receptor-alpha (hRAR–α). N.A.R. 18:6799-6806

Brand N, Petkovich M, Krust A, Chambon P, de Thé H, Marchio A, Tiollais P, Dejean A (1988) Identification of a second human retinoic acid receptor. Nature 332:850-853

De Luca LM (1991) Retinoids and their receptors in differentiation, embryogenesis and neoplasia. FASEB J. 5:2924-2933

de Thé H,Vivanco Ruiz MM,Tiollais P,Stunnenberg H, Dejean A (1990) Identification of a retinoic acid responsive element in the retinoic acid receptor β gene. Nature 343:177-180

Dollé P,Ruberte E,Leroy P,Morriss-Kay GM, Chambon P (1990) Retinoic acid receptors and cellular retinoid binding proteins. I. A systematic study of their differential pattern of transcription during mouse organogenesis. Development 110:1133-1151

Duester G,Shean ML,McBride MS, Stewart MJ (1991) Retinoic acid response element in the human alcohol dehydrogenase gene *ADH3* : implications for regulation of retinoic acid synthesis. Mol. Cell. Biol. 11:1638-1646

Fleischhauer K,Park JH,Disanto JP,Marks M,Ozato K, Yang SY (1992) Isolation of a full length cDNA clone encoding a N-terminally variant form of the human retinoid X receptor β. N. A. R. 20:1801

Giguère V,Ong ES,Segui P, Evans RM (1987) Identification of a receptor for the morphogen retinoic acid. Nature 330:624-629

Giguère V,Shago M,Zirngibl R,Tate P,Rossant J, Varmuza S (1990) Identification of a novel isoform of the retinoic acid receptor gamma expressed in the mouse embryo. Mol. Cell. Biol. 10:2335-2340

Green S,Issemann I, Sheer I (1988) A versatile in vivo and in vitro eukaryotic expression vector for protein engineering. Nucleic Acids Res. 16:369

Hoffmann B,Lehmann JM,Zhang X,Hermann T,Husmann M,Graupner G, Pfahl M (1990) A retinoic acid receptor-specific element controls the retinoic acid receptor β promoter. Mol. Endo. 4:1727-1736

Kastner P,Krust A,Mendelsohn C,Garnier JM,Zelent A,Leroy P,Staub A, Chambon P (1990) Murine isoforms of retinoic acid receptor-γ with specific patterns of expression. Proc.natn.Acad.Sci.U.S.A. 87:2700-2704

Krust A,Kastner P,Petkovich M,Zelent A, Chambon P (1989) A third human retinoic acid receptor, hRAR-γ. Proc. Natl. Acad. Sci. U. S. A., 86:5310-5314

Lafyatis R,Kim SJ,Angel P,Roberts AB,Sporn MB,Karin M, Wilder RL (1990) Interleukin-1 stimulates and all-trans-retinoic acid inhibits collagenase gene expression through its 5' AP-1 binding site. Mol. Endrocrinol. 4:973-980

Lamb P, L. MS (1991) Diversity and specificity in transcriptional regulation: the benefits of heterotypic dimerization. TIBS 16:417-422

LaRosa GJ, Gudas LJ (1988) Early retinoic acid induced F9 teratocarcinoma stem cell gene ERA-1: alternate splicing creates transcripts for a homeobox containing protein and one lacking the homeobox. Molec.cell.Biol. 8:3906-3917

Laudet V, Stehelin D (1992) Nuclear receptor of the thyroid hormone receptor and retinoic acid receptor subfamily bind some DNA response elements in combination with auxiliary proteins, which turn out to be also nuclear receptors. Current Biol. 2:293-295

Leid M,Kastner P,Lyons R,Nakshatri H,Saunders M,Zacharewsky T,Chen J-Y,Staub A,Garnier J-M,Mader S, Chambon P (1992) Purification, cloning and RXR identity of the HeLa cell factor with which RAR or TR heterodimerizes to efficiently bind target sequences. Cell (in press)

Leroy P,Krust A,Zelent A,Mendelsohn C,Garnier JM,Kastner P,Dierich A, Chambon P (1991) Multiple isoforms of the mouse retinoic acid receptor α are generated by alternative splicing and differential induction by retinoic acid. EMBO J. 10:59-69

Mangelsdorf DJ,Borgmeyer U,Heyman RA,Zhou JY,Ong ES,Oro AE,Kakizuka A, Evans R (1992) Characterization of three RXR genes that mediate the action of 9-cis-retinoic acid. Genes and Dev. 6:329-344

Mangelsdorf DJ,Ong ES,Dyck JA, Evans RM (1990) Nuclear receptor that identifies a novel retinoic acid response pathway. Nature 345:224-229

Marks MS,Hallenbeck PL,Nagata T,Segars JH,Appella E,Nikodem VM, Ozato K (1992) H-2RIIBP (RXR-β) heterodimerization provides a mechanism for combinatorial diversity in the regulation of retinoic acid and thyroid hormone responsive genes. EMBO J 11:1419-1435

Nagpal S,Zelent A, Chambon P (1992) RAR-β4, a retinoic acid receptor isoform is generated from RAR-β2 by alternative splicing and usage of a CUG initiator codon. Proc. Natl. Acad. Sci. U.S.A. 89:2718-2722

Nicholson RC,Mader S,Nagpal S,Leid M,Rochette-Egly C, Chambon P (1990) Negative regulation of the rat stromelysin gene promoter by retinoic acid is mediated by an AP-1 binding site. EMBO J. 13:4443-4454

Okazawa H,Okamoto K,Ishino F,Ishino-Kaneko T,Takeda S,Toyoda Y,Muramatsu M, Hamada H (1991) The *oct-3* gene, a gene for an embryonic transcription factor, is controlled by a retinoic acid repressible enhancer. EMBO J. 10:2997-3005

Petkovich M,Brand NJ,Krust A, Chambon P (1987) A human retinoic acid receptor which belongs to the family of nuclear receptors. Nature 330:444-450

Rowe A,Eager NSC, Brickell PM (1991) A member of the RXR nuclear receptor family is expressed in neural crest derived cells of the developing chick peripheral nervous system. Development 111:771-778

Ruberte E,Dollé P,Chambon P, Morriss-Kay G (1991) Retinoic acid receptors and cellular retinoid binding proteins II. Their differential pattern of transcription during early morphogenesis in mouse embryos. Development 111:45-60

Ruberte E,Dollé P,Krust A,Zelent A,Morriss-Kay G, Chambon P (1990) Specific spatial and temporal distribution of retinoic acid receptor gamma transcripts during mouse embryogenesis. Development 108:213-222

Ruberte E,Friederich V,Morriss-Kay G, Chambon P (1992) Differential distribution patterns of CRABP I and CRABP II transcripts during mouse embryogenesis. Development 115:973-987

Ruberte E,Friederich V,Chambon P, Morriss-Kay G (1993) Retinoic acid receptors and cellular retinoid binding proteins III, Their differential transcript distribution during mouse nervous system development. Development 118:267-282

Smith WC,Nakshatri H,Leroy P,Rees J, Chambon P (1991) A retinoic acid response element is present in the mouse cellular retinol binding protein I (mCRBP I) promoter. EMBO J 10:

Sukov HM,Murakami KK, Evans RM (1990) Characterization of an autoregulated response element in the mouse retinoic acid receptor type ß gene. Proc. Natl. Acad. Sci. U.S.A . 87:5392-5396

Yu VC,Deisert C,Andersen B,Holloway JM,Devary O V.,Näär AM,Kim SY,Boutin JM,Glass CK, Rosenfeld MG (1991) RXRβ : a corregulator that enhances binding of retinoic acid, thyroid hormone, and vitamin D receptors to their cognate response elements. Cell 67:1251-1266

Zelent A,Krust A,Petkovich M,Kastner P, Chambon P (1989) Cloning of murine α and β retinoic acid receptors and a novel receptor γ predominantly expressed in the skin. Nature 339:714-717

Zelent A,Mendelsohn C,Kastner P,Krust A,Garnier JM,Ruffenach F,Leroy P, Chambon P (1991) Differentially expressed isoforms of the mouse retinoic acid receptor β are generated by usage of two promoters and alternative splicing. EMBO J. 10:71-81

The Role of Retinoids in Normal Development and Retinoid-Induced Malformations

G. Morriss-Kay, S. Ward and N. Sokolova
Department of Human Anatomy, University of Oxford, Oxford OX1 3QX, UK

The retinoids form a family of compounds structurally and functionally related to vitamin A (all-*trans*-retinol). Their name derives from the role of vitamin A in visual function in the vertebrate retina. The natural retinoids are isoprenoid compounds with a six-membered carbocyclic ring and an eleven-carbon side chain (Fig. 1), although many of the synthetic compounds with biological retinoid activity differ somewhat from this basic plan (Sporn and Roberts 1985).

all-trans-retinol (vitamin A)

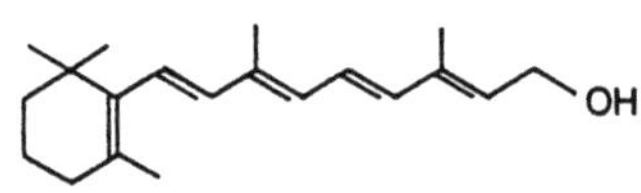

all-trans-retinoic acid

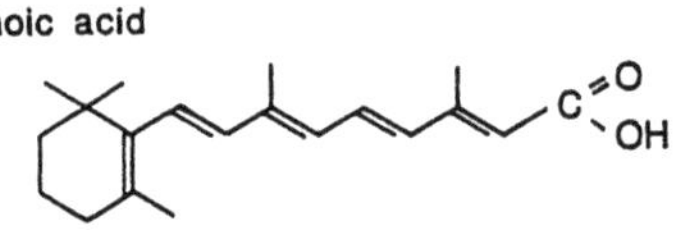

13-cis-retinoic acid

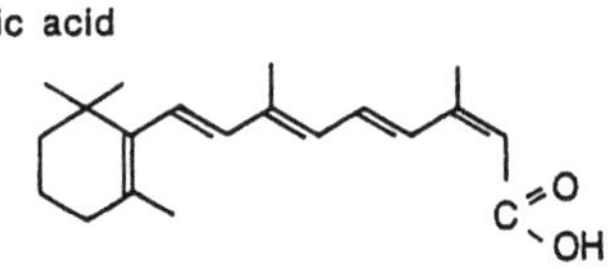

all-trans-retinyl palmitate

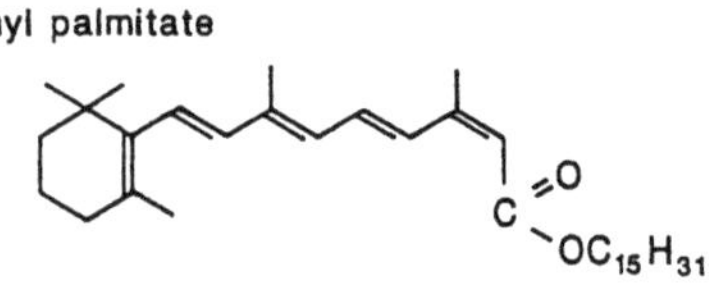

9-cis-retinoic acid

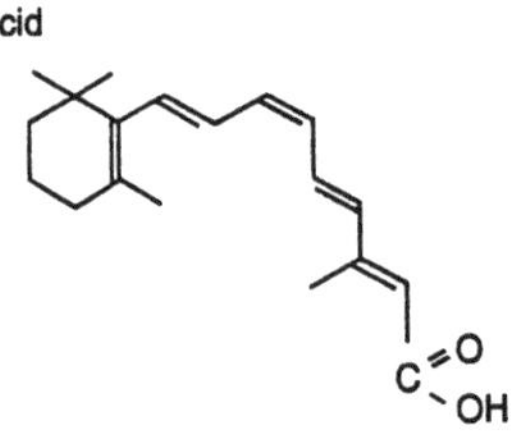

Fig. 1. Chemical structure of the natural retinoids discussed in the text.

Dietary sources of vitamin A include retinyl esters from animal sources (especially liver and kidney), and β-carotene from plants. β-carotene has no intrinsic vitamin A activity, but in the intestinal mucosa and liver is converted by enzymic cleavage at its centre to yield two molecules of retinol. As described in the previous article, the effects of retinoids on gene expression are mediated mainly through conversion to all-*trans*-retinoic acid (RA) and 9-*cis*-retinoic acid; these are ligands for the nuclear retinoic acid receptors (RARs) and the related retinoid "X" receptors respectively, which together act as heterodimeric transcription factors. The roles of retinoic acid in initiation (or suppression) of transcription of those genes with a retinoic acid response element (RARE) in the promotor region are central to its roles in normal and abnormal development.

Normal development depends on a hierarchical programme of gene expression, resulting in an unfolding of the normal sequence of events of morphogenesis and differentiation. These events include the synthesis and secretion of specific extracellular matrix molecules, which play important roles in tissue interactions, and of growth factors, which influence development in a variety of ways through both autocrine and paracrine actions. Synthesis of some of the extracellular matrix molecules known to be essential for normal development (e.g. chondroitin and heparan sulphate proteoglycans) are altered by raised levels of RA, both during early morphogenesis (Morriss-Kay and Mahmood, 1992) and during the differentiation of cartilage (Kistler 1986). The growth factor TGF-β1 also shows altered levels of expression correlated with abnormal morphogenesis when RA levels are raised (Mahmood et al. 1992).

Abnormal mammalian development can result from both vitamin A deficiency and retinoid excess. The malformations due to vitamin A deficiency are different from those induced by retinoid excess, suggesting that different alterations of gene expression are involved. This difference is seen even when the same organ or system is involved, such as the heart. Thus there are specific embryonic cells and tissues that are most vulnerable to vitamin A deficiency, while others are most vulnerable to retinoid excess. These observations imply that when retinoid supply to the embryo is within the normal range, the normal programme of development involves relatively high levels of RA activating the nuclear receptors in those tissues that are most affected by vitamin A deficiency, while relatively low levels of RA (or perhaps none at all) reach the nuclear receptors in those tissues whose development is altered by RA excess. Or, to put it another way, the embryo has a mechanism for differential quantitative access of RA to the nuclear receptors. In the next section we will discuss the evidence that this mechanism is mediated by the cytoplasmic retinoid binding proteins.

Cytoplasmic Retinoid Binding Proteins

Three closely related retinoid binding proteins are present in the cytoplasm of embryonic cells during early morphogenesis and organogenesis stages of development. One of these, cellular retinol binding protein type I (CRBP I) specifically binds retinol; two RA binding proteins (CRABP I and CRABP II) specifically bind RA (see Chytil and Ong 1987 for details and references). A second cellular retinol binding protein (CRBP II), which is involved in the intestinal absorption and metabolism of dietary vitamin A (see Ong 1993 for details), is not expressed in the embryo before day 14 in the liver and day 15 in the gut (H. Nakshatri and P. Chambon, personal communication).

In the early postimplantation stage mouse embryo, in which nutrition is mediated by the yolk sac placenta, CRBP I is expressed in the absorptive cells (visceral endoderm) of the yolk sac, where it is well placed to bind retinol from the maternal blood (Ruberte et al. 1991). Transfer to the embryo probably involves plasma RBP and transthyretin, both of which are expressed at this stage, and are known to be involved in the transport of retinol (Makover et al. 1989). CRBP I is actively involved in RA synthesis through binding retinol and then by direct protein-protein interactions with the dehydrogenases that mediate conversion of retinol to retinal and then (irreversibly) to RA (Napoli 1993). Although these processes have not been observed in embryonic cells, the evidence from metabolic studies strongly suggests that embryonic cells expressing CRBP I are those in which RA is synthesized, and therefore are those in which retinoic acid levels are highest. The good correlation between sites of expression of CRBP I and embryonic tissues that are vulnerable to abnormal development in conditions of vitamin A deficiency (see below) indicates that RA is synthesized precisely where it is required at relatively high levels in order to mediate the normal pattern of RAR/RXR-mediated gene expression.

Within the embryo, CRBP I shows specific spatiotemporal patterns of expression: at early (gastrulation) stages, it is present in the primitive streak, where cells are converting from an epithelial to a migratory mesenchymal phenotype (Ruberte et al. 1991). At later stages (Dollé et al. 1990) it is frequently observed to be co-expressed with RAR-β, which is itself a RA-inducible gene. This co-expression is to be expected if RA is synthesized in the CRBP I-expressing cells.

CRABP I binds all-*trans*-RA with high affinity, but also binds other forms of RA at lower affinities; it mediates RA catabolism by direct protein-protein interaction with appropriate enzymes, having the effect of sequestering it in the cytoplasm and then destroying it (Napoli 1993). Evidence that embryonic cells expressing CRABP I are those in which little or no RA is able to activate the nuclear receptors also comes from a study on F9 cells overexpressing CRABP I: RA-induced gene expression required much higher levels of RA to be supplied than to unaltered F9 cells (Boylan and Gudas 1991).

In embryos, exposure to excess RA alters gene expression and morphogenesis specifically in those regions in which CRABP I is expressed. We interpret this correlation to mean that the normal programme of gene expression in these cells depends on little or no RA reaching and activating the nuclear receptors, so that when RA levels are sufficiently raised to exceed the binding capacity of CRABP I, RA gains access to the nuclear receptors and thereby alters the normally RA-independent programme of gene expression. Alterations in gene expression in cells in which CRABP I is expressed is indeed observed (see below).

CRABP II also has specific spatiotemporal patterns of expression in embryos (Ruberte et al. 1992). These patterns show some overlaps with the expression of both CRABP I and CRBP I, and have no obvious correlation with the effects of either vitamin A deficiency or retinoid excess. CRABP II has a lower affinity for RA than CRABP I has (Bailey and Siu); its function in embryos is unknown.

Vitamin A deficiency and embryonic development

The developmental effects of maternal vitamin A deficiency have been described for pig, rat, rabbit and monkey (Hale 1937, Wilson et al. 1953, Harrington and Newberne 1970, O'Toole et al. 1974). We have begun a study of the effects of maternal vitamin A deficiency in the mouse, in order to relate the morphological effects on specific tissues with the expression of binding proteins and RARs. The lungs and heart are both abnormal in later embryos and neonates.

The lungs develop from an epithelial diverticulum of the embryonic oesophagus that first divides into left and right bronchial branches, then into lobar branches, then into smaller and smaller branches to form the bronchioles and finally alveoli. Normal branching morphogenesis depends on an interaction between the epithelium of the branching tubes and the adjacent pulmonary mesenchyme (Wessels 1970). CRBP I RNA transcripts are at first present in both the bronchial epithelium and pulmonary mesenchyme, but by day 14.5 are restricted to the mesenchyme (Dollé et al. 1990), suggesting that RA is synthesized in the pulmonary mesenchyme during formation of the bronchioles and alveoli. In the lungs of normal neonatal mice, air-filled alveoli are evenly distributed throughout the tissue of each lobe; in neonates from vitamin A-deficient mothers, we have observed collapsed alveoli surrounded by hyperplastic mesenchymal stroma.

Heart development in vitamin A-deficient embryos shows a similar correlation with the type of abnormality and the distribution of CRBP I transcripts. CRBP I is expressed in the endocardial cushion tissue, from which the upper portion of the interventricular septum is formed, and in the outer (growing) layer of the ventricular wall (Dollé et al. 1990). In the vitamin A-deficient neonates, we observed cysts in the outer layer of the ventricular wall, and an open defect in the upper part of the interventricular septum.

Effects of excess retinoic acid on embryonic development

Administration of 100,000 i.u. retinyl palmitate to pregnant rats, or 10-12 mg/kg RA to pregnant mice, around the stage of neural induction (formation of the neural plate) leads to alterations in morphology of the hindbrain (Morriss, 1972; Morriss-Kay et al. 1991). Later stages of development of the retinoid-exposed rat embryos show shortening of the rostral (upper) hindbrain and craniofacial abnormalities related to abnormalities of the structures derived from the neural crest that originates in the affected part of the hindbrain (Morriss-Kay 1993). Exposure of human infants to 13-*cis*-retinoic acid at early stages of gestation has similar effects on craniofacial development (Lammer et al. 1985). In the mouse study, the pattern of gene expression in the rostral hindbrain was altered: in particular, the segmental segregation of different genes was lost, and the expression domain of the homeobox gene *Hox-B1* (formerly *Hox-2.9*) was shifted in position and more extensive than normal.

CRABP I is expressed at a high level in the hindbrain and in the craniofacial neural crest (Ruberte et al. 1991); in contrast, the RA-inducible gene RAR-β is present only in the most caudal (lower) part of the hindbrain. At the time of neural induction, CRABP I is present in the mesenchyme beneath the neural plate rather than in the neural plate itself, and the altered pattern of gene expression induced by RA excess may be mediated by altered gene expression in the mesenchyme prior to any direct effects on the neuroepithelium.

At later stages of development (day 11 in the mouse and day 12-13 in the rat), retinoid excess results in abnormal development of the limbs (Kochhar 1973, Newall and Ward 1990, and our unpublished results). Using high pressure liquid chromatography (HPLC), we have found that oral administration of a teratogenic dose of RA to rats on day 12 of pregnancy results in peak embryonic levels of all-*trans*-RA and 13-*cis*-RA after 2 hours, and a return to normal levels by 4 hours. Transplacental exposure of the embryos therefore involves only a short pulse of raised RA. Long-term effects include not only altered morphogenesis but also altered patterns of gene expression, some of which can still be detected after 48 hours. These include upregulation of RAR-β (our unpublished observations) and altered expression of TGF-βs (Mahmood et al. 1992).

Conclusions

The interactions between retinoids, gene expression and embryonic development are complex. In this article we have emphasised the role of the cytoplasmic retinoid binding proteins in regulating the access of RA to the nucleus for activation of the nuclear receptors. CRBP I binds retinol, and is involved in retinoic acid synthesis. Correlations between the domains of expression of CRBP I in embryos, and the tissues that develop abnormally in conditions of vitamin A deficiency, supports the idea that cells expressing CRBP I are those in which the normal patterns of gene expression, morphogenesis and differentiation depend on relatively high levels of RA binding to the nuclear receptors. In contrast, CRABP I binds RA and mediates its breakdown to inactive metabolites; correlations between the domains of expression of CRABP I and the tissues that are susceptible to the effects of RA excess support the idea that in the embryo, the presence of CRABP I indicates cells in which the normal pattern of gene expression and development depends on little or no RA reaching the nucleus. Excess RA and vitamin A deficiency therefore affect different populations of developing cells, and alter the developmental programme in different ways.

References

Bailey JS, Siu C-H (1988) Purification and partial characterization of a novel binding protein for retinoiç acid from neonatal rat. Proc Natl Acad Sci USA 87:6233-6237

Boylan JF, Gudas L (1991) Overexpression of the cellular retinoic acid binding protein I (CRABP-I) results in a reduction in differentiation-specific gene expression in F9 teratocarcinoma cells. J Biol Chem 112:965-979

Chytil F, Ong DE (1987) Intracellular vitamin A-binding proteins. Ann Rev Nutrition 7:321-325

Harrington DD, Newberne PM (1970) Correlation of maternal blood levels of vitamin A at conception and the incidence of hydrocephalus in newborn rabbits: an experimental animal model. Lab Anim Sci 20:675-680

Kistler A (1986) Hypervitaminosis A: side effects of retinoids. Biochem Soc Trans 14:936-939

Kochhar DM (1973) Limb development in mouse embryos. I. Analysis of teratogenic effects of retinoic acid. Teratology 7:289-298

Lammer EJ, Chen DT, Hoar RM, Agnish ND, Benke PJ, Braun JT, Curry CJ, Fernhoff PM, Grix AW, Lott IT, Richard JM, Sun SC (1985) Retinoic acid embryopathy. N Eng J Med 313:837-841

Mahmood R, Flanders KC, Morriss-Kay GM (1992) Interactions between retinoids and TGF βs in mouse morphogenesis. Development 115:67-74

Makover A, Soprano DR, Wyatt ML, Goodman DS (1989) An in-situ hybridization study of the localization of retinol binding protein and transthyretin messenger RNAs during fetal development in the rat. Differentiation 40:17-25

Morriss GM (1972) Morphogenesis of the malformations induced in rat embryos by maternal hypervitaminosis A. J Anat 113:241-250

Morriss-Kay GM (1993) Retinoic acid and craniofacial development: molecules and morphogenesis. BioEssays 15:9-15

Morriss-Kay GM, Murphy P, Hill R, Davidson D (1991) Effects of retinoic acid excess on expression of *Hox-2.9* and *Krox-20* and on morphological segmentation of the hindbrain of mouse embryos. EMBO J 10:2985-2995

Morriss-Kay GM, Mahmood R (1992) Morphogenesis-related changes in extracellular matrix induced by retinoic acid. In: Morriss-Kay GM (ed) Retinoids in normal development and teratogenesis. OUP, Oxford, pp 165-180

Napoli J (1993) Biosynthesis and metabolism of retinoic acid: roles of CRBP and CRABP in retinoic acid homeostasis. J Nutr 123:362-366

Newall DR, Ward SJ (1990) Relationship between stage of development and type of malformation in retinoic acid-induced teratogenesis of the rat forelimb. Teratology 42(2):24A

Ong D (1993) Retinoid metabolism during intestinal absorption. J Nutr 123:351-355

O'Toole BA, Fradkin R, Warkany J, Wilson JG, Mann GV (1974) Vitamin A deficiency and reproduction in rhesus monkey. J Nutr 104:1513-1524

Ruberte E, Dollé P, Chambon P, Morriss-Kay G (1991) Retinoic acid receptors and cellular binding proteins: II Their differential pattern of transcription during early morphogenesis in mouse embryos. Development 111:45-60

Sporn MB, Roberts AB (1985) Introduction: what is a retinoid? In: Nugent J, Clark S (eds) Retinoids, differentiation and disease. Ciba Foundation Symposium 113, Pitman, London, pp 1-5

Wessels NK (1970) Mammalian lung development: interactions in formation and morphogenesis of tracheal buds. J Exp Zool 175:455-466

Retinoid Teratogenesis: Toxicokinetics and Structure-Specificity

H. Nau
Institute for Toxicology and Embryopharmacology, Free University Berlin, Gary Str. 5, D-14195 Berlin, Germany

It becomes increasingly clear that the dose is often an unsatisfactory correlate to the toxic effect produced by a drug or environmental agent. The effect in a particular target organ is strictly dependent on the concentration-time relationships of the drug and/or its active metabolite(s) in the target tissue, reactive metabolites formed or specific binding phenomena (Monro, 1992). Thus, an effect is dependent on the chemical structure of the xenobiotic (intrinsic activity), the exposure of the target tissue, and the susceptibility or sensitivity toward a particular insult (pharmacogenetics). This general concept will be discussed here with regard to drug teratogenesis, where it is particularly helpful to explain and overcome species differences, which are so extreme in this area of toxicology.

Species Differences

Species differences is still one of the most important problems in toxicology, and in particular teratology. Thalidomide was shown to produce characteristic limb defects in human and subhuman primates, but not in the rat and the mouse, even with doses several orders of magnitude above human therapeutic doses; the rabbit appears to be responsive to thalidomide, at least at high dosing (Neubert et al., 1988; Sterz et al., 1987; Schmahl et al., 1989). 13-Cis-retinoic acid (isotretinoin), a retinoid used for treatment of severe acne and other skin conditions, showed similar, if not as extreme species differences: The human was very sensitive toward this drug, followed by the monkey and the rabbit; the mouse and rat were again relatively insensitive (for recent reviews cf. Nau, 1990; 1993).

The reason behind the species difference of thalidomide teratogenesis is still unclear and pharmacokinetic studies have not yet added to our knowledge. Pharmacokinetic studies are well on the way to explain most, if not all of the major species differences in regard to 13-cis-retinoic acid teratogenicity.

Vitamin A and Retinoids

Retinoids are a family of substances which are structurally and functionally related to retinol (vitamin A) (cf. Blomhoff et al., 1992) (Fig. 1). They are required for a great range of biological processes including vision, reproduction, growth, differentiation and maintenance of epithelial tissues. All-trans-retinoic acid, formed intracellularly via oxidation of dietary vitamin A, is a principal regulatory retinoid which can substitute for retinol for nearly all vitamin A-related functions except vision and certain aspects of reproduction. Indeed, all-trans-retinoic acid is much more active than retinol in the control of differentiation and morphogenesis. The latter process appears to be of fundamental importance in the regulation of the pattern formation of the developing embryo. It is hypothesized that concentration gradients of all-trans-retinoic acid established within the developing limb bud and central nervous system are crucial morphogenetic events defining the three-dimensional structure of the developing embryo (Thaller and Eichele, 1987; Durston et al., 1989).

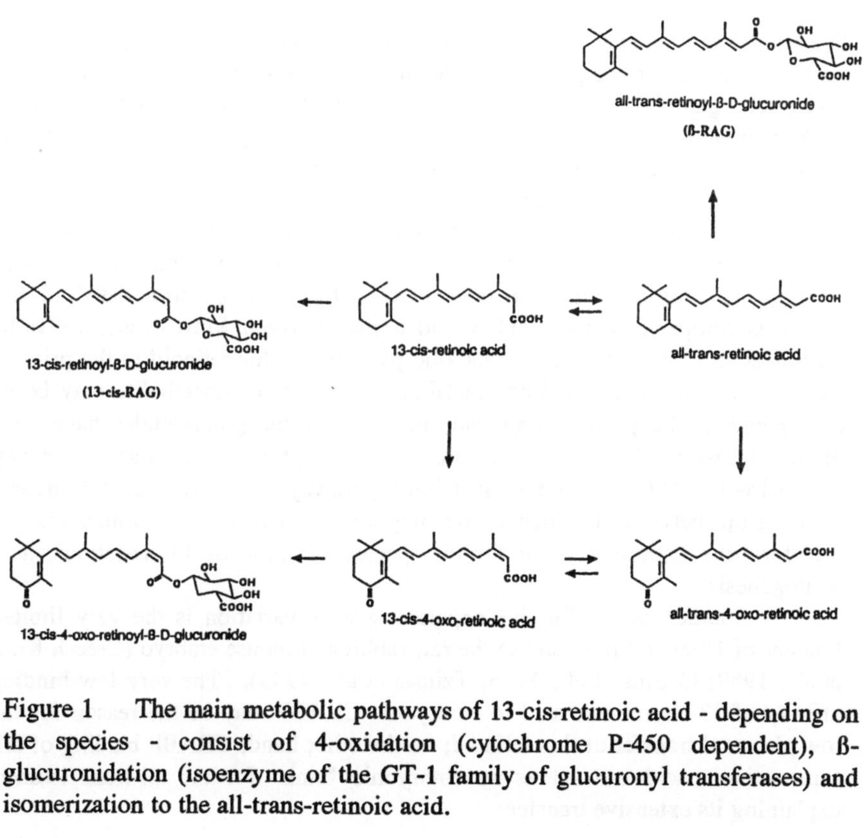

Figure 1 The main metabolic pathways of 13-cis-retinoic acid - depending on the species - consist of 4-oxidation (cytochrome P-450 dependent), ß-glucuronidation (isoenzyme of the GT-1 family of glucuronyl transferases) and isomerization to the all-trans-retinoic acid.

The function of the retinoids has been linked to the presence of cellular binding proteins for retinol (CRBP) and all-trans-retinoic acid (CRABP) (Maden et al., 1991; Kawamura et al., 1993; Gustafson et al., 1993). The regulation of gene expression may be mediated through nuclear receptors for all-trans-retinoic acid (RAR's α,ß,γ) (Hashimoto and Shudo, 1991; Ruberte et al., 1993); the newly discovered 9-cis-retinoic acid may be the endogenous ligand for another set of nuclear receptors (RXR's α,ß,γ) (Levin et al., 1992; Heymann et al., 1992).

Retinoids are important clinical drugs used in the treatment of dermatological disease such as acne and psoriasis, for treatment of various forms of cancer and as chemopreventive agents (Larsen et al., 1992; Bollag and Holdener, 1992). However, all retinoids with proven clinical efficacy also induce severe adverse effects in the developing embryo and fetus (Teelmann, 1989; Nau, 1993); the high teratogenic potency established for 13-cis-retinoic acid (used against acne) and etretinate (used against psoriasis) greatly limit the usefulness and applicability of this class of compounds.

Retinoid teratogenesis and toxicokinetics

The teratogenic potency of 13-cis-retinoic acid (Figure 1) shows a great species variation and the principles discussed above are useful for overcoming this interspecies gap (for reviews, cf. Nau, 1990, 1993). Teratogenic daily doses in the human were below 1 mg/kg/day, in the monkey 5 mg/kg/day, and in the rat 75 mg/kg/day. The rabbit and hamster were more sensitive animal species toward 13-cis-retinoic acid than the rat.

The reasons behind the drastic species difference of 13-cis-retinoic acid are in great part due to differences in toxicokinetics, placental transfer and metabolism. 13-cis-retinoic acid is more extensively glucuronidated in the rodent as compared to the monkey and human (Creech Kraft et al., 1991a,b; Collins et al., 1993); the main metabolic pathway of this retinoid in the primate species leads to the 13-cis-4-oxo-metabolite. The 4-oxo-metabolite may be as teratogenic as the parent compound; in contrast, the glucuronides have very limited access to the embryo and presumably exhibit low teratogenic potency (see below). The variation in metabolic pathways, in particular the inverse relationship between teratogenic potency and extent of glucuronication, is therefore one important factor for the species variation of 13-cis-retinoic acid teratogenesis.

Another reason for this apparent species variation is the very limited transfer of 13-cis-retinoic acid to the rat, rabbit and mouse embryo (Creech Kraft et al., 1989; Collins et al., 1993; Tzimas et al., 1993). The very low binding affinity of 13-cis-retinoic acid to embryonic CRABP may be the reason for the low placental transfer of this retinoid; on the other hand, CRABP binding of all-trans-retinoic acid can act as an embryonic "sink" for the all-trans isomer, explaining its extensive transfer.

It is interesting to note, that the transfer of 13-cis-retinoic acid is more extensive in the monkey as compared to the rodent (Hummler, Hendrickx and Nau, submitted for publication). Differences in placental structure may be important here: the yolk sac placenta is more prominent in the rodent at early stages of organogenesis, when the most drastic teratogenic effects are induced; in the monkey, the hemochorial placenta develops at earlier gestational stages as compared to the rodent species. The human placenta is similar to the monkey placenta in regard to structure and function.

The third reason for the relatively low teratogenicity of 13-cis-retinoic acid in the rodent is the high clearance of this drug in this species - due to rapid glucuronidation as discussed above. The half life of this drug in the mouse is approximately 1h, in the human 10-20h. Thus, the AUC of a particular dose in the mouse is expected to be about 10-20 times lower than that in the human, due to a direct proportionality of AUC and half life values (in the absence of saturation kinetics).

If all factors - metabolism, placental transfer, maternal clearance - are taken together, the high sensitivity of the human and the monkey, and the low sensitivity of the rodent species can entirely be explained by toxicokinetics: a particular AUC of an active retinoid - in this case 13-cis-retinoic acid and its 4-oxo-metabolite - must be reached in the embryo for induction of a frank teratogenic response.

Activation to teratogenic metabolites

It is not known whether retinol or its metabolites are the proximate or ultimate teratogenic agents (Figure 1). We have investigated this problem in the mouse, the rat, the monkey and the human (Eckhoff et al., 1989; 1991a,b; Collins et al., 1993). In the mouse and the rat, the oxidation of retinol cannot account for the entire teratogenic response observed, and implicates also retinol and/or the retinyl esters as a possible proximate or as ultimate teratogens. The same conclusion was reached when the teratogenicity and the metabolism of retinol was compared in control and methylpyrazol-pretreated mice: although methylpyrazol greatly inhibited the formation of retinoic acids from retinol, the teratogenic activity was only slightly reduced - again implicating also retinol as teratogenic entity. These results are compatible with direct binding of all-trans retinol to nuclear retinoic acid recptors (Repa et al., 1993).

In the monkey, proportionally larger amounts of retinoic acids are formed after application of Vitamin A. Also, 13-cis-retinoic acid and 13-cis-4-oxo-retinoic acid are formed in addition to all-trans-retinoic acid, particularly at lower doses.

In the human, relatively large amounts of retinoic acid compounds are formed from vitamin A, and these retinoids accumulate to levels several-fold higher than corresponding endogeneous levels. The human may therefore be more at risk with "megadoses" of Vitamin A or liver consumption during pregnancy; the usual liver meal may contain up to 100,000 IU Vitamin A (corresponding to 30 mg retinol equivalents) which is more than one order of magnitude higher than the Vitamin A contents of most vitamin supplements.

Detoxification pathways to β-D-Glucuronides

β-Glucuronides are major plasma metabolites following pharmacological dosing with retinoids. 13-cis-retinoyl-β-glucuronide actually was the major metabolite of 13-cis-retinoic acid in the rodent, while the glucuronide of all-trans-retinoic acid was more prominent in the monkey. Following retinol administration, β-glucuronides of all-trans-retinoic acid, retinol, 13-cis-retinoic acid and 13-cis-4-oxo-retinoic acid were identified (Figure 1). The inverse relationship between the extent of β-glucuronidation and teratogenic potency of 13-cis and al-trans-retinoic acid suggests that β-glucuronidation is a detoxification pathway in regard to teratogenesis.

Development of Novel Retinoids with Low Teratogenic Potency

Great efforts have been made during the past decade to develop retinoids with high efficacy but low teratogenic potential. Such compounds will be invaluable for the study of the significance of retinoids in normal and abnormal development, would contribute to our knowledge about retinoid molecular mechanism of action, and the potential clinical application of such compounds would be a breakthrough.

The main effort was hitherto mainly directed toward the synthesis of retinoids with particularly high efficacy. It turned out, that these compounds also exhibited high intrinsic teratogenic activity (Bechter et al., 1992) and little was gained in regard to the therapeutic ratio.

There appear to be two promising ways to develop active retinoids with low teratogenic potential: (1) Selective ligands can be developed which bind to and activate one or a limited number of nuclear receptor(s). Such selective agonists have been developed and their binding and gene activation was studied in vitro (Tamura et al., 1990; Martin et al., 1992; Lohnes et al., 1993). So far, however, a direct link between the binding and activation of a particular receptor and the teratogenic outcome has not yet been established. This is indeed a very difficult and complex situation because of the multitude of the retinoid receptors and their isoforms, the demonstrated antagonism as well as cooperation between some, and particularly because of the apparent redundancy of other receptors. Also the complex spatio-temporal expression of the receptors and binding proteins during development will make it very difficult to unambiguously link a specific binding event to the teratogenic effect. Possibly the development of selective retinoid antagonists may be of some help here, although the use of a second compound makes the situation even more complex, particularly in regard to pharmacokinetics.

Selective ligands of RA-receptors
Antagonists

↑

Retinoids with low teratogenic potency

↓

Limited placental transfer

A second way is proposed here to establish retinoids with low teratogenic potency: to limit the access of the retinoids to the embryo. This goal appears to be very realistic because we have shown during the recent years that the placental transfer of retinoids during the sensitive stages of gestation is a specific process (Nau, 1990; 1993).

While all-trans-retinoic acid is efficiently transferred to the embryo of all species investigated in this regard, the access of 13-cis-retinoic acid to the mouse, rat and rabbit embryo is very limited (embryo/maternal plasma ratios below 0.1) (Creech Kraft et al., 1989; Collins et al., 1993, Tzimas et al., 1993). 13-cis-retinoic acid exhibits a higher placental transfer in the monkey (Hummler, Hendrickx and Nau, submitted) and probably human, explaining the higher teratogenic potency of this retinoid in the primate as compared to the rodent.

Glycoconjugates of Retinoids

The ß-glucuronides of retinol and retinoic acid have been found to be metabolites of vitamin A present in bile, liver and kidney (Miller and DeLuca, 1986). Both retinyl- as well as retinoyl-ß-glucuronide (ß-ROG and ß-RAG, respectively) are capable of promoting growth of vitamin A-deficient rats (Barua and Olson, 1987). In addition, ß-RAG can induce the differentiation of HL-60 cells (Zile et al., 1987; Janick-Buckner et al., 1991) as well as of vaginal epithelium cells (Mehta et al., 1991). It was also found that ß-RAG was less toxic in these in vitro systems than all-trans retinoic acid. Thus it was concluded that ß-RAG may be an interesting retinoid drug with lower toxicity. In a recent study a comparison was attempted between the teratogenic potency of ß-RAG and that of all-trans retinoic acid (Gunning et al., 1993). Oral application was used in that study and ß-RAG was essentially not absorbed. No conclusion on the intrinsic teratogenic activity of ß-RAG is thus possible.

ß-Glucuronides of retinoids are major metabolites, but exhibit limited placental transfer

We have recently added to this knowledge in two ways: (1) the ß-glucuronides of retinol, all-trans retinoic acid and 13-cis-retinoic acid are major metabolites following the treatment with the respective retinoids. In the mouse, rat and rabbit, ß-glucuronidation of 13-cis-retinoic acid is much more extensive than ß-glucuronidation of the all-trans-retinoic acid (Creech Kraft et al., 1991a; Collins et al., 1993; Tzimas et al., 1993b); (2) the reverse situation was found in the primate species: here, ß-glucuronidation of the all-trans-retinoic acid predominates and the 13-cis-isomer was mainly oxidized to the 4-oxo-metabolite (Eckhoff et al., 1990; Creech Kraft et al., 1991b; Eckhoff et al., 1991a).

And the most important findings: these ß-glucuronides of retinol, all-trans-retinoic acid and 13-cis-retinoic acid were transferred to the embryo - during the sensitive stages of organogenesis - to a very minor degree only. We have now confirmed these findings in extensive studies in four species: mouse, rat, rabbit and monkey.

These studies indirectly suggest that, because of their limited access to the embryo, the ß-glucuronides and possibly other glycoconjugates as well, may be active retinoids with low teratogenic potential. The reasons behind these exiting findings are not clear and should be investigated. What enzymes are responsible for the formation of ß-glucuronides of retinoids? Why do these conjugates exhibit such a limited placental transfer, while they apparently have ready access to other tissues as demonstrated by their vitamin A-like action? Can these compounds be valuable retinoid drugs? Can such studies provide clues toward the mechanism of retinoid action in normal and abnormal development?

In cultures of mouse palatal shelves, ß-RAG and all-trans retinoic-acid were comparatively tested in regard to their potency to induce cleft palate (Rühl et al., 1993). The dose-response relationships show that the potency of the ß-RAG was about 10-fold lower than the potency of the all-trans-retinoic acid in inhibiting palate fusion. The chemical analyses show isomerization of the all-trans-retinoic acid as well as glucuronidation. ß-RAG was also isomerized to the 13-cis-RAG as well as hydrolysed to the all-trans-retinoic acid. Most importantly, both all-trans-retinoic acid and ß-RAG reached higher concentrations in the tissues as compared to the culture medium at the end of the culture (76h). Thus, ß-RAG reaches high concentrations in the target tissues, but exerts much lower effects than all-trans-retinoic acid; these results indicate a much lower intrinsic activity of ß-RAG as compared to all-trans-retinoic acid. It is therefore likely that the hydrolysis product all-trans-retinoic acid is responsible for most of the effects observed in these in vitro experiments.

The ß-RAG may therefore constitute a water-soluble transport form for all-trans-retinoic acid. This hypothesis suggests that β-RAG would reach significant and persistent concentrations in target tissues of retinoid action (except the embryo), releasing all-trans-retinoic acid in low and persistent amounts for interaction with cellular binding proteins and nuclear receptors or other crucial factors involved in the mechanism of retinoid action. Future studies must show if these kinetic and metabolic properties of β-RAG can be translated into a therapeutic advantage.

Acknowledgements

The original work described here from our laboratory was supported by the Deutsche Forschungsgemeinschaft (Sfb 174, C6) and the Bundesgesundheitsamt Berlin (ZEBET). The preparation of this manuscript by M. Fleminger and of the artwork by U. Schwikowski is gratefully acknowledged.

REFERENCES

Barua AB, Olson JA (1987) Chemical synthesis and growth-promoting activity of all-trans-retinyl ß-D-glucuronide. Biochem J 224:231-234

Bechter R, Terlouw GDC, Tsuchiya M, Tsuchiya T and Kistler A (1992) Teratogenicity of arotinoids (retinoids) in the rat whole embryo culture. Arch Toxicol 66:193-197

Blomhoff R, Green MH and Norum KR (1992) Vitamin A: physiological and biochemical processing. Annu Rev Nutr 12:37-57

Bollag W and Holdener EE (1992) Retinoids in cancer prevention and therapy. Annal Oncol 3:513-526

Collins MD, Tzimas G, Bürgin, H Hummler H and Nau H (1993) Comparative transplacental pharmacokinetics of all-trans-retinoic acid, 13-cis retinoic acid and retinyl palmitate in the rat following daily administrations of equipotent teratogenic doses. Teratology (Abstract P11) 47:415

Creech Kraft J, Löfberg B, Chahoud I, Bochert G, Nau H (1989) Teratogenicity and placental transfer of all-trans-, 13-cis-, 4-Oxo-all-trans-, and 4-Oxo-13-cis-retinoic acid after administration of a low oral dose during organogenesis in mice. Toxicol Appl Pharmacol 100:162-176

Creech Kraft J, Eckhoff C, Kochhar DM, Klug S, Bochert G, Chahoud I, Nau H (1991a) Isotretinoin (13-cis-retinoic acid), metabolism, cis-trans isomerization and transfer to the embryo: consequences for teratogenicity. Terat Carcinogenogen Mutag 11: 21-30

Creech Kraft J, Slikker W, Bailey JR, Roberts LG, Fischer B, Wittfoht W, Nau H (1991b) Plasma pharmacokinetics and metabolism of 13-cis and all-trans-retinoic acid in the cynomolgus monkey and the identification of 13-cis and all-trans-retinoyl-ß-glucuronides. Drug Metabol Disp 19: 317-324

Durston AJ, Timmermans JPM, Hage WJ, Hendriks HFJ , de Vries NJ, Heideveld M, Nieuwkoop PD (1989) Retinoic acid causes an anteroposterior transformation in the developing central nervous system. Nature 340:140-144

Eckhoff C, Löfberg B, Chahoud I, Bochert G, Nau H (1989) Transplacental pharmacokinetics and teratogenicity of a single dose of retinol (vitamin A) during organogenesis in the mouse. Toxicol Lett 48: 171-184

Eckhoff C, Bailey JR, Collins MD, Slikker WJ, Nau H (1991a) Influence of dose and pharmaceutical formulation of vitamin A on plasma levels of retinyl esters and retinol and metabolic generation of retinoic acid compounds and ß-glucuronides in the cynomolgus monkey. Toxicol Appl Pharmacol 111:116-127

Eckhoff C, Collins MD, Nau H (1991b) Human plasma all-trans-, 13-cis-, and 13-cis-4-oxoretinoic acid profiles during subchronic vitamin A supplementation: Comparison to retinol and retinyl ester plasma levels. J Nutr 121:1016-1025

Eckhoff C, Slikker W, Wittfoht W, Nau H (1990) Characterization of oxidized and glucuronidated metabolites of retinol in monkey plasma by thermospray liquid chromatography / mass spectrometry. Biomed Mass Spectrom 19:428-433

Gunning DB, Barua AB, Olson JA (1993) Comparative teratogenicity and metabolism of all-trans retinoic acid, all-trans retinoyl ß-glucose, and all-trans retinoyl ß-glucuronide in pregnant Sprague-Dawley rats. Teratology 47:29-36

Gustafson AL Eriksson U, Dencker L (1993) Retinoid teratogenesis: is a well regulated endogenous retinoid metabolism disturbed? Pharmacol Toxicol Abstract P13/02:129

Hashimoto Y, Shudo K (1991) Retinoids and their Nuclear Receptors. Cell Bio Rev 25:209-230

Heyman RA, DJ Mangelsdorf, JA Dyck, RB Stein, G Eichele, RM Evans and C Thaller (1992) 9-Cis Retinoic Acid is a High Affinity Ligand for the Retinoid X Receptor. Cell 68:397-406

Janick-Buckner D, Barua AB and Olson JA (1991) Induction of HS-60 cell differentiation by water-soluble and nitrogen-containing conjugates of retinoic acid and retinol. FASEB J 5:320-325

Kawamura K, Kara K, Fujiwara S (1993) Developmental role of endogenous retinoids in the determination of morphallactic field in budding tunicates. Development 117:835-845

Larsen FG, Nielsen-Kudsk F, Jakobsen P, Weismann K, Kragballe K (1992) Pharmacokinetics and therapeutic efficacy of retinoids in skin diseases. Clin Pharmacokin 23:42-61

Levin AA, Sturzenbecker LJ, Kazmer S, Bosakowski T, Huselton C, Allenby G, Speck J, Kratzeisen Cl, M Rosenberger Cl, Lovey A, Grippo JF (1992) 9-Cis retinoic acid stereoisomer binds and activates the nuclear receptor RXRα. Nature 355:359-361

Lohnes D, Kastner P, Dierich A, Mark M, LeMeur M, Chambon P (1993) Function of retinoic acid receptor γ in the mouse. Cell 73:643-65

Maden M, Summerbell D, Maignan J, Darmon M Shroot B (1991) The respecification of limb pattern by new synthetic retinoids and their interaction with cellular retinoic acid-binding protein. Differentiation 47:49-55

Martin B, Bernardon JM, Caven MT, Bernard B, Carlavan I, Charpentier B, Pilgrim WR, Shroot B, Reichert U (1992) Selective Synthetic ligands for human nuclear retinoic acid receptors. Skin Pharmacol 5:57-65

Mehta RG, Barua AB Olson JA, Moon RC (1991) Effects of retinoid glucuronides on mammary gland development in organ culture. Oncology 48:505-509

Miller DA, DeLuca HF (1986) Biosynthesis of retinoyl-ß-glucuronide, a biologically active metabolite of all-trans-retinoic acid. Arch Biochem Biophys 244:179-186

Monro A, (1992) What is an appropriate measure of exposure when testing drugs for carcinogenicity in rodents? Toxicol Appl Pharmacol 112:171-181

Nau H (1990) Correlation of transplacental and maternal pharmacokinetics of retinoids during organogenesis with teratogenicity. Meth Enzymol 190:437-448

Nau H (1993) Teratogenesis, transplacental pharmacokinetics, and metabolism of some retinoids in the mouse, monkey, and human. In: Livrea M, Packer L, (eds) Retinoids: Progress in research and clinical applications. Marcel Dekker, Inc., New York, pp 599-616

Neubert D, Heger W, Merker HJ, Sames K, Meister R (1988) Embryotoxic effects of thalidomide derivatives in the non-human primate callithrix jacchus, II. Elucidation of the susceptible period and of the variability of embryonic states, Arch. Toxicol. 61:180-191

Repa JJ, Hanson KK, Clagett-Dame M (1993) All-trans-retinol is a ligand for the retinoic acid receptors. Biochem 90:7293-7297

Ruberte E, Friederich V, Chambon P, Moriss-Kay G (1993) Retinoic acid receptors and cellular retinoid binding proteins III Their differential transcript distribution during mouse nervous system development. Development 118:267-282

Rühl R, Klug S, Sass JO, Kastner U, Nau H (1993) Comparitative investigations of the influence of all-trans-retinoic acid and all-trans-retinoyl-ß-D-glucuronide on the fusion of mouse palatel shelves and of the pharmacokinetics with HPLC-analysis. New Trends in Research and Clinical Applications, Abstract in press.

Schmahl HJ, Heger W, Nau H (1989) The enantiomers of the teratogenic thalidomide analogue EM 12. Toxicol Lett 45:23-33

Sterz H, Nothdurft H, Lex P, Ockenfels H (1987) Teratologic studies on the Himalayan rabbit: new aspects of thalidomide-induced teratogenesis. Arch Toxicol 60:376-381.

Tamura K, Kagechika H, Hashimoto Y, Shudo K, Ohsugi K, Ide H (1990) Synthetic retinoids, retinobenzoic acids Am80, Am580 and Ch55 regulate morphogenesis in chick limb bud. Cell Different Devel 32:17-26

Teelman K (1989) Retinoids: Toxicology and teratogenicity to date. Pharmac Ther 40:29-43

Thaller C, Eichele G (1987) Identification and spatial distribution of retinoids in the developing chick limb bud. Nature 327:625-628

Tzimas G, Burgin H, Collins MD, Hummler H, Nau H (1993) The high senstivitiy of the rabbit to the teratogenic effects of 13-cis-retinoic acid is a consequense of prolonged exposure of the embryo to 13-cis-retinoic acid and not of isomerization to all-trans-retinoic acid. Teratology (Abstract 14) 47:438

Zile MH, Cullum ME, Simpson RU, Barua AB, Swartz DA (1987) Induction of differentiation of human promyelocytic leukemia cell line HL-60 by retinoyl glucuronide, a biologically active metabolite of vitamin A. Proc Natl Acad Sci 84:2208-2212

The Neurochemical Effects of PCB Exposure are Age-Dependent

R.F. Seegal
New York State Department of Health, Wadsworth Center for Laboratories and Research, Box 509, Albany, New York 12201-0509 and
School of Public Health, University at Albany, Two University Place, Albany, New York 12203-3399

Polychlorinated biphenyls (PCBs) are widespread, persistent environmental contaminants that have been associated with neurobehavioral and cognitive dysfunctions in perinatally-exposed humans (Chen et al. 1992; Jacobson et al. 1992) and deficits in learning and memory in perinatally-exposed laboratory rodents and non-human primates (Schantz et al. 1991). However, the underlying biochemical and physiological bases for these changes in behavior have not yet been determined.

In an effort to better understand the direct neurochemical effects of exposure to PCBs, in the absence of developmental changes or alterations in maternal behavior that might occur following perinatal exposure, Seegal et al. (1991a; 1991b) exposed adult rodents and non-human primates to PCBs. These studies demonstrated that exposure of the adult organism to PCBs results in significant decreases in dopamine (DA), an important neurotransmitter involved in the control of locomotor activity, learning and memory (Brozoski et al. 1979; Sawaguchi et al. 1988). Furthermore, the neurochemical changes were limited almost exclusively to DA and did not alter concentrations or activity of other members of the class of neurotransmitters known as biogenic amines (Schwartz, 1981). More recently, we have conducted a series of experiments examining the neurochemical responses seen in rodents perinatally-exposed to Aroclor 1016, a commercial mixture of PCBs consisting of lightly chlorinated, *ortho*-substituted congeners (Erickson, 1986). The purpose of this paper is to compare the neurochemical changes seen in adult PCB-exposed animals with those changes seen following perinatal exposure to PCBs.

Materials and Methods

Animals

Adult animals consisted of laboratory rats (Wistar-derived) and non-human primates (*Macaca nemestrina*). The rats were obtained from the breeding facilities of the New York State Department of Health, weighed between 250 and 300 g, and were orally-exposed to either 500 or 1000 ppm Aroclor 1254-adulterated chow for 30 days. The non-human primates were obtained from either Charles River Primates or from the breeding facilities of the University of Washington and initially weighed between 4 and 6 kg. They were exposed to either Aroclor 1016 or Aroclor 1260 at a dose of 3.2 mg/(kg·day) for a total exposure time of 20 or 66 weeks.

Pregnant rat dams were also obtained from the New York State Department of Health. Dams were exposed to Aroclor 1016 (100 or 300 ppm adulterated chow) during gestational days 8-21, during lactation, or for the combined gestational and lactational periods. Offspring were sacrificed on postnatal days 25, 35 and 60 for determination of regional brain concentrations of biogenic amines and metabolites using high-performance liquid chromatography with electrochemical detection.

High-Performance liquid chromatographic techniques

Following regional dissection of brain tissue from the above-described animals, the tissues were homogenized in 0.2 N perchloric acid containing 100 mg/l of EGTA and the supernatants were analyzed for concentrations of DA, norepinephrine (NE), serotonin (5-HT), and their metabolites (3,4 dihydroxyphenylacetic acid [DOPAC], homovanillic acid [HVA] and 5-hydroxyindoleacetic acid [5-HIAA]) by high-performance liquid chromatography with electrochemical detection using procedures described in Seegal et al. (1986). The resulting chromatograms were analyzed using Maxima chromatography software (Waters Associates, Milford, MA) and data tables were generated for statistical analyses using BMDP software.

Statistical Analyses

The neurochemical data are presented as either ng of neurotransmitter or metabolite per mg wet weight tissue (for the non-human primates and adult rats) or ng neurotransmitter or metabolite per mg protein (for the perinatally-exposed rats). One-way analyses of variance were used to statistically determine the effects of either dose of PCB or period of exposure on the above neurochemical measures. Significant pair-wise t-tests between the exposed groups and the appropriate age-matched controls are indicated in the figures.

Results

Effects of Adult Exposure to Aroclor Mixtures

Exposure of adult male non-human primates to 3.2 mg/(kg·day) of either Aroclor 1016 or Aroclor 1260 for 20 or 66 weeks resulted in significant decreases in DA concentrations in the caudate nucleus, substantia nigra and hypothalamus. However for illustrative purposes, we present only the data from the caudate nucleus (Fig. 1).

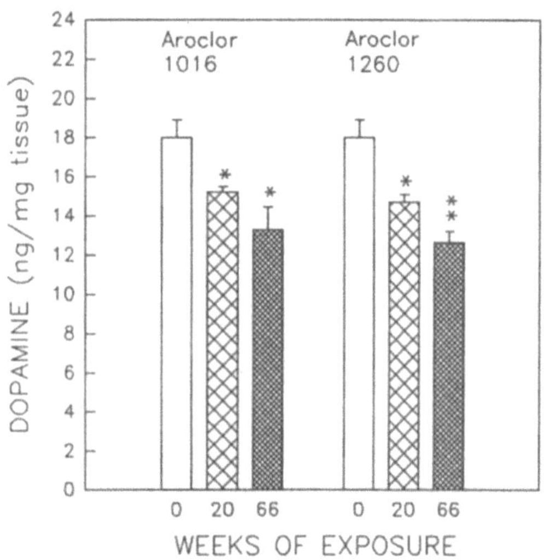

Fig 1. Effects of PCB exposure on dopamine concentrations in the caudate nucleus of adult male non-human primates. Animals were exposed for either 20 or 66 weeks to 3.2 mg/(kg·day) of either Aroclor 1016 or Aroclor 1260. N=4-6 animals per treatment group. Significantly different from control: * ($p \leq 0.05$), ** ($p \leq 0.01$).

These decreases occurred in the absence of either significant changes in body weight or the occurrence of signs of chlorobiphenyl poisoning (Seegal et al. 1991b). Furthermore, there were no significant changes in concentrations of other biogenic amines in any brain regions.

Adult rats, exposed to either 500 or 1000 ppm Aroclor 1254-adulterated chow, were sacrificed after 30 days. PCB exposure resulted in significant decreases in DA, DOPAC and HVA concentrations in the caudate nucleus (C) and significant decreases in DOPAC and HVA concentrations in the lateral olfactory tract (LOT) (Fig. 2).

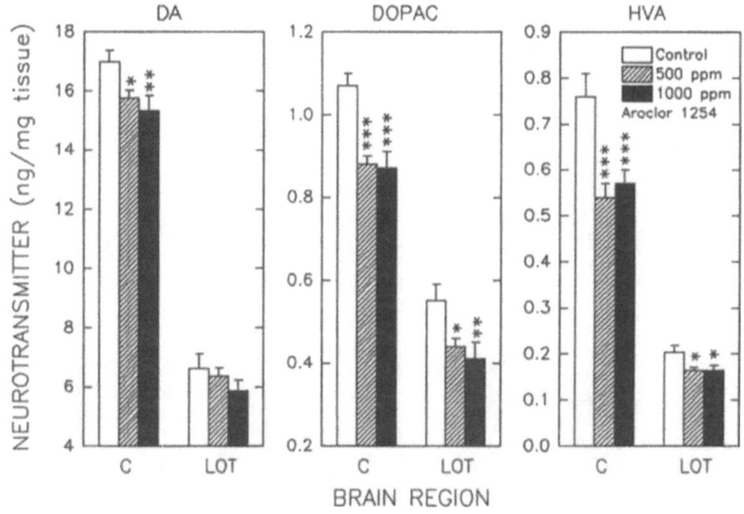

Fig 2. Effects of Aroclor 1254 exposure on dopamine (DA), DOPAC and HVA concentrations in the caudate nucleus (C) and lateral olfactory tract (LOT) of male adult rats. Animals were exposed to either 500 or 1000 ppm adulterated chow for 30 days. N=9 animals per treatment group. Significantly different from control: * ($p \leq 0.05$), ** ($p \leq 0.01$), *** ($p \leq 0.001$).

Effects of Perinatal Exposure to PCBs

Perinatal exposure of rat offspring to either 100 or 300 ppm of Aroclor 1016-adulterated chow during either gestation or lactation, or both periods resulted in significant **increases** in DA concentrations in the substantia nigra, caudate nucleus and nucleus accumbens (Figs. 3, 4), with the majority of significant changes observed in the female offspring. There were no decreases in concentrations of any biogenic amine neurotransmitters (e.g. NE or 5-HT) or metabolites (e.g. 5-HIAA, DOPAC or HVA) following perinatal exposure.

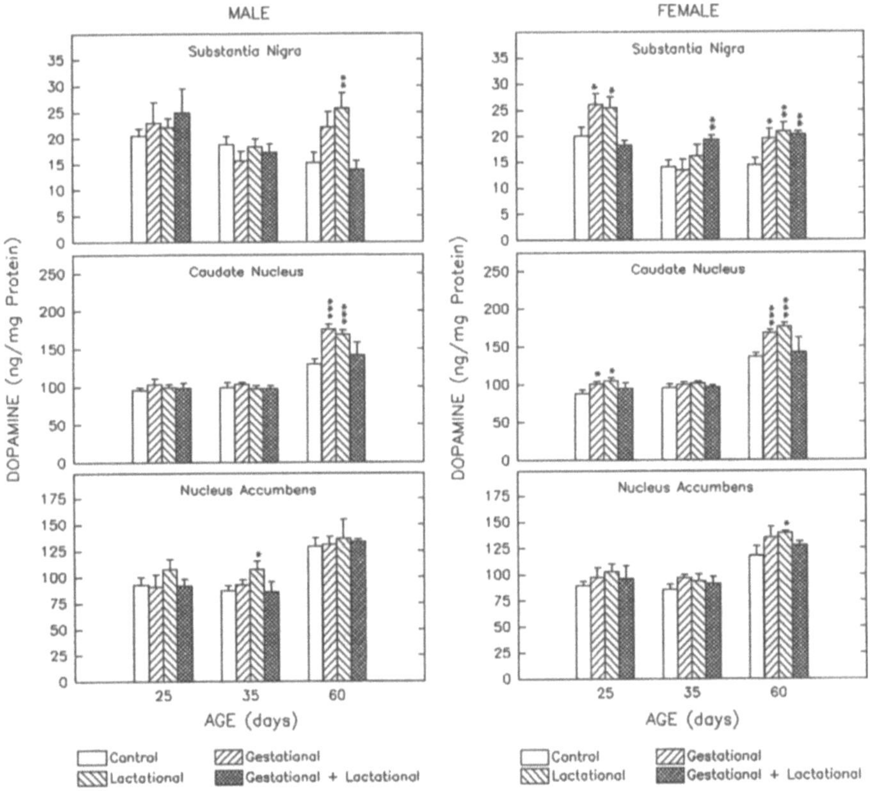

Fig 3. Effects of perinatal Aroclor 1016 exposure on dopamine concentrations in the substantia nigra, caudate nucleus and nucleus accumbens of male and female rats. Dams were exposed to 100 ppm adulterated chow during gestation (GD8-21), lactation (D0-21), or gestation + lactation, and offspring were sacrificed on days 25, 35 or 60 of age. N=6-10 animals per treatment group. Significantly different from control: * ($p \leq 0.05$), ** ($p \leq 0.01$), *** ($p \leq 0.01$).

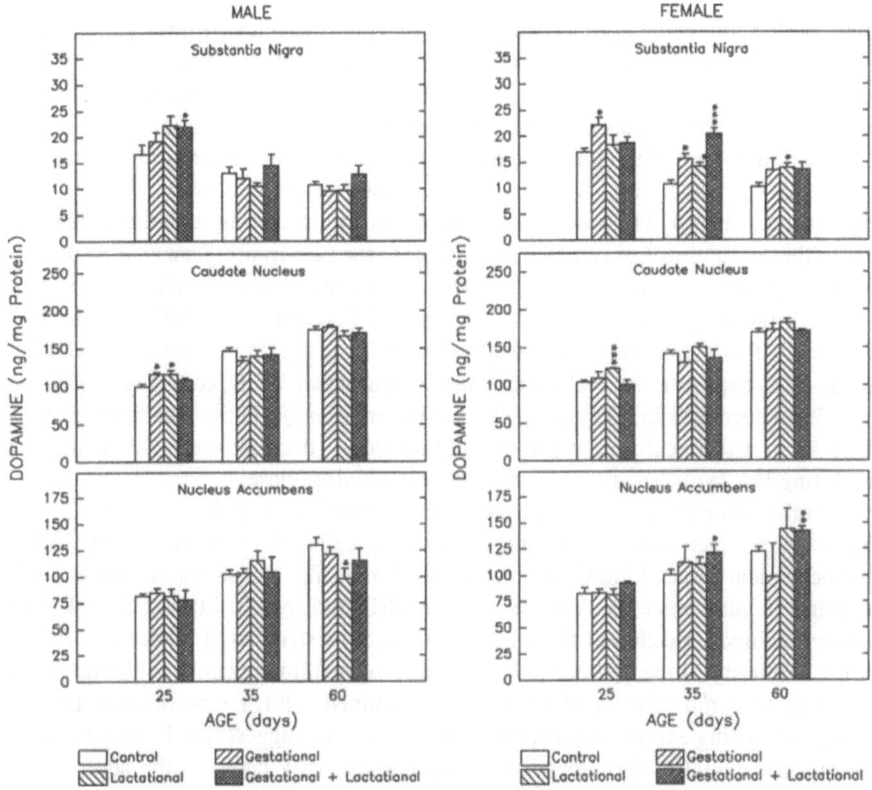

Fig 4. Effects of perinatal Aroclor 1016 exposure on dopamine concentrations in the substantia nigra, caudate nucleus and nucleus accumbens of male and female rats. Dams were exposed to 300 ppm adulterated chow during gestation (GD8-21), lactation (D0-21), or gestation + lactation, and offspring were sacrificed on days 25, 35 or 60 of age. *N*=6-10 animals per treatment group. Significantly different from control: * ($p \leq 0.05$), ** ($p \leq 0.01$), *** ($p \leq 0.01$).

Discussion

Although the elevations in regional brain DA concentrations following perinatal exposure to Aroclor 1016 were not observed across all treatment and dose conditions, the results stand in sharp contrast to the decreases in brain DA concentrations seen in rats and non-human primates exposed to Aroclors 1016, 1254 or 1260 as adults.

Although the mechanisms by which PCBs decrease brain DA concentrations in the adult are not yet fully understood, data that we have obtained in tissue culture using both a rat adrenal-gland-tumor-derived cell line (pheochromocytoma, PC12 cells) and a neuroblastoma cell line (N1E-N115) provide clues to possible mechanisms that may also occur following adult exposure to PCBs. Both Aroclor mixtures and individual PCB congeners decrease PC12 cellular DA concentrations (Shain et al. 1991) without increasing either release of DA into growth medium or the catabolism of DA to its metabolites, DOPAC or HVA (Seegal et al. 1989). These results suggest that, instead, PCBs decrease cellular DA by interfering with the synthesis of DA, either at the level of tyrosine hydroxylase, the rate-limiting enzyme responsible for the synthesis of DA, or at the level of aromatic amino acid decarboxylase (AADC), the enzyme that converts *l*-dopa to DA (Tröster et al. 1991). Using N1E-N115 cells that synthesize large quantities of *l*-dopa, but do not express AADC, we found that exposure to *ortho*-substituted, non-planar PCB congeners results in significant decreases in *l*-dopa (unpublished observations). These results suggest that, at least in tissue culture, *ortho*-substituted PCBs decrease cell DA content by inhibiting DA synthesis at the level of tyrosine hydroxylase.

On the other hand, dioxin-like coplanar congeners (e.g. 3,4,3',4' or 3,4,5,3',4') failed to produce any significant changes in cell DA content. This lack of neurochemical effect following exposure to dioxin-like congeners is similar to that reported in adult mice by Agrawal et al. (1981) and by Chishti and Seegal (1992) following direct injection of these congeners into the striatum of adult rats. Because of the similarity of the neurochemical changes seen following both adult and *in-vitro* exposure to *ortho*-substituted congeners, combined with a lack of neurochemical change following exposure to coplanar congeners, we suggest that PC12 cells are an appropriate model of the biochemical changes that may occur in the adult nervous system following exposure to PCBs.

The situation is considerably more complex in the developing nervous system. For example, both the findings of Tilson et al. (1979) and Agrawal et al. (1981), and our preliminary studies indicate that dioxin-like congeners are active when administered during development, although, as we have previously discussed, dioxin-like congeners are inactive when administered to the adult animal.

There are several possible explanations for the observed differences between adult and perinatal exposure. PCBs, like other known and suspected neuroteratogens, (e.g. alcohol, methyl mercury, drugs of abuse) may alter neuronal migration and neurogenesis (Swaab et al. 1988), which would have profound and long-lasting effects on central nervous system (CNS) function, including regulation of neurotransmitter synthesis. However, because lactational exposure to Aroclor 1016 (see Figs. 3 & 4) results in elevations in brain DA concentrations, when neurogenesis of dopaminergic neurons in the substantia nigra and caudate nucleus has already occurred (Johnston et al. 1988; Herregodts et al. 1990), we must consider other possible explanations.

Although neuronal migration and neurogenesis of dopaminergic neurons have occurred by birth (Johnston et al. 1988), adult levels of neurotransmitters and metabolites are not achieved until approximately day 28 postnatal (Herregodts et al.

1990). Thus, exposure to PCBs during this period of neurochemical maturation may alter receptor development, synaptogenesis and synthetic enzyme activity sufficiently to result in the observed elevations in brain DA concentrations. Experiments are currently underway, using pharmacological techniques, to decrease tyrosine hydroxylase activity during development in order to determine whether similar elevations in brain DA concentrations are observed.

PCB exposure during development has also been shown to alter concentrations and activity of thyroid and steroid (e.g. estrogen) hormonal systems that regulate the activity of biogenic amines and, in particular, DA (Di Paolo et al. 1985; Moore et al. 1980). Perinatal exposure to individual PCB congeners results in significant decreases in circulating T4 levels and histological evidence of excessive TSH stimulation of the thyroid gland (Ness et al. 1993). In turn, decreases in thyroid status affect CNS DA function, although studies in the adult rat suggest decreased dopaminergic function is associated with hypothyroidism. In addition, depending on their structure, PCB congeners have differential effects on estrogen function. *Ortho*-substituted congeners are weakly estrogenic (Korach et al. 1988) while dioxin-like congeners are anti-estrogenic, enhancing the metabolism of estradiol to catecholestrogens (Gierthy et al. 1988), which when directly injected into adult brain, decrease brain DA concentrations by inhibiting the activity of tyrosine hydroxylase (Lloyd, Weisz, 1978). Thus, alterations in estrogenic function induced by perinatal exposure to both *ortho*-substituted and coplanar PCBs congeners, may also produce long-lasting effects on dopaminergic function.

In summary, we have clearly demonstrated a dichotomy in the response of adult and perinatally-exposed animals to exposure to PCB congeners found in the environment. Because the major neurological effects of PCBs are associated with perinatal exposure (Jacobson et al. 1990), it is of great importance to understand the biochemical underpinnings of the cognitive dysfunctions associated with perinatal exposure to PCBs if effective therapeutic strategies are to be developed.

Acknowledgement

Research support for this work was provided in part by NIH grant ESO4913 to Richard F. Seegal.

References

Agrawal AK, Tilson HA, Bondy SC (1981) 3,4,3',4'-Tetrachlorobiphenyl given to mice prenatally produces long term decreases in striatal dopamine and receptor binding sites in the caudate nucleus. Toxicol Lett 7:417-424

Brozoski TJ, Brown RM, Rosvold HE, Goldman PS (1979) Cognitive deficit caused by regional depletion of dopamine in prefrontal cortex of rhesus monkey. Science 205:929-931

Chen Y-CJ, Guo Y-L, Hsu C-C, Rogan WJ (1992) Cognitive development of Yu-Cheng ('oil disease') children prenatally exposed to heat-degraded PCBs. Journal of the American Medical Association 268:3213-3218

Chishti MA, Seegal RF (1992) Neurochemical effects of direct injection of individual PCB congeners into rat brain alters striatal dopamine concentrations. Toxicologist 12:320

Di Paolo T, Rouillard C, Bedard P (1985) 17β-Estradiol at a physiological dose acutely increases dopamine turnover in rat brain. European Journal of Pharmacology 117:197-203

Erickson MD (1986) Analytical chemistry of PCBs. Butterworth, Boston

Gierthy JF, Lincoln DW, Kampcik SJ, Dickerman HW, Bradlow HL, Niwa T, Swaneck GE (1988) Enhancement of 2- and 16α-estradiol hydroxylation MCF-7 human breast cancer cells by 2,3,7,8-tetrachlorodibenzo-p-dioxin. Biochem Biophys Res Commun 157:515-520

Herregodts P, Velkeniers B, Ebinger G, Michotte Y, Vanhaelst L, Hooghe-Peters E (1990) Development of monoaminergic neurotransmitters in fetal and postnatal rat brain: analysis by HPLC with electrochemical detection. J Neurochem 55:774-779

Jacobson JL, Jacobson SW, Humphrey HEB (1990) Effects of in utero exposure to polychlorinated biphenyls and related contaminants on cognitive functioning in young children. J Pediatr 116:38-45

Jacobson JL, Jacobson SW, Padgett RJ, Brumitt GA, Billings RL (1992) Effects of prenatal PCB exposure on cognitive processing efficiency and sustained attention. Dev Psychol 28:297-306

Johnston MV, Barks J, Greenamyre T, Silverstein F (1988) Use of toxins to disrupt neurotransmitter circuitry in the developing brain. Prog Brain Res 73:425-446

Korach KS, Sarver P, Chae K, Mclachlan JA, McKinney JD (1988) Estrogen receptor-binding activity of polychlorinated hydroxybiphenyls: conformationally restricted structural probes. Mol Pharmacol 33:120-126

Lloyd T, Weisz J (1978) Direct inhibition of tyrosine hydroxylase activity by catechol estrogens. J Biol Chem 243:4841-4843

Moore KE, Demarest KT, Johnston CA, Alper RH (1980) Pharmacological and Endocrinological Manipulations of Tuberoinfundibular and Tuberohypophyseal Dopaminergic Neurons. In: Muller EE (ed) Neuroactive Drugs in Endocrinology. Elsevier/North-Holland Biomedical Press, pp 109-121

Ness DK, Schantz SL, Mostaghian J, Hansen LG (1993) Effects of perinatal exposure to specific PCB congeners on thyroid hormone concentrations and thyroid histology in the rat. Toxicol Lett 68:311-323

Sawaguchi T, Matsumura M, Kubota K (1988) Dopamine enhances the neuronal activity of spatial short-term memory task in the primate prefrontal cortex. Neurosci Res 5:465-473

Schantz SL, Levin ED, Bowman RE (1991) Long-term neurobehavioral effects of perinatal polychlorinated biphenyl (PCB) exposure in monkeys. Environ Toxicol Chem 10:747-756

Schwartz JH (1981) Chemical basis of synaptic transmission. In: Kandel ER, Schwartz JH (eds) Principles of neural science. Elsevier/North-Holland, New York, pp 106-120

Seegal RF, Brosch K, Bush B, Ritz M, Shain W (1989) Effects of Aroclor 1254 on dopamine and norepinephrine concentrations in pheochromocytoma (PC-12) cells. Neurotoxicology 10:757-764

Seegal RF, Brosch KO, Bush B (1986) High-performance liquid chromatography of biogenic amines and metabolites in brain, cerebrospinal fluid, urine and plasma. J Chromatogr 377:131-144

Seegal RF, Bush B, Brosch KO (1991a) Sub-chronic exposure of the adult rat to Aroclor 1254 yields regionally-specific changes in central dopaminergic function. Neurotoxicology 12:55-66

Seegal RF, Bush B, Brosch KO (1991b) Comparison of effects of Aroclors 1016 and 1260 on nonhuman primate catecholamine function. Toxicology 66:145-163

Shain W, Bush B, Seegal RF (1991) Neurotoxicity of polychlorinated biphenyls: Structure-activity relationship of individual congeners. Toxicol Appl Pharmacol 111:33-42

Swaab DF, Boer GJ, Feenstra MGP (1988) Concept of functional neuroteratology and the importance of neurochemistry. In: Boer GJ, Feenstra MGP, Mirmiran M, Swaab DF, Van Haaren F (eds) Biochemical basis of functional neuroteratology: Permanent effects of chemicals on the developing brain. Elsevier, Amsterdam, pp 3-14

Tilson HA, Davis GJ, Mclachlan JA, Lucier GW (1979) The effects of polychlorinated biphenyls given prenatally on the neurobehavioral development of mice. Environ Res 18:466-474

Tröster AI, Ruff RM, Watson DP (1991) Dementia as a neuropsychological consequence of chronic occupational exposure to polychlorinated biphenyls (PCBs). Arch Clin Neuropsychol 6:301-318

Risk Estimation and Primary Prevention of Birth Defects

P.W.J. Peters

Unit Teratology, Endocrinology and Perinatal Screening, National Institute for Public Health and Environmental Protection and Medical Faculty, University of Utrecht. PO Box 1, 3720 BA Bilthoven, The Netherlands

Toxicology recognizes several disciplines such as carcinogenicity, mutagenicity and reproductive toxicity. Reproductive toxicology is concerned with possible effects of substances on the reproductive process, i.e. on sexual organs and its functions, endocrine regulation, fertilization, transport of the fertilized ovum, implantation, and embryonic, foetal and postnatal development, until the end-differentiation of the organs is achieved. Reproductive toxicology is divided into areas related with male and female fertility, and developmental toxicology. Developmental toxicology can be further broken down in prenatal and postnatal toxicology.
The aim of this paper is: to stress the complexity of the different stages of normal prenatal development; to mention the principles that are applicable in reproductive and especially prenatal toxicology; to describe the different agents that might act as reproductive toxicants or teratogens through either males or females; to show the broad scope of different effects caused by reproductive toxic agents; and to indicate methods to detect and to recognize causes of developmental defects in a process of risk estimation and in the expectation to primarily prevent these disorders.

General

The estimation of risk for toxicity following a chemical exposure involves both a qualitative evaluation of whether or not a hazard exists (usually based on animal data), and a quantitative evaluation for determining an 'acceptable' level of exposure in humans (Kimmel and Gaylor 1988). This goal is difficult to achieve because of the multiple considerations regarding the qualitative identification of a reproductive hazard both in animals and in man; the interrelationship of these effects and other toxic effects; the interrelationships of the three types of individuals involved (males, females and offspring); the differences in (genetic) sensitivities of the individuals at risk; the association of the (late) effects and the mostly much earlier exposition in man; and the general absence of pharmacokinetic data especially with non-drugs in either animals or humans (Christian 1990). In addition to these scientific considerations, and as is evident from the environmental regulatory and risk communication history, current political and economic considerations additionally confound the interpretation of the data and the identification of appropriate models for use in risk assessment. Even the identification of background incidences of reproductive outcome are subject to political and economic influences. For example, true incidences of spontaneous abortions and infertility are difficult to identify because of the interaction of the socioeconomic background of the population evaluated, the relative availability of data on legal and illegal abortions, the age of the population, and the relative socialization of the medical community (Källén 1988). When setting priorities for different circumstances causing an adverse pregnancy outcome, the risk perception of health professionals in the area of reproduction, policy makers and the public will be very different. Environmental pollution (chemicals in air, soil and water), radiation, and chemical residues in food will get a much higher priority from the public over medicines, infectious diseases, drugs of abuse (alcohol, cocaine, heroin, smoking), maternal (metabolic) diseases, and occupational exposure. It is quite likely that policy makers will follow that public opinion. Professionals involved in the study of the causalities in reproductive and developmental defects, both experimentally and clinically, will have an opposite priority.

Reproductive Cycle

Most prescribers and users of drugs are familiar with the warning against the use of drugs during the first trimester of pregnancy. This phrase has been introduced following the thalidomide disaster that took place some 30 years ago. However, being careful only during the first three months of pregnancy for the effects of drugs and chemicals is short-sighted and in fact impossible. Firstly, because chemicals are able to interfere with all the different stages of development of the reproductive cycle, and secondly at the moment a woman learns that she is pregnant the stage of organogenesis has already progressed a great deal and preventive measures come too late.

The different stages of the reproductive cycle are in fact highlights of a continuum. These stages concern a different developmental time span each with their own sensitivity to a given toxic agent. Let us look at this cycle starting off at the earliest moment of the origination of an individual.

Primordial germ cells are present in the embryo at about one month after the first day of the last menstruation. They originate from the yolksac-entoderm outside the embryo and migrate into the undifferentiated primordia of gonads located at the medioventral surface of the urogenital ridges (Witschi 1948, Kemper and Peters 1987) They subsequently differentiate into oogonia and oocytes, or into spermatogonia. The oocytes in postnatal life are at an arrested stage of meiotic division. This division is restarted much later after birth, shortly before ovulation and is finalized after fertilization with the expulsion of the polar bodies. Thus, all female germ cells develop prenatally, no germ cells are formed after birth and during a female life span approximately 400 oocytes undergo ovulation. This implicates that a woman, whose first day of the last menstruation was 6 weeks ago and who is pregnant of a daughter, is already 'half a grandmother', but also that from far before birth the ova were exposed to endogenic but also exogenic substances. The embryonal spermatogenic epithelium, on the contrary, divides slowly by repeated mitoses, these cells do not differentiate into spermatocytes and do not undergo meiosis in the prenatal period. The onset of meiosis begins at puberty. Spermatogenesis continues throughout (reproductive) life. When the complexity of sexual development and female and male gametogenesis are considered, it becomes apparent that prenatal and postnatal chemical exposure is a special toxicological problem. The specificity of male and female developmental processes also accounts for different reactions to toxic agents in each sex.

After fertilization of the oocyte by one of the spermatozoa in the oviduct there is the stage of cell divisions and transport of the blastocyst into the endocrine-prepared uterine cavity. After implantation the bilaminar stage is formed and embryogenesis starts. These next 7 weeks are a period of finely balanced cellular events, including proliferation, migration, association, differentiation, and programmed cell death; precisely arranged to produce tissues and organs from the genetic information, present in each conceptus. During this period of organogenesis rapid cell multiplication is the rule. Complex processes of cell migration, pattern formation and the penetration of one cell group by another, characterize the later stages.

Final morphological and functional development occurs at different times during foetogenesis and is sometimes completed after birth. Postnatal adaptation characterizes the passage from intra-uterine into extra-uterine life with tremendous changes in, for example, circulatory and respiratory physiology. After puberty the above described reproductive cycle is closed.

Principles of Chemical-Induced Reproductive Toxicity

The determination of whether a chemical has the potential or capability to induce reproductive, including prenatal, toxicity is governed by essentially three established fundamental principles. It can be stated (Wilson 1982), that a prenatal toxic response depends upon the exposure to: a specific substance in a particular dose and duration; a genetically susceptible species; and a conceptus in a susceptible stage of development. Some comments on these three principles follow.

A specific substance in a particular dose and duration. As in other toxicological evaluations, embryo/foetotoxicity is governed by dose-effect relationships; the curve, however, is generally quite steep. The dose-response is of utmost importance in determining if there is a true effect. Moreover, nearly every embryo/foetotoxic chemical that has been realistically tested or that was clinically positive has been shown to have a 'no-effect' level. Another aspect worthy of mentioning here is the sometimes highly specific nature of the substance. For instance, thalidomide is a clear-cut teratogen in the human and in specific species, in contrast to its analogues that were never proven to be developmental toxicants. The effects of a substance might also be very different when such a chemical is used shortly versus chronically due to metabolizing or detoxifying enzyme induction or inhibition.

A genetically susceptible species. Not all mammalian species are equally susceptible or sensitive to the toxic influence of a given chemical. Inter- and intraspecies variability may be manifested in several ways: a chemical that acts in one species may have little or no effects in others; an embryo/foetotoxic agent may produce similar defects in various species, but these defects will vary in frequency; a substance may induce certain developmental disorders in one species that are entirely different from those induced in other species. The explanation is that there are genetic differences that influence the response. This might be further modified by environmental factors.

A conceptus in a susceptible stage of development. For different effects there exists a different sensitive period, i.e. the developmental phase during which originating, proliferating and differentiating cells and organs become susceptible to a given chemical. This period may not be related to critical morphogenetic periods, but may, for example, be related to the appearance of specific receptors. This explains how at an early period of development, malformations are induced by a substance which, at the opposite end of the timetable of development, induces functional disorders such as those of the central nervous system.

Taking these principles into account it seems to be more likely that a embryo/foetotoxic substance induces multiple malformations in a child rather than a single defect.

Characteristics of Chemicals causing Developmental Disorders

The nature and the application of chemicals, including environmental pollutants, that might induce reproductive toxicity are important and must be considered, but also the route by which they can reach the unborn. It has to be realized that a chemical substance nearly always crosses the mother-conceptus barriers, such as the histiotrophic yolksac in early pregnancy and the haematotrophic placenta later. The concentration of the chemical and its metabolite(s) might be even higher in the embryonal or foetal compartment than in the maternal system because detoxification in the embryo or foetus proceeds slower or not at all.

Partners of women who are or want to become pregnant, are rarely, admonished to avoid reproductive toxicants, apart from chemicals known to interfere with fertility. Nevertheless, theories exists suggesting that if males are exposed to reproductive toxic agents before their offspring are conceived, the incidence of birth defects among those offspring may increase. Joffe and Soyka (1982) listed three possibilities: a chemical could damage the sperm genetically; a chemical might act through the semen and impair spermatogenesis or maturation of sperm. Many substances are excreted in semen and are present with sperm before, during but also after conception if intercourse takes place during pregnancy; and the effects of the biological active chemicals could be produced indirectly on the female. However, at present the notion that substances taken by males can contribute to birth defects is not accepted or even considered seriously.

Effects and Manifestations

A wide variety of responses characterizes embryo/foetotoxicity, including teratogenicity (a term restricted for the induction of structural congenital malformations). Infertility, chromosomal and genetic disorders, spontaneous abortion, intrauterine death, prematurity, low birth weight, birth defects and functional disorders are effects of such interference with developmental and reproductive processes. The manifestation of a reproductive toxicant or teratogen can either be seen immediately after exposure or will be expressed at a much later date. Interfering with male or female germ cell development might result in infertility, decreased sperm activity and/or libido and impaired gametogenesis. Effects on the pre-implantation stage will cause early embryonic death, extra-uterine implantation or delayed transport of the fertilized zygote.

A critical phase for the induction of structural malformations usually occurs during the period of organogenesis. In man, this critical period extends from about 20-70 days after the first day of the last menstruation period, or from one week before the missed menstruation until the woman is 44 days overdue. It may be unwise to rely absolutely on this time period.

With physical agents such as X-rays used in laboratory animals, exposure can be limited to a period of minutes to discover the exact sensitive period for inducing a specific disorder. However, in chemicals we are not at all sure about the time courses of absorption, metabolism and excretion. In addition, the actual proximate teratogen may be a metabolite rather than the compound exposed to. If the moment of final differentiation of a particular organ is known with certainty, then a teratogen must have been present prior to that time, if it is supposed to be the causal agent of the malformation. For example the latest sensitive periods for anencephaly and spina bifida are 40, respectively 42 menstrual days, for cleft lip 50 days, and for diaphragmatic hernia, anal atresia and ventricular septal defects 8 weeks.

During the foetal period the manifestations from toxicological interference are growth retardation, some forms of structural malformations, foetal death, functional impairment and transplacental carcinogenesis. The period of organ and system maturation extends beyond the period of organogenesis and even beyond the prenatal period. Therefore, the susceptible period for the induction of insults that may lead to functional deficits is much longer than that for the induction of gross structural defects. Functions that have been shown experimentally to be affected by prenatal and early postnatal exposure to embryo/foetotoxicants include behaviour, reproduction, endocrine function, immune competence, xenobiotic metabolism, and various other physiological functions.

Foetal tissues are intrinsically highly vulnerable to carcinogens because of their high rate of cellular proliferation. This phenomenon has been demonstrated in rats, mice, hamsters, rabbits, pigs, dogs and monkeys with approximately 60 chemicals. Some twenty-three compounds and groups of chemicals and 7 industrial processes have been shown to induce carcinogenic effects in human beings. However, there is convincing epidemiological evidence of transplacental tumour-induction in man for only one chemical, i.e. diethylstilboestrol (DES)(Herbst et al. 1979). Exposure to DES in utero leads to the development of clear-cell adenocarcinoma of the vagina or cervix in about 1 in 1 000 of those at risk. Moreover, DES is now recognised to induce congenital malformations of the female genital tract. The effects of exposure to DES in utero for males are still highly controversial.

Detection and Recognition of Reproductive Toxicants

Different methods exist for assessing the reproductive toxicity, including teratogenicity, of chemical products. With respect to current drugs, pesticides, chemicals in food and in the environment human data are of the greatest value. The risk assessment process of potential reproductive toxic or teratogenic novel chemicals is restricted to experimental studies in laboratory animals.

Qualitative Risk Assessment

Prior to 1960 teratogenicity testing was limited and not standardized, with foetal survival generally being the main parameter. In the wake of the thalidomide disaster guidelines for reproductive studies, which cover also teratogenicity tests, for safety evaluation (Wilson and Warkany 1965, FDA 1966, World Health Organization 1967) appeared. At present, most countries require extensive information on the potential reproductive toxicological effects of a substance before a license to market is approved. Gradually a number of internationally accepted tests have been developed covering not only intra-uterine growth and development, but also gonadal function, mating performance, implantation, parturition and lactation. It has been stressed that thalidomide could not be discovered and identified as a teratogen by the then used test or with the present way of testing. This statement is made because phocomelia, as the structural malformation in the human caused by thalidomide cannot be mimicked in experimental animals. However, the types of effects seen in animal studies do not necessarily have to be the same as those in the human, but might merely be expressed as any of the four endpoints of developmental toxicity (death, weight changes, structural and functional disorders). As mentioned by Palmer (1985), the resorption rate in rats (comparable with abortions in primates) is remarkably high after administration of thalidomide. If at the time before introduction the two-litter test had been used, and if the high resorption rate was recognized as an effect, the animal studies would have predicted the developmental toxicity of that substance.

Although risk and safety assessment in reproductive toxicity should be based upon the identification of reproductive effects as expressed in no-observable-effect-levels (NOELs), birth defects are frequently judged as being out of context of the other effects; just as special endpoints that allow an extrapolation on their own, not taking into account the NOELs. At this stage of testing it has to be realised that the lowest NOEL should be identified with the four general endpoints biologically weighed. There still exist differences in sensitivity between animals and man. Therefore safety factors are applied. Such interspecies differences are explained by the assumption that experimental animals metabolize and excrete substances more rapidly.

Quantitative Risk Assessment

Many investigators attempt to develop models that can be used to assess reproductive and developmental toxicity risks more precisely than with NOELs and safety factors. The use of the NOEL (and especially the no-observable-adverse-effect-level=NOAEL) does not take into consideration the slope of the dose-response curve, the variability of the data, the influence of the number of doses tested or the spacing of the doses. To overcome some of these problems a 'benchmark dose' has been suggested (Kimmel and Gaylor 1988). This benchmark dose allows a reference dose to be established based on an estimated or desired incidence, such as 10%. It can be stated that the benchmark dose is the dose corresponding to the 10% increase in incidence or the lowest effective dose (led_{10}).

In spite of shortcomings animal testing for reproductive toxicity and teratogenicity is the only option when we deal with a novel substance. The danger of disasters, such as thalidomide and DES, may however, be further diminished by close and accurate surveillance of chemicals following their introduction into society and the environment (postmarketing surveillance). It leaves us with the problem of chemical substances already available (a long time) on the market or chemicals produced as waste components.

It can thus be understood that not all teratogenic substances were discovered by laboratory screening methods before they were introduced. With the exception of some drugs and other chemicals, all human teratogens were discovered earlier in man than in animals. Most of these discoveries were made by case reports and not primarily, by epidemiological studies (Goldberg and Golbus 1986). These reports from observant clinicians alert others to look for defects in the babies of women who have been exposed to the same substance(s) in question. If more cases are reported, then specific studies can be designed to test the hypothesis. In general case-control, cohort and - very seldom - clinical studies might provide the final evidence of such hypotheses, but only if the study is designed on biologically plausible principles.

A contribution to test hypotheses is provided by birth defects monitoring studies.

It is worth mentioning here that nearly 20 years ago a collaboration was started among birth defects registries around the world. This International Clearinghouse for Birth Defects Monitoring Systems consists at present of about 30 birth defects surveillance programs. Collaborative research is performed, but the main activity is the exchange of information collected and evaluated in each program. An important aim is a world wide surveillance and in this respect the time factor is important. Preliminary data are collected within three months at the end of each quarter and exchanged, while yearly data are published. A primary goal of the Clearinghouse is to detect changes in the incidence of specific malformations or patterns of malformations that may indicate the presence of teratogens. Follow up studies have to identify causative factors and - if possible - to eliminate them (International Clearinghouse for Birth Defects Monitoring Systems 1991).

Primary Prevention of Developmental Disorders

An essential aim in public health is prevention. Primary prevention of developmental disorders can be defined as the use of methods to prevent the origin of a developmental disorder in contrast to secondary prevention of developmental disorders, which means the prevention of the birth of a child with a developmental defect, mostly by abortion.

The approach for primary prevention of birth defects is most successful when a medicinal product prevents the initiation of a disorder for example in case of rubella vaccination or by correction of an aberrant lifestyle, such as alcohol abuse. Moreover, primary prevention of developmental disorders can be achieved when a chemical or a condition are identified as reproductive toxicant(s) and are not approved, restricted in use or removed from the market or the environment.

When thalidomide was recognized as the causal factor in an epidemic of specific birth defects, removal of the drug from the market resulted in the disappearance of the embryopathy. This event was also accompanied by a transient drastic avoidance of general drug intake by pregnant women. Health care professionals and also pregnant women need to develop a more critical attitude to the use of drugs, and exposure to chemicals during but also before pregnancy, or better to the use or exposition to chemicals during the fertile period, thereby avoiding many unnecessary and unknown risks. These remarks imply that health professionals and couples with child wish or pregnant need to be informed about proven safe drugs, and about the risks of wanted or unwanted exposures to chemicals. Lists exits of 'safe' and 'unsafe' chemicals. However, such a categorization is general and has a 'ready-made' nature.

A more 'tailor-made' procedure to determine the safety of chemicals with respect to reproduction is given by teratology information services. In 1990 two such organizations for collaboration have been established: in the Americas the Organization of Teratology Information Services (OTIS) and in Europe the European Network of Teratology Information Services (ENTIS)(Garbis et al. 1990). A teratology information service provides health professionals and patients with information as related to that specific situation, illness and chemical exposure of the person involved. These services also carry out follow-up studies to learn what happened during the course of pregnancy and about the health of the newborn.

These last examples of activities aimed at primary prevention of chemical-induced birth defects can be characterized as risk-assessment leading to risk-communication. The knowledge gained should lead to a better risk-perception and correspondingly appropriate actions. This applies to the patient, the physician, but also to public health authorities (risk-management).

Conclusions

Some developmental disorders are preventable: not only by secondary prevention, i.e. by preventing the birth of a malformed foetus through abortion, but also by primary prevention, i.e. by avoiding exposure to microorganisms, chemicals or agents that cause developmental disorders. This is not only important for physicians prescribing drugs, midwives monitoring pregnancies, but especially for parents taking the responsibility of having a baby. Moreover, this fact is of importance for public health authorities, approving drugs and chemicals to the market and for teratology information services, giving an individual risk assessment for physicians and patients, since both authorities and institutions have preventive roles to play.

The basic principle in reproductive toxicology and teratology is that the response of an organism to a teratogen depends upon the nature and the dosage of the substance, the stage of development of the embryo/foetus and its genetic make up. Chemicals can induce developmental disorders via the mother and perhaps through the father.

The common misconception that the vulnerable stage of development is only the first trimester of pregnancy is a dangerous one. From oogenesis and spermatogenesis to at least the first years of life the developing organism is specifically susceptible to harmful effects of chemicals. Developmental disorders include not only malformations visible at birth, but also spontaneous abortions, foetal death and functional deficits including behaviourial defects.

Studies both in the human and in laboratory animals make it possible to select chemicals in order to avoid exposure to developmental toxicants which are already on the market or still under development. It is questionable if environmental conditions, such as air pollution, contribute much to the load of birth defects and developmental deficits. The type of substances, the way of exposure, but above all the concentrations of these substances - taking into account the no-effect level principle - make this unlikely. Hence, primary prevention of birth defects through for example controlling air pollution will be doubtful, but indisputably indirect positive effects will be accomplished when the mother with the unborn will have an improved environment without general illnesses, caused by involuntary environmental pollution. The same holds for exposures such as to alcohol and other drugs of abuse; changing personal life styles is a most direct way to primarily prevent birth defects.

This implicates a weighed information: a multi-step procedure leading from risk-evaluation, risk-assessment and risk-communication to risk-perception and the appropriate action (risk-management).

References

Christian MS (1990) Differences in international regulatory needs and approaches. J. Amer. Coll. Toxicol. 9:21-26

FDA-Food and Drug Administration (1966) Guidelines for reproduction studies for safety evaluation of drugs for human use. Washington, D.C.

Garbis JM, Robert E, Peters PWJ (1990) Experience of two information services in Europe. Teratology 42:629-634

Goldberg JD Golbus MS (1986) The value of case reports in human teratology. Am. J. Obstet. Gynecol.154:479-482

Herbst AL, Cole P, Norusis MJ et al. (1979) Epidemiologic aspects and factors related to survival in 384 registry cases of clear cell adenocarcinoma of the vagina and cervix. Am. J. Obstet. Gynecol. 135:876-886

Joffe JM, Soyka LF (1982) Paternal drug exposure effects on reproduction and progeny. Semin. Perinat 6:116-124

International Clearinghouse for Birth Defects Monitoring Systems (1991) Congenital Malformations Worldwide, Elsevier, Amsterdam

Källén BK (1988) Epidemiology of Human Reproduction. CRC Press, Inc. Boca Raton, pp 175-176

Kemper CH, Peters PWJ (1987) Migration and proliferation of primordial germ cells in the rat. Teratology, 36:117-125

Kimmel CA, Gaylor DW (1988) Issues in qualitative and quantitative risk analysis for developmental toxicology. Risk analysis 8:15-20

Palmer AK (1985) Personal communication

Wilson JG (1982) Environment and birth defects. Academic Press, New York

Wilson JG and Warkany J (1965) Teratology: Principles and Techniques. University of Chicago press, Chicago

Witschi E (1948) Migration of the germ cells of human embryos from the yolk sac to the primitive gonadal folds. Contrib. Embryol. 32:67-80 (with nine plates).

World Health Organization (1967) Principles for testing of drugs for teratogenicity; W.H.O. Technical Report Series No. 364

Role of Mechanistic Studies of Toxicity in Risk Assessment

(Chair: S. Orrenius, Sweden)

The role of mechanistic studies in understanding target organ toxicity

Edward A Lock
Research Toxicology Section, Zeneca Central Toxicology Laboratory, Alderley Park, Macclesfield, Cheshire, SK10 4TJ, UK

INTRODUCTION

The approach to understanding the mechanism whereby a chemical produces toxicity has been described in some detail by Aldridge (1986). He discussed the toxic process in terms of a series of related events which occur following the entry of a toxic chemical into the body. These key components include:- (1) all systems that influence the **delivery** of the chemical or its metabolically derived toxic metabolite(s) to its target site(s) of action; (2) the **interaction** of the proximate toxin with critical cellular components which are commonly, but not exclusively, macromolecules; (3) the initiation of a sequence or cascade of **biochemical and physiological events** that follows the earlier primary events(s) and (4) the **consequences** to the target organ that are seen as clinical signs or symptoms which may vary depending on the target organ or organs damaged.

As toxicologists, we are mainly concerned with determining biological **thresholds** of chronic, rather than acute, exposure to chemicals. It is, however, vital to know whether acute exposure leads to chronic toxicity or whether chronic exposure will eventually lead to an accumulation of small effects until clinical signs or symptoms appear. The question of whether a threshold for a particular chemical exists can be defined in more detail if some knowledge of its mode of toxic action is known, so that instead of asking whether a given dose leads to clinical signs of toxicity, one can ask more definitive questions with regard to whether a given degree of reaction with the cellular target leads to clinical signs, or, whether a given reaction with the cellular target causes biochemical, physiological or morphological changes and what relationships there are between these effects and the clinical signs. Identification of the relevant target is an essential component of this type of approach to understanding the basis for target organ toxicity and is a rate-limiting step in the progress of research. Once some understanding of the interaction of the chemical with its target is known, comparison of the properties of this site in experimental animals with the properties of the target site in humans (using post mortem tissue) is also possible. This approach then enables a better assessment of the likely threshold in humans.

In this article I will illustrate with two selected examples from my own experience, how the above approach can help in understanding the biochemical basis for the selective target organ toxicity of chemicals in experimental animals and further help establish a more definitive approach to a threshold for exposure in humans.

1. Renal toxicity following exposure to hexachloro-1,3-butadiene (HCBD)

Some of the very early toxicology conducted with this chemical, which is formed as an unwanted by-product during the manufacture of chlorinated solvents, showed that HCBD caused selective toxicity to the kidney with little or no evidence of hepatic injury. It was also interesting that the renal injury was very selective to just one portion of the nephron, the pars recta (or S_3 segment) of the proximal tubule (Lock and Ishmael, 1979). A number of groups have reported that HCBD primarily damages the kidney, regardless of the route of exposure, or species examined (see reviews by Lock 1988, Dekant et al, 1990). Studies with radiolabelled HCBD showed that radioactivity from the chemical concentrated in the renal cortex at early times after dosing (Nash, et al, 1984) and further, that those cells which concentrated the chemical or its metabolites, where those that subsequently underwent necrosis. The primary route of metabolism of HCBD is via glutathione conjugation, this occurring to a significant extent in the liver. The conjugate formed in the liver is then secreted into bile, where the enzymes γ-glutamyltranspeptidase and other peptidases afford the cysteine conjugate. Following absorption from the gastrointestinal tract and N-acetylation in the liver, the mercapturic acid can enter renal cells via a probenecid sensitive transport system. This complex intra-organ pathway of metabolism appears to be a major route in the rat. However, several other fates do exist for glutathione conjugates formed in hepatocytes, for example, the conjugate may leave via the sinusoidal membrane and then enter renal cells via a glutathione transport system. Some direct formation of a glutathione conjugate in renal cells is also possible. The co-operation between the liver and kidney in this type of metabolism and toxicity has been recently reviewed (Stevens and Jones, 1989; Silbernagl and Heuner, 1993). The important aspect of this pathway of metabolism is that it leads to toxicity. Chemical synthesis of the glutathione, cysteine and mercapturic acid conjugates and their administration to rats caused selective toxicity to the kidney (Ishmael and Lock 1986). The basis for the selective toxicity to the pars recta is believed to be due to the transport of the mercapturate into renal cells on a probenecid-sensitive organic anion transport system (Lock and Ishmael 1985). Subsequently deacetylation affords the cysteine conjugate (Vamvakas et al, 1987; Pratt and Lock, 1988), which is a substrate for the renal enzyme cysteine conjugate β-lyase (Green and Odum, 1985). This enzyme cleaves the C-S bond, liberating ammonia, pyruvic acid and a reactive thiol moiety. Elegant studies by Dekant and Anders have shown that the proximate acylating agent is probably the thioketene (Dekant et al, 1993). Generation of this reactive electrophile in renal tubular cells leads to the acylation of proteins, glutathione, phospholipids and DNA, extensive acylation of critical proteins in membranes, especially those in the mitochondria, are thought to lead to cytotoxicity (Jones et al, 1986; Wallin et al, 1987; Groves et al, 1991).

Many groups have extended and built on these findings and those of Schultze and coworkers (1959) and Parker (1965) such that we now have a good working knowledge on the mechanism for the selective toxicity of a whole series of haloalkenes to the kidney (Commandeur J N M and Vermeulen N P E, 1990; Dekant et al, 1993).

How can we utilise this knowledge gained on the metabolism, disposition and activation to help assess the likely response in humans? Several key steps in the metabolism of HCBD can be examined in vitro using human tissue, and the findings compared with those seen in rodents. For example, the rate of conjugation of HCBD with glutathione by human liver enzymes, the rate of metabolism of the cysteine conjugate by human renal cysteine conjugate β-lyase and the cytotoxicity of these conjugates to human renal proximal tubular cells in culture can be assessed. To date, only limited data with HCBD conjugates is available with human tissue, although some studies are available with other haloalkene cysteine conjugates (eg. Green et al, 1990).

In rat liver HCBD conjugation with glutathione is mediated by both cytosolic and microsomal glutathione S-transferases, the lipophilic nature of HCBD making it a better substrate for the microsomal enzyme (Wolf et al 1984). Human liver cytosolic and microsomal fractions will metabolise HCBD to form a glutathione conjugate with a forty-fold greater activity in the microsomal fraction (Oesch and Wolf, 1989). Both the human and rat liver microsomal glutathione S-transferase responsible for HCBD conjugation have been isolated and purified (Morgenstern and DePierre, 1983; McLellan et al, 1989). The specific activity with HCBD as the substrate was 10 and 6 nmol/min/mg protein for the rat and human enzyme, respectively. Thus, in both species the enzyme is able to conjugate HCBD with glutathione indicating no major difference in this metabolic step.

Cysteine conjugate β-lyase has been isolated and purified from rat kidney cytosol (Stevens et al, 1986; McFarlane et al, 1989) and human kidney cytosol (Lash et al, 1990). Rat kidney cytosol will readily cleave S-1,2,3,4,4,-pentachlorobuta-1,3-dienyl-L-cysteine (PCBC) the cysteine conjugate of HCBD (Green and Odum, 1985) although detailed kinetic studies with this substrate have not been reported. Some studies are available with the human renal enzyme (Lash et al, 1990; Buckberry et al, 1990) although neither of these groups have studied PCBC. Findings with the substrates S-2-benzothiazoyl-L-cysteine (BTC) and S-1,2-dichlorovinyl-L-cysteine (DCVC) indicated that the activity of human kidney cytosol was only 10% of that present in rat kidney cytosol. We have examined a number of samples of human renal cytosol for β-lyase activity with a series of substrates including PCBC and found the production of pyruvic acid to be at least 50-fold lower in human tissue than in rat renal cytosol (McCarthy et al, 1992). In fact, the activity with five different samples was at or close to the limits of detection of about 5nmol pyruvic acid formed/mg protein/120 min. Thus, the ability of human renal cortical β-lyase to metabolise PCBC appears to be many-fold lower than that of the rat renal enzyme.

Only one study has so far been reported with human proximal tubular cells in culture, where a number of nephrotoxic haloalkene cysteine conjugates were examined (Chen et al, 1990). PCBC at 500μM caused cytotoxicity over 24 hour which could be prevented by inhibition of β-lyase with aminooxyacetic acid. Thus, under these conditions sufficient metabolism of PCBC can occur to cause cytotoxicity. PCBC is also toxic to rat or rabbit renal cells, however, at much lower doses and within a shorter time frame, for example, 25μM PCBC will produce marked toxicity to rabbit renal cells within 4hr (Groves et al, 1991). This species difference in sensitivity presumably reflects the rate of formation of the toxic species, as shown by the relative activities of the enzyme β-lyase, and any species difference that may exist in other pathways of detoxification such as S-oxidation, S-methylation and N-acetylation.

Based on our knowledge of the role of metabolism in the production of nephrotoxicity it appears that humans have the ability to form the glutathione conjugate and can metabolise PCBC via β-lyase to form a toxic metabolite. However, the rate of metabolism particularly via the β-lyase enzyme appears to be many-fold lower than that of rat tissue.

2. Heptatic Protoporphyria

Recent studies have shown the chemical 1-[4-(3-acetyl-2,4,6-trimethylphenyl)-2,6-cyclohexanedionyl]-O-ethyl propionaldehyde oxime (for short ATMP) produces hepatic protoporphyria in the mouse following a single dose and further that these changes are not seen in similarly treated rats, guinea pigs, hamsters or in chick embryo cultures (Frater et al, 1993). Hepatic protoporphyria is a disorder of liver haem metabolism characterised by inhibition of the enzyme ferrochelatase, the last enzyme in the pathway, and a consequent accumulation of protoporphyrin, the substrate for the inhibited enzyme. When this disorder is marked, protoporphyrin accumulates in high concentrations in the biliary system, leading to cholestasis and biliary cirrhosis. Two drugs known to induce such a disorder are 3,5-diethoxycarbonyl-1,4-dihydrocollidine (DDC) and griseofulvin (GF). Both these drugs are thought to be metabolised by cytochrome P-450 as suicide substrates, so that a fragment of drug (a methyl group) becomes attached to the haem moiety of the enzyme producing N-methylprotoporphyrin, a powerful inhibitor of ferrochelatase (see reviews by DeMatteis et al 1987; Marks et al, 1988).

In order to understand in more detail the basis for the species difference in response seen with ATMP studies were undertaken to ascertain the mechanism whereby ATMP, which is structurally unrelated to either DDC or GF produces protoporphyria. A single dose of ATMP to the mouse produced a marked accumulation of hepatic porphyrins within 24 hr which was associated with a 70% inhibition of liver ferrochelatase. The two drugs DDC and GF produce a similar effect, associated with an accumulation in the liver of green pigments which inibit ferrochelatase (DeMatteis and Gibbs, 1980). This inhibitory pigment has been identified as N-methylprotoporphyrin (DeMatteis et al, 1980; Oritz de Montellano et al, 1981a), the N-methyl group originating from the drug both in the case of DDC (DeMatteis et al, 1981; Oritz de Montellano et al, 1981b) and possibly also with GF (Holley et al, 1991) Studies with ATMP showed that a green pigment did accumulate in the liver after dosing mice and the isolated pigment markedly inhibited the enzyme ferrochelatase in vitro (Frater et al, 1993). The green pigment was identified by its electronic absorption spectrum and chromatographic properties on HPLC as N-methylprotoporphyrin. Two inhibitors of hepatic cytochrome P-450 namely SKF 525A and piperonyl butoxide when given to mice before ATMP, afforded protection against ATMP-induced porphyria and the production of N-methylprotoporphyrin suggesting a role for cytochrome P-450 in the production of this metabolic disorder.

These findings indicate that ATMP requires metabolism in the mouse to produce a reactive species which in turn alkylates the haem moiety of liver cytochrome P-450, thus producing N-methylprotoporphyrin. This inhibits ferrochelatase and as a secondary response protoporphyrin accumulates. As indicated previously studies in a number of other laboratory species showed no accumulation of porphyrins in the liver or inhibition of mitochondrial ferrochelatase (Frater et al, 1993). Hence the likely basis for the species difference in protoporphyria is presumably due to the absence or reduced activity of the pathway of metabolism required to form the reactive metabolite of ATMP which can alkylate the haem moiety.

It is possible to model in cultured hepatocytes the key pathways in haem metabolism associated with protoporphyria following the drugs DDC and GF (Brady and Lock, 1992). Addition of either of these drugs to cultured mouse hepatocytes produces a marked inhibition of the enzyme ferrochelatase and consequent accumulation of porphyrins, predominantly protoporphyrin in the culture medium. Similar studies with DDC and GF using rat liver hepatocytes, showed a marked inhibition of mitochondrial ferrochelatase and accumulation of protoporphyrin with DDC but not with GF (Brady et al, 1993). This species difference in response with GF is similar to that seen in vivo where the mouse shows a marked hepatic protoporphyria (DeMatteis and Rimington, 1963) while the other species examined, rats, guinea pigs, rabbits or chickens were refractory (Weston Hurst and Paget, 1963; Racz and Marks, 1969). Thus GF like ATMP appears to show a species specific protoporphyria. Recent studies with DDC and GF using primary cultures of human hepatocytes has shown inhibition of ferrochelatase with DDC but not with GF (Brady AM personal communications). The likely basis for the species difference in protoporphyria with GF is presumably due to a reduced activity of the pathway of metabolism required to form the reactive metabolite which can alkylate the haem moiety. This approach of understanding the basis for the hepatic protoporphyria, enables studies to be conducted in vitro with hepatocytes from humans help to assess the likely risk of drugs such as GF and ATMP producing protoporphyria.

CONCLUSION

The two examples illustrate how by understanding in some detail the mechanism underlying the basis for the target organ toxicity, it is possible to obtain a better assessment of the likely risk of these chemicals to humans. In both cases metabolism of the chemical to generate a reactive species is a critical step in the toxic process, the key targets being renal mitochondrial function and inhibition of hepatic haem metabolism. In both examples there is a marked species difference in the rate of metabolism of the chemical between rodents and humans. With HCBD the key metabolism via β-lyase appears to be considerably lower in human kidney while with GF and ATMP the pathway of metabolism appears to be absent or possess very low activity in all species examined except the mouse. By understanding the mechanism of target organ toxicity it is possible to make a more realistic assessment of the risk of exposure to a chemical and determine the likely threshold for toxicity.

REFERENCES

Aldridge WN (1986) The biological basis and measurement of thresholds. Ann Rev Pharmacol Toxicol 26: 39-58

Brady AM and Lock EA (1992) Inhibition of ferrochelatase and accumulation of porphyrins in mouse hepatocyte cultures exposed to porphyrinogenic chemicals. Arch Toxicol 66: 175-181

Brady AM, Hasmall RL and Elcombe BM (1993) Comparison of the effects of griseofulvin and dihydropyridines on ferrochelatase activity and porphyrin accumulation in primary cultures of mouse and rat hepatocytes. Toxicol in vitro in press

Buckberry LD, Blagbrough IS, Bycroft BW and Shaw PN (1990) Human renal C-S lyases: two cytosolic isoenzymes. Toxicol Letts 53: 253-255

Chen JC, Stevens JL, Trifillis AL and Jones TW (1990) Renal cysteine conjugate β-lyase mediated toxicity studies with primary cultures of human proximal tubular cells. Toxicol Appl Pharmacol 103: 463-473

Commandeur JNM and Vermeulen NPE (1990) Molecular and biochemical mechanisms of chemically induced nephrotoxicity: a review. Chem Res Toxicol 3: 171-194

Dekant W, Anders MW and Monks TJ (1993) Bioactivation of halogenated xenobiotics by S-conjugate formation. In. Renal Disposition and nephrotoxicity of xenobiotics. Ed Anders MW, Dekant W, Henschler D, Oberleithner H and Silbernagl S. Acaemic Press London pp187-215

Dekant W, Vamvakas S and Anders MW (1990) Bioactivation of hexachlorobutadiene by glutathione conjugation. Fd Chem Tox 28: 285-293

De Matteis F and Rimington C (1963) Disturbance of porphyrin metabolism caused by griseofulvin in mice. Br J Dermatol 75: 91-104

De Matteis F and Gibbs AH (1980) Drug-induced conversion of liver haem into modified porphyrins: evidence for two classes of products. Biochem J 187: 285-288

De Matteis F, Gibbs AH and Holley AE (1987) Occurrence and biological properties of N-methyl protoporphyrin. Ann NY Acad Sci 514: 30-40

De Matteis F, Gibbs AH, Farmer PB and Lamb JH (1981). Liver production of N-alkylated porphyrins caused in mice by treatment with substituted dihydropyridines evidence that the alkyl group on the pyrrole nitrogen atom originates from the drug. FEBS Lett 129: 328-331

De Matteis F, Gibbs AH, Jackson AH, Weerasinghe S (1980) Conversion of liver haem into N-substituted porphyrins or green pigments. Nature of the substituent at the pyrrole nitrogen atom. FEBS Lett 119: 109-112

Frater Y, Brady A, Lock EA and De Matteis F (1993) Formation of N-methyprotoporphyrin in chemically-induced protoporphyria Arch Toxicol 67: 179-185
Green T and Odum J (1985) Structure/activity studies of the nephrotoxic and mutagenic action of cysteine conjuates of chloro- and fluoroalkenes. Chem Biol Interact 54: 15-31
Green T, Odum J, Nash JA and Foster JR (1990) Perchloroethylene-induced rat kidney tumours: an investigation of the mechanisms involved and their relevance to humans. Toxicol Appl Pharmacol 103: 77-89
Groves CE, Schnellmann RG, Sokol PP, Steffens TG and Lock EA (1991). Pentachlorobutadienyl-L-cysteine (PCBC) toxicity: The importance of mitochondrial dysfunction. J Biochem Toxicol 6: 253-260
Holley A, Frater YA, Gibbs AH, De Matteis F, Lamb JH, Farmer PB, Naylor S (1991) Isolation of two N-monosubstituted protoporphyrins being either the whole drug or a methyl group on the pyrrole nitrogen atom, from liver of mice given griseofulvin. Biochem J 274: 843-848
Ishmael J and Lock EA (1986) Nephrotoxicity of hexachlorobutadiene and its glutathione-derived conjugates. Toxicol Pathol 14: 258-262
Jones TW, Wallin A, Thor H, Gerdes RG, Ormstad K and Orrenius S (1986) The mechanism of pentachlorobutadienyl-glutathione nephrotoxicity, studies with isolated rat renal epithelial cells. Arch Biochem Biophys 251: 504-513
Lash LH, Nelson RM, Van Dyke RA and Ander MW (1990) Purification and characterisation of human kidney cytosolic cysteine conjugate β-lyase activity. Drug Metab Dispos 18: 50-54
Lock EA (1988) Studies on the mechanism of nephrotoxicity and nephrocarcinogenicity of halogenated alkenes. CRC Crit Rev 19: 23-42
Lock EA and Ishmael J (1979) The acute toxic effects of hexachloro-1,3-butadiene on the rat kidney. Arch Toxicol 43: 47-57
Lock EA and Ishmael J (1985) Effect of the organic acid transport inhibitor probenecid on renal cortical uptake and proximal tubular toxicity of hexachloro-1,3-butadiene and its conjugates. Toxicol Appl Pharmacol 81: 32-42
McCarthy R, Hawksworth GM and Lock EA (1992) Subcellular localisation of C-S lyase activity in human renal cortex. 13th European workshop Drug Metabolism. Sept 21-25, Bergamo, Italy
McFarlane M, Foster JR Gibson GG, King LJ and Lock EA (1989) Cysteine conjugate β-lyase of rat kidney cytosol: characterisation, immunocytochemical localisation and correlation with hexachlorobutadiene nephrotoxicity. Toxicol Appl Pharmacol 98: 185-197
McLellan LI, Wolf RC and Hayes JD (1989) Human microsomal glutathione S-transferase Biochem J 258: 87-93
Marks GS, McCluskey SA, Mackie JE, Riddick DS and James CA (1988) Disruption of hepatic heme biosynthesis after interaction of xenobiotics with cytochrome P-450. FASEB J 2: 2774-2783

Morgenstern R and DePierre JW (1983) Microsomal glutathione transferase, purification in unactivated form and further characterisation of the activation process, substrate specificity and amino acid composition Eur J Biochem 134: 591-597

Nash JA, King LJ, Lock EA and Green T (1984). The metabolism and disposition of hexachloro-1,3-butadiene in the rat and its relevance to nephrotoxicity. Toxicol Appl Pharmacol 73: 124-137

Oesch F and Wolf CR (1989) Properties of the microsomal and cytosolic glutathione transferases involved in hexachloro-1,3-butadiene conjugation. Biochem Pharmacol 38: 353-359

Ortiz de Montellano PR, Beilan HS, Kunze KL (1981a) N-Methyl protoporphyrin IX: chemical synthesis and identification as the green pigment produced by 3,5-diethoxycarbonyl-1,4-dihydro-collidine treatment. Proc Natl Acad Sci USA 78: 1490-1494

Ortiz de Montellano PR, Beilan HS, Kunze KL (1981b) N-Alkyl-protoporphyrin IX: formation in 3,5-dicarbethoxy-1,4-dihydro-collidine-treated rats. Transfer of the alkyl group from the substrate to the porphyrin. J Biol Chem 256: 6708-6713

Parker VH (1965) A biochemical study of the toxicity of S-dichlorovinyl-L-cysteine. Fd Chem Toxicol 3: 75-84

Pratt IS and Lock EA (1988) Deacetylation and further metabolism of the mercapturic acid of hexachloro-1,3-butadiene by rat kidney cytosol, in vitro. Arch Toxicol 62: 341-345

Racz WJ and Marks GS (1969). Drug-induced porphyrin biosynthesis-II. Simple procedure for screening drugs for porphyria-inducing activity. Biochem Pharmacol 18: 2009-2018

Schultze MO, Klubes P, Perman V, Mizuno NS, Bates FW and Sautter JH (1959) Blood dyscrasia in cattle induced by S-(dichlorovinyl)-L-cysteine. Blood 14: 1015-1025

Silbernagl S and Heuner A (1993) Renal transport and metabolism of mercapturic acids and their precursors, the S-conjugates of glutathione and cysteine. In. Renal Disposition and nephrotoxicity of xenobiotics. Ed. Anders MW, Dekant W, Henschler D, Oberleithner H and Silbernagl S. Academic Press, London pp135-154

Stevens JL and Jones DP (1989) The mercapturic acid pathway: biosynthesis, intermediary metabolism and physiological disposition. In: Glutathione: chemical, biochemical and medical aspects. Eds D Dolphin, R Poulson and O Avramovic, Wiley, New York, pp46-84

Stevens JL, Robbins JD and Byrd RA (1986) A purified cysteine conjugate β-lyase from rat kidney cytosol. J Biol Chem 261: 15529-15537

Vamvakas S, Dekant W, Berthold K, Schmidt S, Wild D and Henschler D (1987). Enzymatic transformation of mercapturic acids derived from halogenated alkenes to reactive and mutagenic intermediates. Biochem Pharmacol 361: 2741-2748

Wallin A, Jones TW, Vercesi AE, Cotgreave I, Ormstad K and Orrenius (1987). Toxicity of S-pentachlorobutadienyl-L-cysteine studies with isolated rat renal cortical mitochondria. Arch Biochem Biophys 258: 365-372

Weston Hurst E and Paget GE (1963) Protoporphyrin, cirrhosis and hepatomata in the livers of mice given griseofulvin. Br J Dermatol 75: 105-112

Wolf CR, Berry PN, Nash JA, Green T and Lock EA (1984) The role of microsomal and cytosolic glutathione-S-transferases in the conjugation of hexachloro-1,3-butadiene and its possible relevance to toxicity. J Pharmacol Exp Ther 228: 202-208

Rational Species Extrapolation of Toxic Effects

F. Oesch and B. Diener

Institute of Toxicology, University of Mainz,
Obere Zahlbacher Str. 67, D-55131 Mainz, Germany

One of the main problems in toxicology and pharmacology is the extrapolation from one species to another, particularly from laboratory animals to man or in ecotoxicology from laboratory animals to free living species. Many toxicological data were raised only in one or a few species of laboratory animals. In the last decades more and more experimental data were obtained from in vitro-systems. Therefore extrapolations from in vitro to in vivo and from species to species are required. Species differences in uptake, distribution, metabolism, site of action, elimination and accumulation have to be taken into account. Most foreign compounds are metabolically transformed by diverse xenobiotica metabolizing enzymes. The metabolism can lead to detoxication and elimination but metabolism can also be responsible for toxication. For most foreign compounds the metabolism is predominantly controlled by hepatic metabolism. Here the extrapolation from in vitro to in vivo and between species is discussed exemplary for the liver.

Metabolizing cellular in vitro systems

More and more in vitro-systems are used for toxicological investigations. The reasons are manifold: reduction of the in vivo complexity to more simple in vitro-systems improving the possibilities for the establishment of causal relationships; time saving; cost reduction; ethical reasons; and in some cases there will be a better extrapolation to man, namely if in vitro systems of human origin are included.

The liver is the organ with the highest activities of most xenobiotic metabolizing enzymes in vertebrates. Today some in vitro-systems representing liver metabolism with different complexity are available. Most complex and closest to the in vivo situation is the isolated perfused liver followed by liver slices. In both systems the structure of the liver is maintained to a greater or lesser extent, respectively. Since the introduction of the two step collagenase isolation of liver parenchymal cells (hepatocytes) (Berry & Friend 1969) isolated hepatocytes are a widely used in vitro-system. The disadvantage of these isolated cells is that they die within one day in suspension. They survive for a longer period when they are taken into cell culture but in this case they rapidly dedifferentiate and lose many of the activities of their xenobiotic metabolizing enzymes. The co-culture of the hepatocytes with a cell line (Guguen-Guillouzo et al. 1983, Utesch & Oesch 1992[a]) or the culture of hepatocytes on extracellular matrix compounds (Reid et al. 1986, Utesch & Oesch 1992[b]) leads to a stabilization of the differentiated status of the hepatocytes to some extent. With the progress in cell and molecular biology immortalized cell lines (hybrid cell lines) (Utesch et al. 1992[a]) and cell lines transfected with genes coding for xenobiotic metabolizing enzymes (Döhmer et al. 1988) are additionally used.

Isolated and cryopreserved hepatocytes

Since most foreign compounds are predominantly metabolized in the liver, in most cases there is a good correlation between the metabolism of a foreign compound in vivo and in isolated hepatocytes of the same species. This was clearly demonstrated for the two drugs Alinidin and Nebracetam (Ruß 1992).

Freshly isolated hepatocytes from different species in suspension can be used as an easy screening system to compare the hepatic metabolism of a xenobiotic compound for different species. If freshly isolated human hepatocytes are also available it will be possible to make predictions for the hepatic metabolism in man. For further investigations the species with the metabolism closest to man can be choosen. This will help to get information for a convincing extrapolation at an early time point of the investigation.

One major obstacle for this procedure is the limited access to human liver. To overcome this a method for the successful cryopreservation of isolated hepatocytes was developed (Utesch et al. 1992[b], Diener et al. 1993), the two main points being to allow sufficient time at zero degree for thorough equilibration of the entire sample at low temperature and to decrease the temperature from the freezing point extremely fast in order to prevent the damage to the cells caused by the crystallization heat. Freshly isolated hepatocytes of various species were frozen with this computer-controlled cryopreservation protocol. After thawing the hepatocytes showed satisfactory maintainance of viability and xenobiotic metabolizing enzyme activities. The activities of microsomal epoxide hydrolase (styrene 7,8-oxide = STO), phenol sulfotransferase (2-naphthol) and of the UDP-glucuronosyl transferase (1-naphthol) were maintained in the same range as in freshly isolated hepatocytes, whereas there was a decrease in the activities of the cytosolic epoxide hydrolase (trans-stilbene-oxide = TSO) and the glutathione S-transferase (1-chloro-2,4-dinitrobenzene = CDNB) (Table 1). This decrease correlated with the viability of the isolated hepatocytes. The viability was judged by the trypan blue exclusion test.

In conclusion, cryopreserved liver parenchymal cells from different species including man can be a helpful in vitro tool for rational species extrapolation. Thereafter the species with the metabolism pattern of the compound in question closest to man will be the best choice for an in vivo investigation of toxicity.

Hepatocytes in culture

Hepatocytes which retain their differentiated status in culture for sufficient time can be used for toxicological investigations such as genotoxicity e.g. unscheduled DNA-synthesis; cytotoxicity (Jover et al. 1992); detection of tumor promotors which act on the cell to cell communication (Ruch et al. 1990); investigations of regulatory mechanisms e.g. proliferation of peroxisomes

Table 1: Activities of xenobiotic metabolizing enzymes: comparison between freshly isolated and cryopreserved human hepatocytes from two isolations.

Donor		% activity of freshly isolated cells				
No.	Viab.	mEH	cEH	GST	UGT	ST
1	76	117	45	34	87	102
2	91	104	89	64	nt	106

mEH = microsomal epoxide hydrolase (STO); cEH = cytosolic epoxide hydrolase (TSO); GST = glutathione S-transferase (CDNB); ST = phenol sulphotransferase (2-naphthol); UGT = UDP-glucuronosyl transferase (1-naphthol); nt = not tested.

(Chatterjee et al. 1987); enzyme induction and repression studies. In all cases it is necessary that the cultured hepatocytes retain their differentiated status more or less.

Hepatocytes in conventional primary culture show a drastic decrease in the activities of xenobiotic metabolizing enzymes with culture time. Therefore it is proposed that hepatocytes of the rat in primary culture may be used only for a short period for in vitro studies where an intact metabolizing system is necessary. It was reported (Guillouzo 1986) that human hepatocytes are more stable in culture. In our laboratory it was possible to keep human hepatocytes for up to four weeks in conventinal culture.

Hepatocytes can be stabilized in co-culture with some cell lines for long-term culture (Gugen-Guilluozo et al. 1983, Utesch & Oesch 1992[a]). In dependency of the used cell line various xenobiotic metabolizing enzymes are stabilized to a different extent (Figure 1). The use of C3H mouse embryo fibroblasts instead of non parenchymal rat liver epithelial cells (NEC) has the advantage that they are freely available as a commonly used cell line and that they are stabilizing several xenobiotic metabolizing enzymes better than rat liver epithelial cells (Utesch & Oesch 1991, Utesch & Oesch 1992[a]).

Figure 1:Activities of xenobiotic metabolizing enzymes in 7 day-old hepatocyte monoculture and coculture with different cell lines.

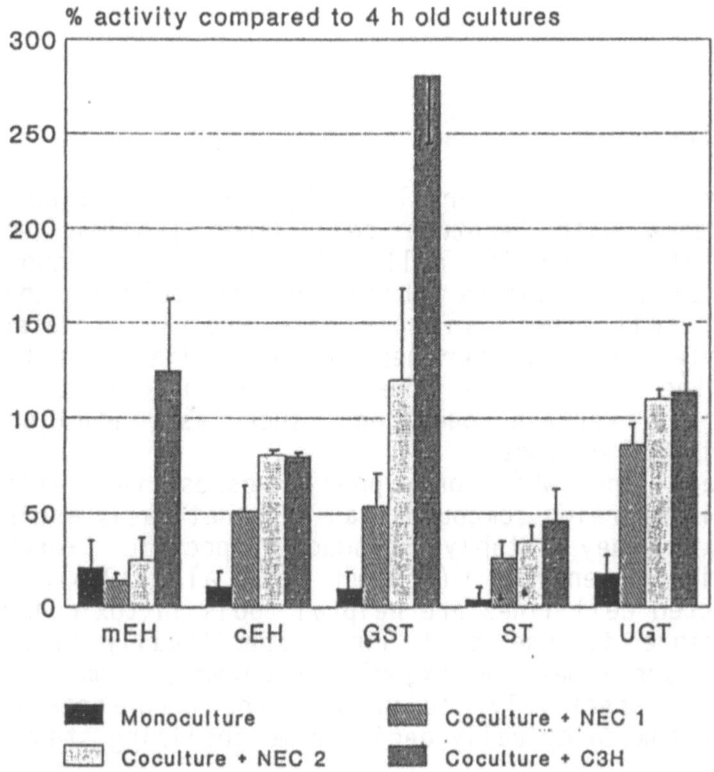

mean of at least 3 independent experiments with standard deviation

mEH = microsomal epoxide hydrolase (substrate: styrene oxide); cEH = cytosolic epoxide hydrolase (substrate: trans-stilbene oxide); GST = glutathione S-transferase (substrate: 1-chloro-2,4-dinitrobenzene); ST = phenol sulfotransferase (substrate: 2-naphthol); UGT = UDP-glucuronosyl transferase (substrate: 1-naphthol). NEC = non parenchymal rat liver epithelial cell line, no. 1 from Utesch (Mainz), no. 2 from Guillouzo (Rennes). C3H = mouse embryo fibroblast C3H cell line.
The activities are compared in terms of activities per culture dish to those of 4 h old cultures.

We and others (Corlu et al. 1991, Diener, Traiser, Oesch manuscript in preparation) showed that for the stabilization of the hepatocytes in co-culture plasma membrane components of the cell lines are mainly responsible.

Cell lines

Hepatocytes can be immortalized by virus infection or by cell fusion with hepatoma cell lines (Föllman et al. 1989). Those hybrid cell lines possess xenobiotic metabolizing activities which are in between the two parental cell types, in this case freshly isolated rat hepatocytes and a Reuber hepatoma cell line. A potential use of the hybrid cell line in toxicological studies is limited to certain questions such as induction and repression of enzymes.

Cell lines which only poorly possess the ability to metabolize foreign compounds can be genetically engineered so that they stably produce specific xenobiotic metabolizing enzymes (Doehmer et al. 1988). These transfected cell lines are helpful tools in toxicology. It is possible to use cell lines specifically expressing defined isoenzymes for example isoenzymes from different species, respectively, to predict the occurrence of one distinct toxicologically decisive metabolizing step.

Conclusions

Since for most foreign compounds the metabolism is predominantly controlled by hepatic metabolism, isolated hepatocytes predict the pattern of the overall metabolism of a given compound quite well for most chemical compounds. Methods were developed for cryopreserving isolated hepatocytes from man and other species with satisfactory maintenance of foreign compound metabolizing enzyme activities. The installation of a bank of cryopreserved hepatocytes from different species is possible and may be used for rational species extrapolation. For some toxicological questions it is necessary to have cultured hepatocytes which retain their differentiated status in culture for a time period sufficient for the problem in question. This might be achieved by the model of co-culturing hepatocytes with diverse cell lines. But there are differences in the pattern of stabilization from cell line to cell line. Questions on metabolic action of individual isoenzymes can be answered by the use of genetically engineered cell lines.

All mentioned in vitro-systems, especially such wich contain differentiated human cells or at least human isoenzymes are helpful in the rational species extrapolation of toxic effects from animal to man.

References

Berry MM, Friend DS (1969) High-yield preparation of isolated rat liver parenchymal cells. J Cell Biol 43:506-520

Chatterjee B, Murty RCV, Olson MJ, Roy AK (1987) Cloning and expression of rat liver cDNA for peroxisomal enoyl-CoA hydratase, 3-hydroxyacyl-CoA dehydrogenase in GT11. Eur J Biochem 166:273-278

Corlu A, Kneip B, Lhadi C, Leray G, Glaise D, Baffet D, Guguen-Guillouzo C (1991) A plasma membrane protein is involved in cell contact-mediated regulation of tissue-specific genes in adult hepatocytes. J Cell Biol 115:505-515

Diener B, Utesch D, Beer N, Dürk H, Oesch F (1993) A method for the cryo-preservation of liver parenchymal cells for studies of xenobiotics. Cryobiology 30:116-127

Doehmer J, Dogra S, Friedberg T, Monier S, Adesnik M, Glatt H, Oesch F (1988) Stable expression of rat cytochrome P-450IIB1 cDNA in Chinese hamster cells (V79) and metabolic activation of aflatoxin B_1. Proc Natl Acad Sci USA 85:5769-5773

Föllmann W, Immenschuh S, Gebracht U, Eigenbrodt E, Katz N, Petzinger E (1989) Morphology and enzymatic pattern of rat hepatocytoma cells. Naunyn Schemiedebergs Arch Pharmacol Suppl II 340:R71

Guguen-Guilluozo C, Clement B, Baffet G, Beaumont C, Morel-Chany E, Glaise D, Guillouzo A (1983) Maintenance and reversibility of active albumin secretion by adult rat hepatocytes co-cultured with another liver epithelial cell type. Exp Cell Res 143:47-54

Guillouzo A (1986) Use of isolated and cultured hepatocytes for xenobiotic metabolism and cytotoxicity. In: Guillouzo A, Guguen-Guillouzo C (eds) Isolated and cultured hepatocytes. John Libbey Eurotext, London, pp 313-332

Jover R, Ponsonda X, Castell JV, Gomez-Lechon MJ (1992) Evaluation of cytotoxicity of ten chemicals on human cultured hepatocytes: predictability of human toxicity and comparison with rodent cell culture systems. Toxic in Vitro 6:47-52

Reid LM, Narita M, Fujita M, Murray Z, Liverpool C, Rosenberg L (1986) Matrix and hormonal regulation of differentiation in liver cultures. In: Guillouzo A, Guguen-Guillouzo C (eds) Isolated and cultured hepatocytes John Libbey Eurotext, London, pp 225-258

Ruch RJ, Fransson R, Flodstrom S, Warngard L, Klauning JE (1990) Inhibition of hepatocyte gap junctional intercellular communication by endosulfan, chlordane and heptachlor. Carcinogenesis 11:1097-1101

Ruß H (1992) Entwicklung und Charakterisierung eines Hepatozyten-Zellkultur-Modells für in vitro-Metabolismus- und Enzyminduktionsstudien an Arzneistoffen. PhD thesis Frankfurt Germany

Utesch D, Molitor E, Platt KL, Oesch F (1991) Differential stabilization of cytochrome P-450 isoenzymes in primary cultures of adult rat liver parenchymal cells. In Vitro Cell Dev Biol 27A:858-863

Utesch D, Oesch F (1992[a]) Dependency of the in vitro stabilization of differentiated functions in liver parenchymal cells on the type of cell line used for co-culture. In Vitro Cell Dev Biol 28A:193-198

Utesch D, Oesch F (1992[b]) Phenol sulfotransferase activity in rat liver parenchymal cells cultured on collagen gels. Drug Metab Disp 20:614-615

Utesch D, Arand M, Thomas H, Petzinger E, Oesch F (1992[a]) Xenobiotic- metabolizing enzyme activities in hybrid cell lines established by fusion of primary rat liver parenchymal cells with hepatoma cells. Xenobiotica 22:1451-1457

Utesch D, Diener B, Molitor E, Oesch F, Platt KL (1992[b]) Characterization of cryopreserved rat liver parenchymal cells by metabolism of diagnostic substrates and activities of related enzymes. Biochem Pharmacol 44:309-315

Predicting the toxic dose or concentration. Is there a difference ?

R. Jochemsen
Institut de Recherches Internationales Servier, 6 Place des Pléiades,
92415 Courbevoie Cedex, France.

Introduction.

The ultimate objective of conducting toxicity studies is to be able to effectively treat patients with relative safety. Traditionally, safety margins have been considered to be the ratio between the no observed toxic effect dose levels in animal and the therapeutic dose level. The underlying "believes" of this practice are that toxic effects or absence of them can be extrapolated from animal to man and that effects are related to doses. More recently, the latter has been refined to the hypothesis that effects are related to total exposure to the compound and that exposure is better determined by plasma concentrations than by dose (Collins et al 1990).

When administering a dose of a compound to animal or man, the concentration levels obtained will depend on a number of factors, the most important of which are :

1) the bioavailability of the compound in the given formulation. This includes aspects of the extent of the absorption as well as of first pass metabolism.

2) the rate of absorption

3) the clearance of the compound in the species studied.

Important parameters for defining concentration levels and profiles, are maximum concentration (Cmax) and area under the concentration-time profile (AUC). Under linear pharmacokinetic conditions, Cmax and AUC are proportionally related to dose and bioavailability, (clearance is constant). Often, particularly in the high dose range used in toxicity studies, the situation is more complex. The metabolism of a compound can become saturated and in that case, AUC increases much more rapidly than predicted on the basis of linearity with dose. For compounds with high clearance values the first pass extraction can become saturated, also leading to more than proportional increase of AUC with dose. This is shown for a compound in actual development in figure 1. On the other hand for poorly soluble compounds, absorption may become dissolution rate limited and the AUC increases less rapidly than the dose or even does not increase anymore at all with dose.,

In figure.2, the example of another compound under development is shown. The compound was very insoluble and dissolution rate limited absorption occured at higher doses in all species (shown for the monkey). To counteract this problem, the dose was divided into 2 given with a 6 hours interval. This significantly increased the total absorption and thus exposure. The importance of this can be seen in figure 3, where the relationship between AUC and dose is shown for different species. In man, the AUC increased more rapidly with the dose than in the toxicologic species, in particular the monkey. Clearly, insufficient exposure for a correct safety margin would have been obtained with a once a day administration only.

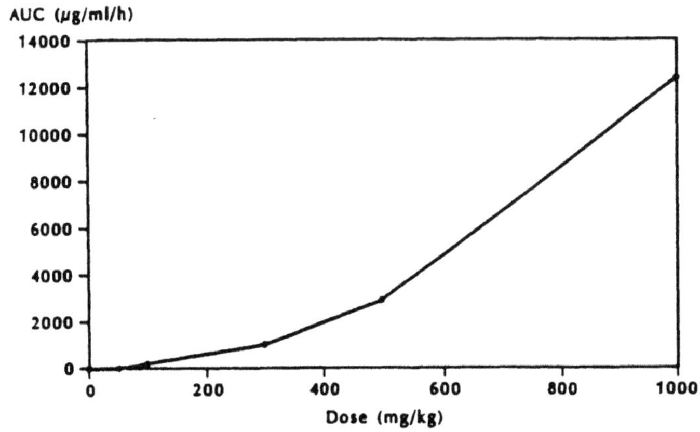

Fig.1 : Dose-AUC relationship in the monkey for a compound with a saturable first pass-effect.

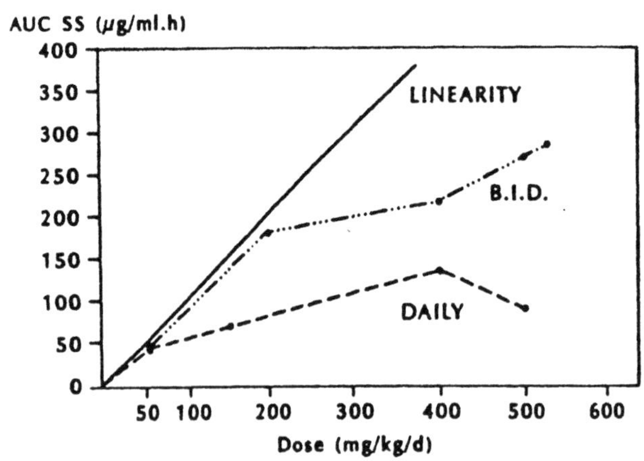

------ AUC when dose is given in one administration per day
—···— AUC when dose is given in 2 administration with a 6 hour interval
—— AUC if kinetics would have been linear from first dose (50mg/kg/d) onwards

Fig. 2 : Dose-AUC relationship in the monkey for a compound with dissolution rate limited absorption

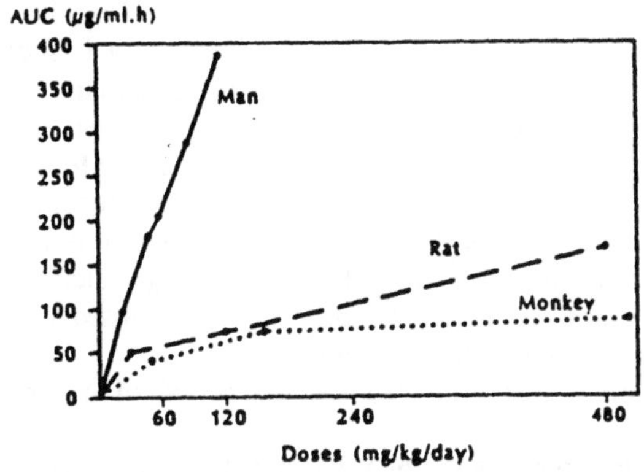

Fig. 3 : Dose-AUC relationship in different species for the same compound as in figure 2

These examples show the problems that can be encountered when attempting to predict toxic effects across species based on dose only. Measurement of concentrations and determination of exposure will contribute to obtain better predictibility but are not sufficient. Other important factors to be considered are protein binding which can be very different between species and metabolism. Differences in metabolic profiles, both qualitative and quantitative often are also important determinants of interspecies differences in (toxic) effects.

Another problem to be considered is, that not for all type of effects similar effects are obtained for similar concentrations across species, even in the absence of differences between protein binding and metabolism. The example of carcinogenic effects was discussed recently in a paper by Monro (1992), where he stated that determination of the concentrations of the test substance cannot be used for interspecies comparisons of carcinogenetic risk. Even within one species tumoriginecity is better correlated with cumulative amounts absorbed and metabolised than with plasma concentrations. Nevertheless, in this type of situations, estimations of areas under the curves are useful in assessing relative bioavailability between diet and gavage administration and for exploring sex and age differences.

Determination of AUC in diet studies.

Determination of areas under the curves can be achieved relatively easily in studies when the drug is administered by gavage. In contrast when the substance is administered with food, the problem is more difficult since the substance is ingested mainly at night and highest plasma concentrations will be obtained at that time.

To increase the feasibility of determining AUC_{24} routinely in dietary studies, we wanted to evaluate the possibility of calculating this parameter using only a limited number of samples. We decided therefore to perform a dietary study using compounds with different kinetic behaviour.

Fischer rats of stabilised body weight (6 to 10 months) were fed, for 3 weeks, a diet mixed with one of the following 3 compounds : paracetamol, 100mg.kg^{-1} (short half-life compound), antipyrine, 100 mg.kg^{-1} (medium half-life compound) or phenylbutazone, 50 mg.kg^{-1} (long half-life compound). It had been shown in a preliminary study that these compounds at these dose levels did not influence feeding behaviour. At the end of the 3 weeks, 5 rats were sampled every 3 hours during 24 hours, startind and finishing at 19.00h (time of lights off) and concentrations were measured using HLPC. The area under the curve over 24 hours (AUC_{24}) calculated using all 9 concentrations was considered to be the true AUC_{24}. Subsequently estimates of this parameter were made using different combinations.

Food consumption of each animal was evaluated for the period of 4 hours preceding the blood sample.

A gavage study was performed to determine the pharmacokinetic parameters and in particular the elimination half-life of these compounds in the respective same strain and age rats used in diet study.

Results and discussion

In the gavage study, elimination half-lives of 1,3h, 3,1h and 4,8h were found for paracetamol, antipyrine and phenulbutazone respectively. Absorption was relatively rapid with peak times of less than 2 hours. In the feeding study, plasma concentrations were relatively stable during dark hours (Fig 4), with fluctuation of less than a factor 2. In all cases, the highest concentrations were found at 7.00h (lights on). Subsequently levels decreased until 13.00h (paracetamol, antipyrine) or 16.00h (phenylbutazone) and increased gradually from then on showing that the animals started eating again in the afternoon.

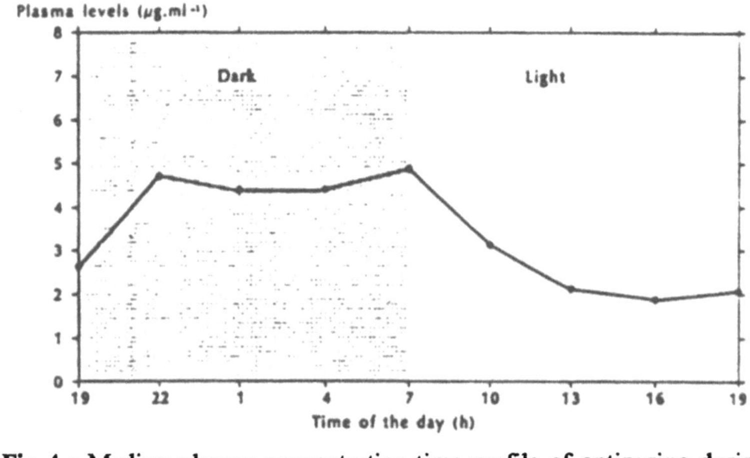

Fig 4 : Median plasma concentration-time profile of antipyrine during continuous exposure to the compound in the diet (taken from R. Jochemsen et al, 1993 with permission)

In fact, the results on food consumption were very similar to those found in literature for mature rats (Armstrong 1978 ; Zucker 1970), about 70% of the daily food intake was consumed during the dark hours and 30% during the light hours. In table 1, AUC_{24} concentrations calculated using all data points as well as estimated values using only 3 selected time points for the calculations are presented.

Table 1 : Caculated an estimated AUC_{24} values of paracetamol, antipyrine and phenylbutazone in the feeding study with daily doses of 100 $mg.kg^{-1}$, 100 $mg.kg^{-1}$ and 50 $mg.kg^{-1}$ respectively

AUC_{24} (µg.ml-1h)	Paracetamol	Antipyrine	Phenylbutazone
all 9 time points	**18.4**	**81.3**	**296.3**
7-10-16 h	**19.2** (+4.5%)	**74.6** (-8.2%)	**309.8** (+4.4%)
7-10-19 h	**20.5** (+11%) **18.4** (+0%)	**91.8** (+12.9%) **81.6** (+0.2%)	**275.3** (-7.2%) **283.2** (-4.6%)

Percentage of deviation from "true" value are given in brackets

It was found that using the concentrations at 7.00h (lights on), at 10.00h and at 16.00h and 19.00h and the assumption that at steady state the concentration for a time point on the subsequent day will be the same as that measured on the previous day, an error of less than 15% was found for all 3 compounds used. This accuracy can be considered as quite acceptable for this type of study. Using the food consumption profile as input function, and the pharmacokinetic parameters from the gavage study, a good simulation of the concentrations obtained during the feeding study was obtained, showing the importance of the diurnal feeding pattern on the concentration profile (figure 5).

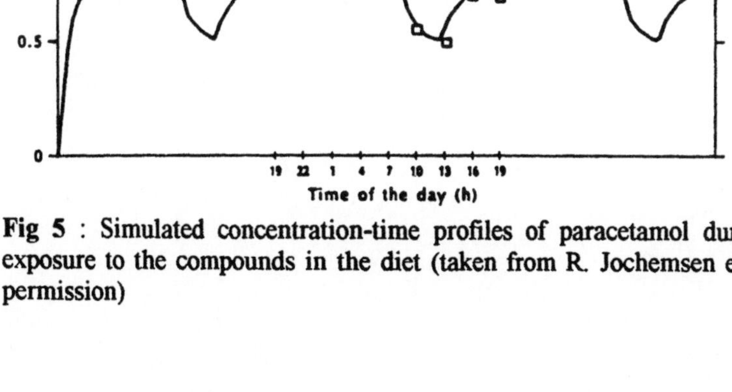

Fig 5 : Simulated concentration-time profiles of paracetamol during continuous exposure to the compounds in the diet (taken from R. Jochemsen et al, 1993 with permission)

Many compounds have a shorter half-life in rats than 1,3h and therefore it could be argued that the conclusions of the study can not be generalised to such compounds. We simulated the concentration profile to be expected in a diet study for a hypothetical compound at a dose of 100mg/kg with an elimination half-life of 0.5h and a volume of distribution of 1l.h-1, using the average food consumption profile.

The theoretical AUC_{24} was then compared with the AUC calculated using only the concentration at 7.00h, 10.00h and 16.00h. An error of 2% was found suggesting that these sampling times are also valid for compounds with a very short elimination half-life.

Therefore we think that for compounds with relatively rapid absorption and elimination, half-lives of up to 5 hours exposure in indiet studies can be estimated with acceptable accuracy using only 3 concentrations taken during day time. During the sampling day, food consumption should be measured over 24 hours divided in two times 12 hours (light and dark).

CONCLUSIONS

It was shown that often concentrations are not in a predictable way related to dose in particular in the high dose range used in toxicity studies. Calculations of safety margins can therefore be very different whether they are based on dose or on concentration. Since it is thought that safety margins based on (free) concentration will generally give the highest predictibility in extrapolation to man, it is important to measure concentrations (AUC) in toxicity studies. Where effects are more related to total absorbed dose than to concentrations of unchanged drug, AUC measurements contribute to the determination of total absorbed drug.

A feasable method to determine AUCs in indiet studies using only 3 concentrations during day time was presented.

References

Armstrong, S (1978) A chronometric approach to the study of feeding behaviour. Neuroscience and Behavioural Reviews 4 : 27-53.

Collins, JM, Grieshaber, CK, Chabner, BA (1990). Journal of the National Cancer Institute 82 : 1321

Jochemsen, R, Bazot, D, Brillanceau, MH, Lupart, M (1993) Assessment of drug exposure in rat dietary studies. Xenobiotica (In press)

Monro, A (1992) Contemporary issues in toxicology : what is an appropriate measure of exposure when testing drugs for carcinogenicity in rodents ? Toxicology and Applied Pharmacology 112 : 171-181.

Zucker, I (1970) Light-dark rhythmes in rat eating and drinking behaviour. Physiology and Behaviour 6 : 115-126.

The Role of Investigative Toxicology in Pharmaceutical Industry

E.S. Harpur and F.W. Bonner
Sterling Winthrop Research Centre, Willowburn Avenue, Alnwick, Northumberland, NE66 2JH, U.K.

Within the context of drug development in pharmaceutical industry what do we mean by investigative toxicology? We include, among other things, the development of new procedures for assessing toxicity, the investigation of mechanisms of toxicity and the application of mechanism-based screens with the overall goal of enhancing the effectiveness of drug discovery and development. Although investigative toxicology has an important role to play in drug discovery by aiding the selection of compounds for development, through, for example, the study of structure-toxicity relationships, this article will concentrate on the role of investigative toxicology in drug development. Here the principal goals could be stated as assisting rational decision making and improving the risk-benefit assessment process, thus facilitating regulatory review.

In an ideal world it would be possible to demonstrate, in all of the plethora of animal studies which constitute the pre-clinical safety evaluation of a drug, the absence of any toxicological potential of the candidate compound. However, this is a comparatively rare occurrence and more often than not we are faced with the need to decide whether or not to continue to develop a compound which has produced abnormal pathological and/or biochemical findings in one or more animal models. In this situation an understanding of the mechanism of the effect can be of considerable help. It should be borne in mind, however, that mechanistic information is only one facet of understanding species differences in toxicological response and the relevance of findings in animals for humans. Zbinden (1987a) has suggested a number of 'conversion factors' for predicting from animal studies to humans (Table 1).

Table 1. Conversion Factors for Predicting From Animal Studies to Humans

Bioavailability	Receptor Sensitivity
Pharmacokinetics	Anatomical Characteristics
Protein Binding	Physiological Characteristics
Metabolism	Repair Mechanisms
Dose and Time	Species - Specific Response
Concentration	Mechanisms of Action

Adapted from Zbinden (1987a)

Many of these factors (those listed in the left-hand column) are essentially related to differences in the handling of zenobiotics by the various species but intrinsic differences between species in anatomy and physiology, as well as mechanistic information, must also be considered (right-hand column). While pharmacokinetic differences both between species and between compounds form an important dimension to the two examples which are cited below, this article will focus on how an understanding of mechanisms can in some cases increase confidence that one is not dealing with an effect of toxicological significance for humans.

The starting point for a mechanistic study is often an observation of an effect in a toxicological study of standard design. Existing toxicological knowledge can be valuable initially in helping to formulate a hypothetical mechanism and also, ultimately, in assessing the risk of a similar occurrence in patients.

Non-genotoxic carcinogenesis is one area which has been the subject of much investigative activity and the data generated have been accepted by regulatory authorities. The phenoxyisobutyrates or 'fibrates', such as clofibrate, ciprofibrate, fenofibrate, bezafibrate and gemfibrozil are a class of hypolipidaemic compounds which are non-genotoxic but nevertheless have a variable proclivity to induce tumours in the liver, thyroid, pancreas, stomach and testis of rodents. Such a spectrum of carcinogenic potential in rodents presents a risk assessment challenge. The value of mechanistic information can be demonstrated by reference to the stomach and thyroid lesions induced by ciprofibrate.

Neuroendocrine Cell Hyperplasia and Gastric Carcinoids

In a two-year study in rats ciprofibrate was found to produce carcinoid tumours in the glandular stomach at the top dose (10 mg/kg/day). The incidence of tumours was low; 5/59 males, 1/60 females and none were seen in a chronic mouse study. In subsequent chronic studies in the rat the propensity of ciprofibrate to produce dose-related neuroendocrine cell hyperplasia and, at high doses, a low incidence of carcinoid tumours was confirmed.

Literature reports that clofibrate was associated with a gastric tumour of similar morphology in the rat (Svoboda and Azarnoff, 1979) and possessed gastric antisecretory activity (Lippmann and Seethaler, 1976; Rheault et al, 1982) suggested a possible mechanism for the induction of gastric carcinoids by ciprofibrate. It is widely reported that long-term administration of potent inhibitors of gastric acid secretion is associated with the formation of gastric carcinoid tumours in rodents (Betton et al 1988; Ekman et al 1985; Hirth et al 1988; Poynter et al 1986a and 1986b; Streett et al 1988). This led to the hypothesis that the gastric carcinoids produced by ciprofibrate in the rat could be explained if ciprofibrate also possessed gastric antisecretory activity. Since ciprofibrate has a very long half-live in the rat (in excess of 70 hours) it was speculated that any antisecretory activity would probably be prolonged, resulting in the hypergastrinaemia which has been shown to be an essential stimulus to the hyperplastic growth of neuroendocrine cells, leading eventually to formation of a gastric carcinoid. A series of studies were, therefore, conducted to test this hypothesis.

It was shown that ciprofibrate, as well as clofibric acid and bezafibrate, produced a dose related inhibition of gastric acid secretion in the rat following either oral or intraduodenal administration (Eason et al 1988a). A study of the time course of this inhibition revealed that ciprofibrate produced a long acting effect, acid output being still significantly depressed 8-12 hours following a single dose (Eason et al 1988a). It was further shown that repeated administration of ciprofibrate resulted in a sustained elevation of plasma gastrin levels in the rat (Eason et al 1988b). In contrast, single administration of ciprofibrate did not inhibit gastric acid secretion in the mouse (Eason et al 1988a), nor did ciprofibrate produce any sustained elevation of plasma gastrin concentration in the mouse, even when administered at a dose of 20 mg/kg/day for 26 weeks (Eason et al 1988b). This lack of change in plasma concentration of gastrin in response to chronic administration of ciprofibrate was also apparent in another species, the marmoset (Eason et al 1988b). These species differences in the changes in plasma gastrin concentrations were paralleled by species differences in the morphological response of the stomach to ciprofibrate administration in sub-chronic studies conducted for 8 or 26 weeks (Eason et al 1988b). Dose-related changes in the parietal cells, consistent with inhibition of secretion, and progressive neuroendocrine cell hyperplasia were seen in the rat but not in the mouse or the marmoset. Indeed, no morphological changes have been observed in the stomach of marmosets administered doses of up to 80 mg/kg/day for 9 months or in rhesus monkeys administered doses up to 50 mg/kg/day for 12 months. Thus the rat seems to be singularly sensitive to the induction and progression of neuroendocrine cell hyperplasia since no such response was seen in the mouse or in two non-human primate species. This suggests a low risk of any such occurrence in man, an assessment which is supported by data which show that plasma gastrin levels are not elevated in patients receiving long-term ciprofibrate treatment (unpublished data). This risk assessment is facilitated because many of the arguments in support of the indirect mechanism have been advanced and accepted when considering the risk of gastric cancers being induced in patients receiving therapy with the potent and long-acting pharmacological anti-secretory agents which also induce gastric carcinoids in rodents by a similar mechanism.

Thyroid Hyperplasia and Neoplasia

Clofibrate and fenofibrate have been shown to produce morphological changes in the thyroid of rats when administered for periods up to 18 months (Price et al 1984). These changes, decreased follicular size, more intense basophilic staining of the colloid and ultrastructural alterations in the follicular cells were interpreted as being indicative of a hyperactive gland. In a two-year study which we have conducted in mice, ciprofibrate produced dose-related follicular hyperplasia in the thyroid and, only in the high dose group (10 mg/kg/day), a low incidence of follicular cell adenomas; 3/50 males and 6/50 females. Subtle changes indicative of a hyperactive gland, detectable only using electronmicroscopy, have been seen with ciprofibrate, fenofibrate and bezafibrate after 3 and 28 doses in mice. Chronic administration of ciprofibrate to rats produced morphological changes in the thyroid, diffuse follicular hyperplasia and increased basophilic casts in the colloid, first seen in a premature decedent after 13 weeks. Qualitatively similar, but generally less marked, changes were seen with bezafibrate at equivalent exposure. Subtle morphological changes in the thyroid, indicative of hyperactive gland, have been seen in the rat after as few as five daily doses of ciprofibrate at a dose of 30 mg/kg.

Serum concentrations of thyroid hormones have been measured in chronic studies in the rat after a minimum of six weeks administration. Thyroxine (T_4) was found to be decreased whereas T_3 was unchanged or decreased and TSH unchanged. We developed a hypothesis that fibrates produce a hyperactive thyroid indirectly by increasing T_4 clearance resulting in decreased serum T_4 concentrations and, through a reduction of the negative feedback on the pituitary, increased TSH secretion. This would provide an increased trophic drive to the thyroid resulting in a hyperactive gland. Existing knowledge which supported this hypothesis included the fact the metabolic clearance of T_4 has been shown to be increased by clofibrate (Ruegamer et al 1969; Harland and Orr 1975; Hinton et al 1986), fenofibrate (Hinton et al 1986) and nafenopin (Kaiser et al 1988). Clofibrate has also been shown to stimulate biliary clearance of T_4 (Harland and Orr 1975) and nafenopin strongly stimulates the faecal clearance of T_4 (Kaiser et al 1988). Finally, fibrates, including ciprofibrate, are known to induce hepatic bilirubin-UDPGT (Burchell and Coughtrie 1989 and Boiteux-Antoine et al 1989).

By analogy with other compounds, an H_2 receptor antagonist (Brown et al, 1987) and phenobarbital (McClain et al, 1989) it was considered possible that a measurable elevation of TSH would be present only transiently, early in the course of administration of the drug. Therefore, a series of studies were conducted to determine the effects of short-term ciprofibrate administration to the rat on plasma concentrations of thyroid hormones, on plasma clearance of T_4 and on hepatic metabolism and disposition of thyroid hormones. Ciprofibrate, 30mg/kg/day was administered to male Fischer 344 rats by gavage for between 1 and 14 days. As expected, administration of this dose of ciprofibrate for 14 days produced a significant increase in liver weight (two-fold). Serum concentrations of T_4 were profoundly and persistently decreased, whereas T_3 levels were initially decreased and then increased. There was a small but significant increase in serum TSH concentration when this hormone was measured 6 hours post-dose at a time close to the peak in the circadian variation in serum TSH concentration. Paralleling the decrease in serum T_4 concentration it was demonstrated that ciprofibrate produced a progressive increase in T_4 clearance to 3.8x control after 14 doses. Deiodination and glucoronidation are the major routes of metabolism of thyroid hormones in the rat. After 14 doses, hepatic deiodinase activity (measured as conversion of T_4 to T_3) was significantly decreased, whereas hepatic UDPGT activities for T_3 and particularly T_4 were increased. Taking account of the increase in liver weight, total hepatic deiodinase activity was significantly decreased to 0.44x control while T_3- and T_4-UDPGT activities were significantly increased to 3.2x control and 5.8x control respectively. A study was made of cumulative elimination of radioactivity derived from ^{125}I-T_3 or ^{125}I-T_4 in urine (from deiodinative metabolism) and faeces (from glucuronidation) over 72 hours following 1 or 7 doses of ciprofibrate. A single dose of ciprofibrate had no effect on urinary and faecal disposition of ^{125}I-T_3, whereas there was a small but significant increase in excretion of both faecal and total radioactivity derived from ^{125}I-T_4. After 7 doses of ciprofibrate, excretion of radioactivity derived from T_3 was significantly increased in faeces and decreased in urine compared to controls at most time points over the 72 hr collection period. The respective changes were similar in magnitude so that there was no net change in cumulative total excretion of radioactivity at any timepoint. However, in the case of T_4 there was a greater increase in the faecal excretion of radioactivity compared with the decrease in urinary excretion resulting in a net increase in the cumulative total excretion of radioactivity. These results are consistent both with the measured changes in deiodinase and UDPGT activities in liver and also with the observation that there was a profound and sustained decrease in serum T_4 concentration while the effect on serum T_3 concentration was transient. These results support the hypothesis that the thyroid hyperplasia produced by administration of ciprofibrate in the rat is secondary to altered thyroid hormone status.

No thyroid pathology was detected in rhesus monkeys following administration of ciprofibrate at doses up to 50mg/kg/day for one year, nor in marmosets treated for 9 months at doses up to 80mg/kg/day. In assessment of the risk of thyroid hyperplasia occurring in patients it has to be borne in mind that the rodent thyroid is particularly sensitive to disturbances in thyroid hormone metabolism. The particular sensitivity of the rodent may be explained at least in part by the fact that rodents lack thyroxine binding globulin (Dohler et al, 1979, Zbinden, 1987b) resulting in a short half-life of T_4, 12-24 hours in rats compared with 5-9 days in humans (Dohler et al, 1979). Consequently, any induction of T_4 metabolism in rats, even transiently, leads to a rapid depletion of the hormone and an immediate regulatory response by the pituitary with an increased secretion of TSH. It is also possible that the responsiveness of the thyroid to TSH varies amongst species. For example in cell culture TSH alone does not induce proliferation of porcine, ovine or human thyroid cells, but does stimulate growth of cells from rat and dog thyroids (Westermark et al, 1985). Thus the lack of thyroid changes in the one year monkey study and the 9 month marmoset study are probably of greater relevance to man than the effects observed in rodents. It was concluded from these findings that ciprofibrate is unlikely to produce a significant effect on thyroid function in humans.

Summary

In summary, the benefits of a mechanistic toxicology approach include increased understanding of the processes underlying primary pathology, the development of predictive capability, enhanced decision making and improved responsiveness to regulatory agencies. The overall objective of toxicology in the pharmaceutical industry is risk benefit assessment in order to support product registration. This requires both that studies be conducted in accordance with regulatory guidelines and that the biological processes underlying pathology be understood. These should be regarded as inseparable components of the risk assessment process.

References

Betton GR, Dormer CS, Wells T, Pert P, Price CA and Buckley P (1988) Gastric ECL-cell hyperplasia and carcinoids in rodents following chronic administration of H_2-antagonists SK&F 93479 and oxmetidine and omeprazole. Toxicol. Pathol. **16**, 288-298

Boiteux-Antoine AF, Magdalou J, Fournel-Gigleux S and Siest G (1989) Comparative induction of drug-metabolizing enzymes by hypolipidaemic compounds. Gen. Pharmacol **20**, 407-412

Brown, CG, Harland, RF and Atterwill, CK (1987) Increased thyroxine clearance in rats treated with high doses of a histamine H_2-antagonist, SKF 93479. Arch. Toxicol., Suppl. **11**, 253-256

Burchell B and Coughtrie MWH (1989) UDP-glucuronyltransferases. Pharmacol Ther **43**, 261-289

Dohler, KD, Wong, CC and Von Zur Muhlen, A (1979) The rat as a model for the study of drug effects on thyroid function: consideration of methodological problems. Pharmac. Ther., **5**, 305-318

Eason, CT, Pattison, A, Howells, DD, Spencer, AJ and Bonner, FW (1988a) Assessment of Gastric Antisecretory Effects of Phenoxyisobutyrate Derivatives in the Rat and the Mouse. Scand J Gastroenterol,23, 1063-1071

Eason, CT, Spencer, AJ, Pattison, A, Howells DD, Henry DC and Bonner FW (1988b) Species Variation in Gastric Toxicity following Chronic Administration of Ciprofibrate to Rat, Mouse and Marmoset. Toxicol. Appl. Pharmacol. **95**, 328-338

Ekman L, Hansson E, Havu N, Carlsson E and Lundberg C (1985) Toxicology studies on omeprazole. Scand. J. Gastroenterol. **20**, 53-69

Harland, WA and Orr, JS (1975) The effect of clofibrate on thyroxine metabolism. In 'Thyroid Hormone Metabolism' Ed. W.A. Harland and JS Orr. Academic Press, pp 65-87

Hinton, RH, Mitchell FE, Mann A, Chescoe D, Price SC, Nunn A, Grasso P and Bridges JW (1986) Effects of phthalic esters on the liver and thyroid. Environ Health Perspect **70**, 195-210

Hirth RS, Evans LD, Buroker RA and Oleson FB (1988) Gastric enterochromaffin-like cell hyperplasia and neoplasia in the rat: An indirect effect of the histamine H_2 blockade and gastric carcinoid tumours in the rat. Gut **26**, 1284-1295

Kaiser CA, Seydoux J, Giacobino JP, Girardier L and Burger AG (1988) Increased plasma clearance rate of thyroxine despite decreased 5'-monodeiodination: study with a peroxisome proliferator in the rat. Endocrinology **122**, 1087-1093

Lippman W and Sethaler K (1976) Clofibrate: antisecretory antiulcer activity in the rat. J. Pharm. Pharmac. 23, 165

McClain, MR, Levin AA, Posch, R and Downing JC (1989) The effect of phenobarbital on the metabolism and excretion of thyroxine in rats. Toxicol. Appl. Pharmacol., **99**, 216-228

Poynter D, Pick CR, Harcourt RA, Selway SAM, Ainge G, Harman IW, Spurling NW, Fluck PA, and Cook JL (1986a) Association of long lasting unsurmountable histamine H_2 blockade and gastric carcinoid tumours in the rat. Gut 26: 1284-1295.

Poynter D, Selway SAM, Papworth SA and Riches SR (1986b) Changes in the gastric mucosa of the mouse associated with long lasting unsurmountable histamine H_2 blockdage. Gut **27**, 1338-1346

Price, SC, Chesco, D and Hinton RH (1984) Changes in the colloid and lysosomes of the thyroid of rats treated with clofibrate. Biochem. Soc. Trans. **12**, 1034-1035

Rheault MJ, Laberge M, Lelomer J and Tetreault C (1982) The gastric antisecretory effects of clofibrate in the rat. Rev. Can. Biol. Exp. 41, 165

Ruegamer, WR, Ryan NT, Richert, DA and Weshenfield, WW (1969) Effect of p-chlorophenoxyisobutyrate on the turnover rate and distribution of thyroid hormone in the rat. Biochem. Pharmacol., **18**, 613-624

Streett CS,Robertson JL and Crissman JW (1988) Morphologic stomach findings in rats and mice treated with the H_2-receptor antagonist, ICI 125,211 and ICI 162,846. Toxicol. Pathol. **16**, 299-304

Svoboda DJ and Azarnoff DL (1979) Tumours in male rats fed ethyl chlorophenoxyisobutyrate, a hypolipidaemic drug. Cancer Res. 39, 3419-3428

Westermark, K, Karlsson, FA and Westermark, B (1985) Thyrotropin modulates EGF receptor function in porcine thyroid follicle cells. Mol. Cell. Endocrinol., **40**, 17-23

Zbinden, G (1987a) CMR Annual Lecture: Predictive Value of Animal Studies in Toxicology. Published by Centre for Medicines Research

Zbinden, G (1987b) Assessment of hyperplastic and neoplastic lesions of the thyroid gland. TIPS, **8**, 511-514

Mechanisms of Metal Toxicity

(Chair: M. Vahter, Sweden)

Metal-induced autoimmunity

Ph. Druet
INSERM U 28, Hôpital Broussais, 96, rue Didot, 75014 Paris, France.

Introduction.

Several metals have been known for a long time to be associated with immune-mediated pathological effects. These effects may be classified into two categories. Hypersensitivity reactions are the consequence of either cell-mediated or antibody (humoral)-mediated reactions against the metal which behave as a hapten. These manifestations represent an exageration of a normal immune response against an exogenous antigen. Autoimmune reactions represent quite different situations where the immune response, again either cell- or antibody-mediated, is directed against an autoantigen. In this review the second situation will be mainly considered. Before that the basis of the normal immune response will be summarized.

Basis of the immune response

Considerable advances have been achieved during the last 10 years which have clarified to a great extent our understanding of the normal immune response. Two signals are essential for the induction of an immune response against an exogenous antigen.

The specific signal

An exogenous antigen entering the body will be first taken in charge by an antigen presenting cell (APC) and processed. There are two different kinds of APC, the common hallmark of which is the presence of class II major histocompatibility complex (MHC)-encoded molecules at their surface. The first (macrophages, dentritic cells, Langerhans cells ...) are able to internalize non-specifically any kind of antigen. B lymphocytes represent the second category of APC ; they internalize the antigen they have recognize through their IgM membrane receptor. A given B cell will only pick up the antigen it has recognized through its membrane receptor. Whatever the cell, the antigen that has been internalized is degraged into peptides in the endosomal compartment of the cell. Class II molecules of the MHC are produced by APC and transferred from the endoplasmic reticulum to the endosomes where they take in charge the peptides. The complex class II molecule plus peptide is then conveyed to the membrane cell surface. Class II molecules (HLA DP, DQ or DR in the human, RT1-B or -D in the rat) are made of two α and ß glycoproteic chains with a monomorphic portion (C terminal) and a polymorphic portion (N terminal) which has been shown by crystallography to realize a groove where peptides (about 12 aminoacids) are found. The complex will then be recognized by the CD4+ T cell clone which has the T cell receptor specific for the peptide (Schwartz 1985). The peptide is only recognized in the context of class II molecules which are histocompatible with the T cell. This phenomenon is called class II restriction. Following this recognition, the T cell is then activated.

The non-specific signals

The trimolecular complex (T cell receptor of the CD4+ cell, class II plus peptide) is essential for T cell activation but other non specific signals are also crucial. Some are delivered through adhesion molecules and others through soluble mediators named cytokines.

Various molecules are present at the T cell surface and APC cell surface which are required for the T cell to be activated : CD4+ molecules bind to a monomorphic determinant of the class II molecules ; other molecules present at the T cell surface : CD28, CD2, ICAM1 bind respectively to B7, LFA1 and LFA3 molecules present at the cell surface of APC (Springer 1990).

In addition cytokines which are produced following T cell activation are required for the CD4+ T cell to proliferate and differentiate. Two different CD4+ T cell subsets have been described during the past few years which are mainly characterized by the pattern of cytokine they produce (Mosmann et al. 1991). Th1 CD4+ T cells produce mainly IL-2 (the main growth factor for this subset) and interferon gamma (IFN gamma). Th2 CD4+ T cells produce mainly IL-4 (the main growth factor for this subset), IL-10, IL-5 and IL-6. Th2 cells also proliferate in response to IL-2. Activated Th1 cells express receptors for IL-2 while activated Th2 cells express receptors for IL-2 and IL-4. These T cell subsets have other important properties. First, they do not exist as such but both come from a common CD4+ Th precursor that differentiates in either T subset probably depending on the APC and on the costimulatory signals they receive. It is likely that Th1 cells are generated when the APC is a conventional one (such as dentritic cell) and when IFN gamma is present in the environment while Th2 cells are generated when the APC is a B lymphocyte and if IL-4 and/or IL-10 are present in the environment of the T cell. Second, the effector function of these subsets seems to differ : Th1 cells are mainly responsible for cell-mediated reactions (type IV hypersensitivity and cell-mediated autoimmunity) while Th2 cells mainly collaborate with B cells and provide help for antibody (IgE, IgG1, IgA) production. This latter subset is responsible therefore for antibody (humoral)-mediated reactions (type I hypersensitivity and antibody-mediated autoimmunity). Third, these two T cell subsets are reciprocally controlled through the cytokines they produce : IFN gamma inhibits Th2 cells while IL-10 inhibits Th1 cells.

Tolerance

One of the main characteristics of the immune system is its ability to differentiate between self and non-self. Several non exclusive mechanisms explain the tolerance state. First, when T cells mature in the thymus during development, those potentially autoreactive T cells are eliminated by apoptosis (Kisielow et al. 1991) if peptides derived from autoantigens are presented to them by class II positive cells of the thymus (epithelial cells). The same holds true probably for B cells. These informations have been obtained from works using transgenic models. It is however clear that many autoantigens are not presented to T cells in the thymus. Autoreactive T and B cells are therefore present in the periphery (Kroemer and Carlos-Martinez 1992). The second mechanism that explains tolerance is that autoreactive cells in the periphery would be rendered anergic which means that although they may be activated by a self peptide they do not proliferate, for example because they are not sensitive to IL-2. Finally, it is more and more likely that at least some autoreactive cells present in the periphery do not respond to autoantigens because they are negatively controlled. This negative control may be achieved by T cells with a suppressor effect, through the idiotypic network or due to cytokines.

Hypersensitivity and autoimmunity

Hypersensitivity (HS) reactions are the consequence of a normal immune response to an exogenous antigen but, for a reason or another, the intensity of the immune response is such that pathological manifestations occur. These manifestations are often due to the production of an excess of mediators by cells of the immune system. It is classical to distinguish 4 different types of HS. The first three are mediated by antibodies : type I or immediate HS is mediated by IgE, type II or cytotoxic HS is mediated by IgG antibodies while type III would be mediated by immune complexes. The importance of type III HS is controversial. Finally, type IV HS is mediated by CD4+ (Th1) cells.

Autoimmunity is defined, by contrast, as the pathological situations that are due either to autoreactive T cells (cell-mediated) or to autoantibodies. It is important to that point to stress that the presence of autoreactive T cells and/or of autoantibodies associated with a disease does not mean that the disease is autoimmune. To demonstrate the responsibility of T cells or of autoantibodies needs to be able to transfer the disease passively or to be able to reproduce it in vitro. Autoimmune diseases may be classified also in other ways : one may oppose spontaneous to induced- autoimmune diseases or systemic to organ-specific autoimmune diseases.

Since normal individuals are tolerant to autoantigens, it is important to try to identify factors that may lead to autoimmunity. Genetic as well as environmental factors have shown to be important and it is well established that all known autoimmune diseases are multifactorial (Carson 1992). Some of these factors are listed in Table 1.

Table 1. Factors governing the appearance of autoimmune diseases

Genetic factors	
	Major histocompatibility complex - encoded molecules
	T cell receptor or immunoglobulin genes
	Defect in the elimination of autoreactive cells
	Hormonal factors (sexual hormones)
	Enzyme polymorphism (drug metabolism)
Environmental factors	
	Exposure to chemicals
	Infections (virus, parasite, bacteria)
	Stress (production of substances such as heat shock proteins)

Among the genetic factors the best characterized are those which depend upon the MHC. Numerous diseases are associated with a given HLA haplotype and most often with a given HLA DR or DQ. Several explanations have been given for this association, but the most likely one is that some class II molecules are more suitable than others to present a given self-peptide to the T cell. Among the environmental factors, chemicals are those that we are interested with in this review and we will consider in the next chapter how chemicals (and metals) may induce autoimmunity.

Chemical-induced autoimmunity

A chemical or its metabolites may affect the immune system in several ways. First, it may bind to self proteins and may be presented as a hapten to the T cells. In these situations, the pathological situations observed will be a consequence of HS. Second, a drug may bind directly to class II molecules without prior processing by the APC or even to the T cell receptor. In this situation too, HS may be observed as a consequence. Third, reactive drugs may modify self proteins which then become autoantigenic because the T cells that recognize the modified protein will also recognize, due to cross reactions, the native protein. This mechanism has been proposed to explain hepatitis induced by drugs such as tienilic acid or dihydralazine (Beaune et al. 1987). In these situations, reactive metabolites are formed after metabolisation of the drugs by isoforms of cytochrome P-450 which bind to and modify the P-450. This would explain the presence of autoantibodies specific for the P-450 isoform responsible for the drug metabolisation. Finally, as we shall see later, a drug may deregulate in a way or another the immune system leading to a so-called polyclonal activation.

Metal-induced immune reactions in humans and experimental models

Numerous metals may induce pathological immune manifestations in humans (Table 2).

Table 2. Pathological immune manifestations induced by metals.

Metal	Human	Experimental animal
Nickel	Contact dermatitis	?
Chromium	Contact dermatitis	Contact dermatitis (guinea-pig)
Berylium	Contact dermatitis Pneumonitis (granulomas)	Contact dermatitis (guinea-pig)
Gold	Glomerulopathy, dermatitis,Cytopenia, Hepatitis, Pneumonitis	Systemic manifestations (rat, mouse) Glomerulopathy (rat, mouse, guinea-pig)
Mercury	Contact dermatitis, Glomerulopathy	Contact dermatitis (guinea pig) Glomerulopathy (rat, mouse) Sjögren syndrome, skin lesions Vasculitis (rat)
Cadmium	?	Glomerulopathy (?)
Platinum	Asthma	?
Lithium	Glomerulopathy, cardiomyopathy	?

Several of them such as contact dermatitis are due to HS. T cell clones have been derived from patients with Ni-induced dermatitis but also from normal individuals. They produce mainly IFN gamma (Kapsenberg et al. 1991). It is therefore likely that these T cells are Th1 cells. The difference between normal and affected people seems to be a higher frequency of Ni-specific T cells in the latter. It has been shown that Ni binds directly to class II molecules and even to the T cell receptor (Kapsenberg et al. 1985). Gold salts have also been shown to bind class II molecules and gold-specific T cell clones have been derived from gold-treated rhumatoid arthritis patients (Romagnoli et al. 1992). It is interesting to note that gold may inhibit binding of given peptides into the groove of MHC molecules which could also induce immunosuppression (Romagnoli et al. 1991). The presence of metal-specific T cells certainly explains HS observed with these metals but is not likely to explain the occurrence of autoimmune reactions. It must be stressed to that point that none of the immune manifestations observed in exposed humans has been demonstrated to be of autoimmune origin.

Experimental models have been most helpful in that respect although relevance of these models remains to be firmly established. Mercury and gold-induced manifestations have been obtained in susceptible strains of rats (BN : $RT1^n$) and mice ($H2^s$ such as $B\text{-}10^s$) (Goldman et al. 1991) but also in rabbits. Interestingly the disease observed with either metal is quite similar. A series of manifestations appear (Table 3).

Table 3. Manifestations observed in Brown-Norway rats exposed to mercury or gold.

Lymphoproliferation (affecting T CD4+ and B cells)
Hyperimmunoglobulinemia (affecting mainly IgE)
Production of autoantibodies (laminin, collagen IV, tubular basement membrane, ADN, rheumatoid factor, thyroglobulin, myeloperoxidase)
Production of antibodies to non-self antigens (TNP. sheep red blood cells)
Increased production of IL-4 ; decreased production of IFN gamma
Presence of autoreactive (anti-class II plus peptide) T cells
Pathological manifestations (see Table 2).

It is clear that in the rat model the glomerulopathy observed is really autoimmune since anti-laminin autoantibodies have been eluted from the diseased glomeruli whatever the pattern observed (linear in a first step, granular with a typical membranous glomerulopathy later on). The manifestations observed in rats also include skin and mucosal lesions, a Sjögren syndrome, a vasculitis with anti-myeloperoxidase antibodies and wide spread immune complex deposition. It has been shown that these chemicals induce in the susceptible strain of rats a polyclonal activation of B cells that requires the presence of autoreactive anti-class II T cells. They presumably recognize either class II or class II plus peptides as a consequence of an initial modification of the complex by these highly reactive metals. T cells specific for mercury have also been found in mice. The role of autoreactive anti-class II T cells is supported by the fact that such cells are indeed present in both models and that T cells from diseased rats transfer the disease in naive recipients (Pelletier et al. 1988). Finally, T cell lines have been obtained from BN rats injected with gold. These cells have been only stimulated with syngeneic APCs and they are able to induce normal B cells to produce in vitro anti-laminin, anti-DNA and anti-TNP antibodies as well as important amounts of IgE (in preparation). These T cells belong to the Th2 subset. First, treatment of susceptible strains of mice with an anti-IL-4 monoclonal antibody suppresses the hyper production of IgE (Ochel et al. 1991). Second, T cells from susceptible animals induce an increase in class II molecule at the B cell surface, a well-known effect in IL-4. Finally, in BN rats and in mice, a strong message for IL-4 has been detected using an RT PCR. More important, T cells 98 % pure, when cultured in the presence of non toxic amounts of $HgCl_2$ express after 2 hours a strong message pour IL-4 (in preparation). It is therefore likely that the disease results from at least two signals. The first one is non specific and consists in the activation of the IL-4 gene. Other cytokines are affected. A significant decrease in IFN gamma produced has been observed after a single injection of $HgCl_2$ in BN rats (Van der Meide et al. 1993). The second signal would be provided by the autoreactive anti-class II T cells. Finally, although 50 % of the diseased rats die, those who do not die recover and then become resistant to disease reinduction. Evidence has been obtained suggesting that recovery could be due to an activation of Th1 cells (Dubey et al. 1993).

We have mentioned that besides the susceptible BN rat strain, many others are resistant. Susceptibility to Hg-induced disease depends upon 3 or 4 genes one of which is a gene encoding for class II molecules (Druet et al. 1977; Aten et al. 1991).

The disease in the susceptible strains of mice is characterized by the presence of anti-nucleolar autoantibodies and is also genetically-controlled (Mirtcheva et al. 1989).

Conclusion.

Experiments performed using mercury or gold salts in rats or mice have shown that these agents may indeed induce autoimmune disorders in susceptible strains. These metals break tolerance because they induce the appearance of autoreactive T cells to normal class II molecules which explains the B cell polyclonal activation observed in these models. In addition, these metals affect directly the production of cytokines in a way that favors the expansion of Th2 cells. The relevance of these findings to the human situation remains to be established.

References

Aten J, Veninga A, De Heer E, Rozing J, Nieuwenhuis P, Hoedemaeker PJ, Weening JJ (1991) Susceptibility to the induction of either autoimmunity or immunosuppression by mercuric chloride is related to the major histocompati- bility complex class II haplotype. Eur J Immunol 21:611-616.

Beaune Ph, Dansette PM, Mansuy D, Kiffel L, Finck M, Amar C, Leroux JP, Homberg JC (1987) Human anti-endoplasmic reticulum autoantibodies appearing in a drug-induced hepatitis are directed against a human liver cytochrome P-450 that hydroxylates the drug. Proc Natl Acad Sci USA 84:551-555.

Carson DA (1992) Genetic factors in the etiology and pathogenesis of autoimmunity. FASEB J 6:2800-2805.

Druet E, Sapin C, Günther E, Feingold N, Druet P (1977) Mercuric chloride-induced anti-glomerular basement membrane antibodies in the rat. Genetic control. Eur J Immunol 7:348-351.

Dubey C, Kuhn J, Vial MC, Druet P, Bellon B (1993) Anti-interleukin-2 receptor monoclonal antibody therapy supports a role for Th1-like cells in $HgCl_2$-induced autoimmunity in rats. Scand J Immunol 37:406-412.

Goldman M, Druet P, Gleichmann E (1991) Th2 cells in systemic autoimmunity : insights from allogeneic diseases and chemically-induced autoimmunity. Immunol Today 12:223-227.

Kapsenberg ML, Van der Pouw-Kraan T, Stiekema FE, Schootemeijer A, Bos JD (1988) Direct and indirect nickel-specific stimulation of T lymphocytes from patients with allergic contact dermatitis to nickel. Eur J Immunol 18:977-982.

Kapsenberg ML, Wienrenga EA, Bos JD, Jansen HM (1991) Functional subsets of allergen-reactive CD4+ T cells. Immunol Today 12:392-395.

Kisielow P, Swat W, Rochat B, Von Boehmer H (1991) Induction of immunological unresponsiveness in vivo and in vitro by conventional and super-antigens in developing and mature T cells. Immunol Rev 122:69-85.

Kroemer G, Carlos-Martinez A (1992) mechanisms of self- tolerance. Immunol Today 13:401-404.

Van der Meide PH, de Labie MCDC, Botman CAD, Van Bennekom WP, Olsson T, Aten J, Weening JJ (1993) Mercuric chloride down-regulates T cell interferon-gamma production in Brown-Norway but not in Lewis rats ; role of glutathione. Eur J Immunol 23:675-681.

Mirtcheva J, Pfeiffer C, Bruijn JA, Jacquesmart F, Gleichmann E (1989) Immunological alterations inducible by mercury compounds. III. H-2A acts as an immune response and H-2E as an immune "suppression" locus for $HgCl_2$-induced antinucleolar autoantibodies. Eur J Immunol 19:2257-2261.

Mosmann TR, Schumacher JH, Street NF, Budd R, O'Garra A, Fong TAT, Bond MW, Moore KWM, Sher A, Fiorentino DF (1991) Diversity of cytokine synthesis and function of mouse CD4+ T cells. Immunol Rev 123:209-229.

Ochel M, Vohr HW, Pfeiffer C, Gleichmann E (1991) IL-4 is required for the IgE and IgG1 increase and IgG1 autoantibody formation in mice treated with mercuric chloride. J Immunol 146:3006-3011.

Pelletier L, Pasquier R, Guettier C, Vial MC, Mandet C, Nochy D, Bazin H, Druet P (1988)Autoreactive T cells in mercury-induced autoimmunity. Ability to induce the disease. J Immunol 140:750-754.

Romagnoli P, Labhardt AM, Sinigaglia F (1991) Selective interaction of Ni with an MHC-bound peptide. EMBO J 10:1303-1306.

Romagnoli P, Spinas GA, Sinigaglia E (1992) Gold-specific T cells in rheumatoid arthritis patients treated with gold. J Clin Invest 89:254-258.

Schwartz RH (1985) T-lymphocyte recognition of antigen in association with gene products of the major histocompatibility complex. Ann Rev Immunol 3:237-261.

Springer TA (1990) Adhesion receptors of the immune system. Nature 346:425-434.

Genotoxic Effects of Metal Compounds

D. Beyersmann and A. Hartwig
Department of Biology and Chemistry, University of Bremen, D - 28334 Bremen, Germany

Carcinogenicity of Metal Compounds

Genotoxic effects are important indicators of carcinogenicity. The International Agency for Research on Cancer has classified compounds of arsenic, chromium (VI), nickel and cadmium as carcinogenic to humans (IARC 1980, 1990, 1993). Based on animal data, the German Science Foundation (DFG) also has classified antimony trioxide, beryllium and cobalt as mammalian carcinogens (DFG 1992). In addition, certain metals further the generation of tumors in animals depending on their chemical form and route of administration (Sunderman 1984; Lauwerys 1989; Magos 1991). Regarding the potential risk to human by airborn carcinogens, a German Government Commission has ranked seven substances as the most important for the general population. These are (in decreasing order of contribution to the total risk): polycyclic aromatic hydrocarbons, asbestos, benzene and four metals, i.e. chromium, arsenic, cadmium and nickel. The carcinogenic potential of metals is highly dependent on their speciation. Only the chromates (with chromium in its hexavalent state) but not Cr(III) compounds are carcinogens, since the chromates are able to penetrate cell membranes via anion carriers whereas Cr(III) is not (Jenette 1979). Nickel metal and nickel subsulfide are more efficient carcinogens than nickel chloride, since the soluble nickel ion is taken up slowly. At variance, the soluble $CdCl_2$ is a stronger carcinogen than the weakly soluble CdS, since the soluble ion is better bioavailable (Heinrich et al. 1989). With iron and platinum, only special complexes with enhanced bioavailability are carcinogenic to animals. These include the complexes Fe(III)NTA (Ebina et al. 1986) and cis $Pt(II)Cl_2(NH_3)_2$ (Leopold et al. 1979).

Mutagenicity and Comutagenicity

Mutagenicity is an indicator of the potency of a substance to initiate tumors, but also an indicator of a potential risk of hereditary damage in germ cells. Table 1 summarizes carcinogenic and mutagenic properties of metal compounds in bacterial and mammalian cells. With prokaryotic cells, only chromates give positive results in genotoxicity assays. Mammalian cells also detect the other carcinogenic metal compounds albeit with poor potency. Lead ions, which were carcinogenic only in some animal experiments, are not mutagenic in bacterial or mammalian cell assays.

At variance, Fe(III) is rendered mutagenic and clastogenic by complexation with nitrilotriacetate (Nakatsuka et al 1990; Hartwig et al. 1993).

Table 1. Carcinogenicity, mutagenicity and comutagenicity of metal compounds

Compound	Carcinogenic[a] Human Epidem.	Animal Experim.	Mutagenic[b] Bacteria	Mammal. cells	Comutagenic[c] +UV	+Alkylant
Be(II)		+	-	+	n.t.	n.t.
Cr(VI)	+	+	+	+	+	n.t.
Co(II)		+	-	+	+	n.t.
Ni(II)	+	+	-	+	+	+
As(III)	+	-	-	(+)	+	+
Cd(II)	+	+	(+)	+	+	+
Pb(II)		+	-	-	+	+

+: positive, (+): weakly positive, -: negative, n.t.: not tested
[a]Data compiled from IARC 1980, 1990, 1993 and DFG 1992.
[b]Data compiled from Cohen et al. 1991; Magos 1991
[c]Data compiled by Hartwig and Beyersmann 1993

In contrast to their weak mutagenicity, carcinogenic metal compounds with the exception of beryllium are consistently comutagenic in mammalian cells when combined with UV-irradiation or alkylating agents (Table 1). This effect is interpreted by inhibitory actions of the toxic metals on the repair of DNA damage that has been caused by other agents. Ni^{2+}, Co^{2+}, Pb^{2+} and Cd^{2+} also have been shown to enhance the frequency of sister-chromatid exchanges in mammalian cells exposed to UV-light (Hartwig and Beyersmann 1989; Hartwig et al. 1991a; Hartwig 1993). In the case of nickel, also its effect on the cytotoxicity of cis-$Pt(NH_3)_2Cl_2$ and mitomycin C has been investigated, and an increased toxicity was observed (Hartwig 1993). Since the cytotoxity of these agents can be attributed to alkylation or crosslinking of DNA, an enhancement of alkylating damage by Ni^{2+} can be concluded.

DNA Damage Caused by Metals

Metal cations efficiently bind to the negatively charged nucleic acids, if they are not captured by other ligands before they reach the nucleus of a cell (Cohen et al. 1991). However, metal ions do not damage DNA by an immediate reaction with DNA, but by causing the generation of reactive intermediates, often reactive oxygen species. This concept is supported by the observation that many metal compounds cause the formation of hydroxylated nucleic acid bases (Table 2). The actual reactions differ between the individual metal compounds, however. Chromium(VI) is taken up by cells as the chromate anion, it is reduced within cells via chromium(V) to the most stable state of chromium(III). A complex of chromium(V) with the reducing agent glutathione has been isolated from a reaction of Cr(VI) with glutathione, and the complex was shown to cause DNA strand breaks with isolated DNA in vitro (Kortenkamp et al. 1989). Chromate and other metal ions capable of electron transfer react with hydrogen peroxide to more reactive oxygen species like superoxide or the hydroxyl radical (Klein et al.1991).

Table 2. Oxidative DNA Base Damage by Metal Compounds (Modified from: Kasprzak, 1991)

Compound	System	Elevated Products
Ni(II)acetate	Rats i.p. injection	renal 8-OH-dG
Fe(II)NTA	Rats i.p. injection	renal 8-OH-dG
$Ni(II)Cl_2$	Isolated chromatin $+H_2O_2$	8-OH-Purines
$Co(II)Cl_2$	Isolated chromatin $+H_2O_2$	8-OH-Purines
$FeCl_3$	Isolated chromatin $+H_2O_2$	8-OH-Purines
$CuSO_4$	Isolated chromatin $+H_2O_2$	8-OH-Purines
$Na_2Cr_2O_7$	Isolated DNA $+ GSH + H_2O_2$	8-OH-Purines

Other metal ions like Be^{2+} and Cd^{2+} are not capable of participating in redox reactions, and these metals could only promote oxidative damage to DNA by interfering with enzymes detoxifying reactive oxygen species like catalase, glutathione peroxidase, or superoxide dismutase (reviewed by Kasprzak 1991).

Interference with DNA replication and repair

Genotoxic metal ions also can enhance genetic damage by indirectly furthering mutations. One possible mechanism involves the binding of detrimental metal ions to the complex of DNA with DNA polymerase, thereby lowering the frequency of correct base pairing in DNA synthesis and increasing the incorporation of genetic errors (Table 3). For some metal ions cytotoxic concentrations in the millimolar range are required to achieve this effect. With Cd^{2+} a concentration of 40 μM suffices to increase the error frequency in DNA. The low and varying mutagenicity of carcinogenic metal ions points to a still more indirect mechanism. As already indicated by the consistent comutagenicity of metal ions when combined with UV-irradiation and/or alkylating agents, metal ions may interfer with the repair of damaged DNA. Since the enzymic steps in the removal of UV-induced thymine dimers from DNA are well-characterized, the effects of metal ions on UV repair will be taken as an example.

Table 3. Error frequencies with DNA polymerase induced by metal ions (Selected from Sirover and Loeb, 1976 and Tkeshelashvili et al., 1980).

Compound	x fold increase in error frequency	Metal concentration [mM]
$MgSO_4$	1.1	10.0
$BeCl_2$	15.0	10.0
$CrCl_3$	2.2	1.0
$MnCl_2$	3.8	10.0
$CoCl_2$	8.4	4.0
$NiCl_2$	1.9	8.0
$CuCl_2$	2.9	0.1
$ZnCl_2$	1.1	0.4
$CdCl_2$	1.4	0.04
$PbCl_2$	1.5	4.0

Employing the technique of DNA strand break analysis by nucleoid sedimentation it could be shown that several carcinogenic metal ions (Ni^{2+}, Co^{2+}, Pb^{2+}, Cd^{2+}) inhibited the resealing of DNA strand breaks caused by UV-irradiation of human HeLa cells (Hartwig and Beyersmann 1989; Hartwig et al. 1991a; Hartwig 1993). In the case of Co^{2+} it was demonstrated that the metal ion inhibited an early step in DNA repair, i.e. the excision of UV-induced thymine dimers from DNA (Hartwig et al. 1991a). Investigations on the molecular interactions of metal compounds with individual enzymes of DNA repair after UV- and alkylating damage are carried out in the authors' laboratory.

Epigenetic Effects of Metal Ions

Toxic metals are known to exert systemic effects, and it is not possible to assign single endpoints to the individual metals. Besides their role on the level of DNA synthesis and repair, metal ions may interfere with several levels of cellular signalling, and these interactions have been discussed as possible mechanisms of cytotoxicity (Rossi et al. 1991) or of tumor promoting effects of metals (Beyersmann 1991). Cadmium ions, which are not capable of redox reactions under physiological conditions, have recently been investigated as modulators of the inositol triphosphate - calcium - protein kinase C signalling cascade. Cd^{2+} binds to the cell surface of human skin fibroblasts and evokes intracellular inositol triphosphate and calcium mobilization (Smith et al. 1989). Cadmium ions also interfere with the intracellular calcium homeostasis by inhibiting vesicular Ca^{2+} ATPase and Ca^{2+} transport (Hechtenberg, Beyersmann 1991). Furthermore, cadmium ions enhance the translocation to cell nuclei and nuclear activity of protein kinase C (Block et al.

1992). Most relevant to carcinogenesis is the observation that cadmium ions induce transcription of proto-oncogeneses c-jun and c-myc in rat myoblasts (Jin, Ringertz 1990). Hence, the genetic activity may be changed by toxic metal ions in addition to effects on the genetic material itself.

Interactions in Metal Genotoxicity

Due to their multiple interactions with proteins and nuclei acids, metal ions also exert multiple interactions with each other on various levels of cellular metabolism. Metal ions interact on the level of metal uptake, binding to cellular macromolecules and modulation of enzymic activities (Beyersmann 1991). The infidelity in DNA synthesis produced by many toxic metal ions can be interpreted as an interference with the function of the essential magnesium ion which is required for the proper interaction between DNA and DNA polymerases. Also the impairment of DNA repair by toxic metal ions may be due to competition of these ions for magnesium sites in repair enzymes. This view is supported by the observation that Mg^{2+} partially prevents the inhibition by Ni^{2+} of the repair of UV-damage to DNA (Hartwig et al. 1991b). On the other hand, the disturbance of cellular and genetic regulation by genotoxic and/or carcinogenic metals may be interpreted by interactions with calcium- and zinc-dependent processes. Cadmium ions have been shown to interfere with calcium-dependent signal transduction (Smith et al. 1989; Zhang et al. 1990; Hechtenberg, Beyersmann 1991) and with calcium-dependent DNA fragmentation in apoptosis (Lohmann, Beyersmann 1993). Substitution of zinc by cadmium in zinc finger motifs of transcription factors is an attractive hypothesis to interpret the altered gene expression in the presence of cadmium (Sunderman, Barber 1988). These observations and speculations are in accordance with the protective effects of magnesium in experimental carcinogenesis by cadmium, nickel and lead and the protective effect of zinc in experimental carcinogenesis by cadmium (Furst 1987).

Conclusions

The elucidation of mechanisms of metal-induced genotoxicity is important for the understanding of metal-induced carcinogenesis and possible protective measures. There is no single mechanism in metal carcinogenesis, since metals exert multiple systemic effects on all levels of cellular metabolism. There are, however, three emerging main lines of interactions. Firstly, several ions including those of chromium and nickel are capable of catalyzing the formation of reactive oxygen species within cells. This is the probable cause of the oxidative damage to DNA observed in vivo. Secondly, many metal compounds enhance the genotoxicity of other agents including UV-irradiation and alkylating agents. In many instances, inhibition of DNA repair by metal ions has been demonstrated. Thirdly, metal ions interfere with intracellular signal transduction and expression of genetic information. Hence, carcinogenic metal compounds may act via multiple mechanisms within the same cell, including genetic and epigenetic effects. Further research is needed to elucidate which mode of interaction is prevailing for each individual toxic metal.

References

Beyersmann D (1991) The significance of interactions in metal essentiality and toxicity. In Merian E (ed.) Metals and their compounds in the environment. VCH Weinheim, pp. 491 - 509

Block C, Freyermuth S, Beyersmann D, Maliviya A N (1992) Role of cadmium in activating nuclear protein kinase C and the enzyme binding to nuclear protein. J. Biol. Chem. 267: 19824 - 19828

Cohen M, Latta D, Coogan T, Costa M (1991) Mechanisms of metal carcinogenesis: The reactions of metals with nucleic acids. In: Foulkes EC (ed) Biological effects of heavy metals. Vol II. CRC Press, Boca Raton, pp. 19 - 75

DFG (1992) Maximale Arbeitsplatzkonzentrationen und Biologische Arbeitsstoff-Toleranzwerte. VCH, Weinheim

Ebina Y, Hamazaki S, Li J L, Midorikawa O, Ogino F, Okada S (1986): Nephrotoxicity and renal cell carcinoma after use of iron- and aluminium-nitriloacetate complexes in rats. J. Ntl. Cancer Inst. 76: 107 - 113

Furst A (1987) Metal interactions in carcinogenesis. In: Fishbein L, Furst A, Mehlman MA (eds) Genotoxic and carcinogenic metals: Environmental and occupational occurrence and exposure. Princeton Publishing Co., Princeton, pp 279 - 293

Hartwig A (1993) Role of DNA repair inhibition in lead- and cadmium-induced genotoxicity: A review. Environ. Health Persp., in press

Hartwig A, Beyersmann D (1993) mechanisms in metal genotoxicity: The significance of interactions with DNA repair. Toxicol. Letters, in press

Hartwig A, Schlepegrell R, Beyersmann D (1990) Indirect mechanism of lead-induced genotoxicity in cultured mammalian cells. Mutat. Res. 241: 75 - 82

Hartwig A, Snyder R D, Schlepegrell R, Beyersmann D (1991a) Modulation by Co(II) of UV-induced DNA repair, mutagenesis and sister-chromatid exchanges in mammalian cells. Mutat. Res. 248: 177 - 185

Hartwig A, Schlepegrell R, Beyersmann D (1991b) Interactions in nickel mutagenicity: Indirect mechanisms in genotoxicity and interference with DNA repair In: Merian E, Haerdi W (eds) Metal Compounds in Environment and Life 4: Science Reviews, Wilmington, pp 475 - 480

Hartwig A, Klyszcz-Nasko H, Schlepegrell R, Beyersmann D (1993) Cellular damage by ferric nitrilotriacetate and ferric citrate in V79 cells: Interrelationship between lipid peroxidation, DNA strand breaks and sister chromatid exchanges. Carcinogenesis 14: 107 - 112

Hechtenberg S, and Beyersmann D (1991) Inhibition of sarcoplasmic reticulum Ca^{2+} ATPase activity by cadmium, lead and mercury. Enzyme 45: 109 - 115

Heinrich U, Peters L, Ernst H, Rittinghause S, Dasenbrock C, König H (1989) Investigation on the carcinogenic effects of various cadmium compounds after inhalation exposure in hamsters and mice. Exper. Pathol. 37: 1 - 4

IARC (1980) Some metals and metal compounds. IARC Monographs. Vol. 23. International Agency for Research on Cancer, Lyon

IARC (1990) Chromium, nickel and welding. IARC monographs Vol. 49. International Agency for Research on Cancer, Lyon

IARC (1993) Cadmium and cadmium compounds. IARC monographs. International Agency for Research on Cancer, Lyon, in print

Jenette K E (1979) Chromate metabolism in liver microsomes. Biol. Trace Elem. Res. 1: 55 - 62

Jin P. Ringertz N R (1990) Cadmium induces transcription of proto-oncogenes c-jun and c-myc in rat L6 myoblasts. J. Biol. Chem. 265: 14061 - 14064

Kasprzak K (1991) The role of oxidative damage in metal carcinogenicity. Chem. Res. Toxicol. 4: 604 - 615

Klein C B, Frenkel K, and Costa M (1991) The role of oxidative processes in metal carcinogenesis. Chem. Res. Toxicol. 4: 592 - 604

Kortenkamp A, Ozolins Z, Beyermann D, O'Brien P (1989) Generation of PM2 DNA breaks in the course of reduction of chromium(VI) by glutathione. Mutat. Res. 216: 19 - 26

Lauwerys R R (1989) Metals - Epidemiological and experimental evidence for carcinogenicity. Arch. Toxicol. Suppl. 13: 21 - 27

Leopold W R, Miller E C, Miller J A (1979) Carcinogenicity of antitumor cis-platinum(II) complexes in the mouse and rat. Cancer Res. 39: 913 - 918

Lohmann R D, Beyersmann D (1993) Cadmium and zinc mediated changes of the Ca^{2+}-dependent endonuclease in apoptosis. Biochem. Biophys. Res. Commun. 190: 1097 - 1103

Magos L (1991) Epidemiological and experimental aspects of metal carcinogenesis: Physicochemical properties, kinetics, and the active species. Environ. Health Perspect. 95: 157 - 189

Nakatsuka S, Tanaka H, Namba M (1990) Mutagenic effects of ferric nitrilotriacetate (Fe-NTA) on V79 chinese hamster cells and its inhibitory effects on cell-cell communication. Carcinogenesis 11: 257 - 260

Rossi A, Manzo L, Orrenius S, Vahter M, Nicotera P (1991) Modifications of cell signalling in the cytotoxicity of metals. Pharmacology Toxicology 68: 424 - 429

Sirover M A, and Loeb L A (1976) Infidelity of DNA synthesis in vitro: Screening for potential metal mutagens or carcinogens. Science 194: 1434 - 1436

Smith J B, Dwyer S D, Smith L (1989) Cadmium evokes inositol polyphosphate formation and calcium mobilization. J. Biol. Chem. 264: 7115 - 7118

Sunderman F W, Jr. (1984) Recent advances in metal carcinogenesis. Ann. Clin. Lab. Sci. 14: 93 - 122

Sunderman F W, Jr., Barber A M (1988) Fingerloops, oncogenes and metals. Ann. Clin. Lab. Sci. 18: 267 - 288

Tkeshelashvili L K, Shearman C W, Zakour R A, Koplitz R M, and Loeb L A (1980) Effects of arsenic, selenium and chromium on the fidelity of DNA Synthesis. Cancer Res. 40: 2455 - 2460

Zhang G H, Yamaguchi M, Kimura S, Higham S, and Kraus-Friedman N (1990) Effects of heavy metals on rat liver microsomal Ca^{2+} ATPase and Ca^{2+} sequestering. J. Biol. Chem. 265: 2184 - 2189

Toxicokinetics of Drugs

(Chair: D. S. Davies, United Kingdom)

The Importance Of Exposure Measurements In Risk Assessment Of Drugs

D. Hattis
Center for Technology, Environment, and Development, Clark University, 950 Main Street, Worcester, Massachusetts, USA 01610

I am an outsider to the field of drug evaluation, although I have considerable experience with exposure and quantitative risk assessment for a variety of other environmental and occupational chemicals. Below I will first give my outsider's view of the state of the art of "risk assessment" or "risk-benefit assessment" as applied to pharmaceuticals. Then I will survey some ways in which, with the aid of measures of in-use exposures, risk-benefit assessments of drugs might be made more quantitative and potentially helpful for optimizing drug therapeutic regimens.

Current Paradigms of Risk-Benefit Assessment for Already-Marketed Drugs

The journal, *Drug Safety* has a specific category of "Risk-Benefit Assessment" reviews that offers a convenient window on assessment practices for drugs that are already on the market. Table 1 shows my outsider's perspective in analyzing 12 of these papers--all I could find in my local library published between January, 1991 and June, 1993.

In reading the articles I was struck by the qualitative nature of the questions usually addressed, and the almost total reliance on simple summarization of direct clinical observational studies. There was nothing resembling the quantitative risk assessment modeling that is now common in the analysis of environmental and occupational health hazards (Hattis and Smith, 1987; Hattis and Silver, 1993). There was no assessment of distributions of drug "exposures" observed or likely to occur in actual clinical practice, no dose response modeling in the sense of fitting some mechanism-based or empirical equation to available data on either beneficial or adverse effects in relation to dose, and no "risk/benefit characterization" in the sense of quantitatively projecting likely individual and aggregate risks and benefits for specific patient subgroups and estimating the associated uncertainties. Only one paper (Kanto, 1991) found occasion to make use of a mechanism-based equation in the text (in that case, to relate likely dosages to newborns with milk to maternal serum concentrations and milk/serum partition coefficients). Only 4 had any quantitative measures of both beneficial and adverse effects--and in all cases these were based on direct clinical observations of specific study populations, rather than modeling. The restriction to direct observations meant that likely important, serious, but rare adverse responses were often not quantified.

Among the 12 papers there was, however, one outstanding example of the use of empirical dose response data for both beneficial and undesirable consequences of drug use (Lipworth, 1992). This author asks not just whether the risks of current uses of the drugs seem likely to be warranted in relation to the benefits, but he casts a critical eye on whether current recommendations for drugs in his review category are as beneficial as they might be. That is, does current usage strike the most favorable balance of risks and benefits that might be found for the use of those drugs? Since this paper is the only one of the twelve to use dose response

Table 1. Current risk benefit analysis for already-marketed drugs based on a review of 12 papers appearing January 1991- June 1993 in *Drug Safety*

Therapeutic Category	Quantitative Measure of Risk?	Quantitative Measure of Benefit?	Use of Measures of Exposure?	Reference
Anti-coagulants	Yes--incidence of hemorrhage, thrombocyto-penia, etc.--no dose response	Limited--"Potential benefit" quantified as risks without treatment; reductions in risk with treatment	Yes--serum concentra-tions, clotting times	Harrington and Ansell (1991)
Anti-convulsants (potentially pregnant women)	Yes--malformation rates; also some specific maternal effect rates given	No--Prevention of serious complications of epilepsy mentioned, but not quantified	No	Cleland (1991)
Treatment of menstrual disorders	No	Yes--Reduced blood loss; Percentage of women benefiting	No	Higham and Shaw (1991)
Anesthetics used during and after childbirth	No	No	Yes--concentra-tions in maternal and newborn plasma, milk	Kanto (1991)
Tocolytic drugs (to delay early childbirth)	No	Yes--but very indirect--national time trends in perinatal mortality	Yes--fetal concentra-tions relative to maternal	Wischnik (1991)
Antihyper-tensives in pregnancy	No	No	No	Kyle and Redman (1992)
Withdrawal of antihyper-tensive therapy	Yes--recurrence rates of blood pressures above criteria. But consequence for major cardiovascular events (e.g., strokes) not quantified	Limited-- changes in serum potassium and uric acid. However other possible benefits of intermittent treatment (lowering serum lipids, blood sugar, costs) not quantified	No	Schobel et al. (1992)
β_2-Agonists for asthma	Yes--dose response--finger tremor, heart rate, plasma potassium. Arrhythmias, sudden death not estimated.	Yes--dose response--continuous increase in FEV_1 over a 40-fold range of dosage of salbutamol (100-4000 µg). Also see figures 1 and 2.	No	Lipworth (1992)
Vaccines	Yes--adverse effect rates	No--problem of quantifying individual vs social benefit discussed	No	Duclos and Bentsi-Enchill (1993)

Table 1, Continued. Current risk benefit analysis for already-marketed drugs based on a review of 12 papers appearing January 1991- June 1993 in *Drug Safety*

Therapeutic Category	Quantitative Measure of Risk?	Quantitative Measure of Benefit?	Use of Measures of Exposure?	Reference
Tamoxifen for breast cancer	Yes--incidence of life-threatening thromboembolic events and endometrial cancer, plus some others.	Yes--5-year recurrence-free survival; decrease in mortality at 5 and 10 years, contralateral breast cancer, cardiovascular disease, loss of bone	Yes--ranges of steady-state and peak serum concentrations	Catherino and Jordan (1993)
Nicotine preparations in smoking cessation	No	Yes--doubling cited for long term abstinence rates	Yes--blood nicotine levels compared with those produced by smoking	Hughes (1993)
Withdrawal of anti-epileptic therapy after 2 years successful control	Yes--seizure recurrence rates. But no study of risks from dosage reduction, rather than withdrawal, after relapse.	No--long term adverse effects of continued antiepileptic therapy mentioned, but not quantified.	No	Scherokman and Jabbari (1993)

information, it seems worth quoting on the subject; "The aim for bronchodilator therapy should be to produce optimal airway response with minimal systemic adverse effects. This aim can only be achieved with a thorough understanding of dose-response relationships." I should think that similar reasoning would be more generally applicable to other categories of drugs.

Figure 1 shows some basic efficacy dose response data from this paper and indicates why the author thinks current recommendations for the use of bronchodilators are worth reexamining. He comments, "Only 1 patient [of 14] achieved maximal FEV_1 response with the recommended dose [200 µg]...the majority required either 2000 or 4000 µg in order to achieve maximal bronchodilatation."

Figure 2 shows how a dose response modeler like myself would go further to characterize the population distribution of dosage requirements to achieve maximal effect, using a classical log probit analysis (Finney, 1971). The good fit of the line to the points indicates that the assumption of this analysis--that maximal effect concentrations have a lognormal distribution in the population studied--is compatible with the observations. The low probit slope (1.56 standard deviations per 10-fold change in dose) provides a preliminary quantitative estimate of the enormous amount of population heterogeneity in response suggested here (95% of patients' maximal effect dosages would be expected to be spread out over nearly a 400 fold range; from about 20-fold less to 20-fold greater than the median). I would proceed to use this and other distributions characterizing different intensities of benefit and risk to help extend analysis beyond the available data and paint a more complete picture of likely risks and benefits from different protocols for drug

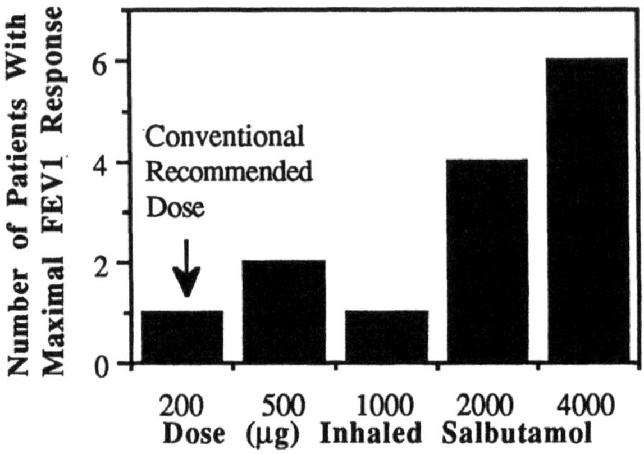

Fig. 1. Histogram of the numbers of asthmatic patients requiring various doses of Salbutamol to achieve maximal FEV_1 response (data of Lipworth et al. (1988).

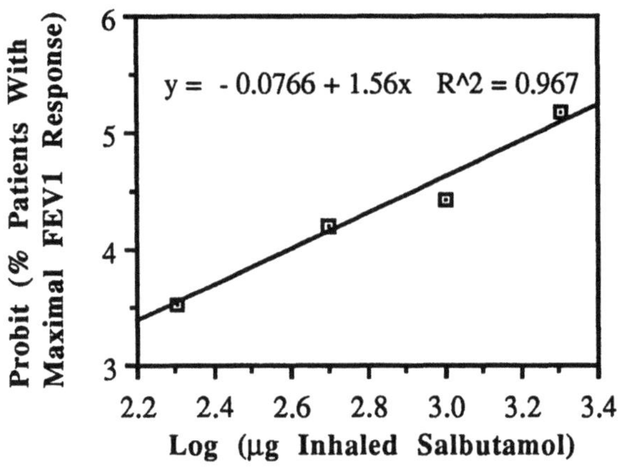

Fig. 2. Classical log probit plot of the dosages needed for maximal FEV_1 responses to Salbutamol (based on the data of Lipworth et al., 1988 in figure 1).

usage (see below), including in this case, protocols for measuring each individual patient's responses in relation to dose.

Practices for Evaluation of New Drugs--One Likely Source of the Problem

Doing the research to develop new drugs, prove that they are safe and effective, and thereby get them on the market is a risky and very expensive undertaking (Bienz-Tadmor et al., 1992; DiMasi et al., 1991). It also has the potential for very large rewards, both in monetary terms and in terms of improvements in health. For this reason it is not surprising that the research needed to initially get new drugs on the market has been a primary focus of attention both in the pharmaceutical industry and in academia. I believe it is likely that the concepts and objectives for this type of research--molded by the pressures of costs and information requirements for "demonstrating" efficacy and safety in a regulatory context--are have strongly influenced the kinds of risk-benefit analysis seen in the papers reviewed in Table 1.

Lai et al. (1988) provide a useful summary of the process of creating the basis for drug approval from the perspective of clinical pharmacokinetics. This perspective is valuable in the context of risk benefit analysis because, as the authors emphasize, pharmacokinetics is a "bridge science"--bridging "efficacy and safety assessments of a new chemical entity (NCE) with dosage and dosing schedule." In the context of this paper, of course, pharmacokinetics is one key category of quantitative information that must be based on quantitative external and/or internal measurements of exposure.

The work of the clinical pharmacokineticist begins in phase I testing, where graded doses of the new drug are administered to relatively small groups (usually <20) of normal healthy adults--generally young males. The primary goal of first single-dose administrations, and later repeated-dose studies, is to establish maximally tolerated doses for both acute and chronic administration. The trials therefore need to be conducted in such a way that at least some early toxic effect concentrations are reached. To be most useful, Lai et al. (1988) emphasize that in the chronic dosing study "The dosing schedule, which includes daily dosages and frequency, dose escalations, and dose tapering, should simulate the regimen to be followed in efficacy testing." Thus the expectation is that even prior to any appreciable efficacy information derived from the use of the drug in patients with the condition to be treated, the pharmacokineticist needs to anticipate key features of the dynamics of drug usage and action that will be incorporated into efficacy testing protocols.

In planning the protocols for the efficacy testing trials (the exploratory Phase II, and the large scale double blind trials in Phase III), the pharmacokineticist uses parameters derived from the limited Phase I studies to "computer-simulate plasma NCE levels for various dosage regimens. In this exercise, the disease and/or physiological states of the test patients (e.g., organ dysfunction as a function of age), concurrent medications (e.g., enzyme inducers or inhibitors), and the safety data obtained earlier must be taken into consideration before arriving at an optimal regimen for inclusion in the protocol." Peck et al. (1992) strongly recommend that this be done based on quantitative pharmacokinetic and dose effect studies extending into Phase II and Phase III.

The main goal of the efficacy testing is to demonstrate that the new drug "works," at the most basic level. Operationally, this is defined as rejecting the "null hypothesis" that it is completely worthless (indistinguishable from a placebo) in a double-blind trial. At the same time, of course, it is important to show that the drug causes no more detectable side effects than are deemed tolerable in the light of the therapeutic benefits realized.

The costs of the Phase III trials to accomplish this are very substantial. And it can be expected that these costs would be multiplied whenever the investigators choose to include either additional dose groups, therapeutic regimen groups (e.g. twice per day vs once per day), or additional subgroups of patients to the program. The incentives therefore are to limit the numbers of these trial categories as much as feasible. Moreover, there is also reason for the experimenters to limit the diversity of patients included within the groups subject to test. Generally, the greater the variability of a tested patient population in characteristics that may affect response, the larger the numbers of people that must be included in treatment and placebo groups in order to demonstrate a difference for any specific degree of statistical confidence.

All of these considerations suggest that an economically optimal research strategy to prove the qualitative efficacy and safety of a drug, and get it on the market as soon as possible, will not be the same as the research strategy one would adopt if one wished to address the quantitative questions involved in deciding how to best use the drug in practice. The latter type of question is best informed by including as much realistic diversity in the study population as is possible and ethical under the circumstances. Rather than viewing diversity/ variability as a nuisance to be rid of, investigators with the latter goal should view it as a basic fact to be preserved and studied in its own right as an important determinant of in-use risks and benefits. These sources of variability need to be used to help determine optimal recommendations for indications of use, patient selection, and monitoring precautions. Unfortunately in the early 1980's, when a prominent official of the U.S. Food and Drug Administration proposed that patients in Phase III trials be routinely monitored to define pharmacokinetic variability and detect subgroups with unusual pharmacokinetics, the response from the industry was so hostile (apparently because of fears of delaying drug approval) that the proposal was abandoned (Sheiner and Benet, 1985). More recently, some use of this opportunity seems to be gaining acceptance (Lai et al., 1988; Peck et al., 1992; Thomson and Whiting, 1992).

Contribution of Exposure Measures to Elucidating Interindividual Variability in Pharmacokinetics

Repeated measures of serum concentrations are, of course, a principal data source for pharmacokinetics. The variability of major pharmacokinetic parameters in normal healthy adults has been summarized previously (Hattis et al., 1987).

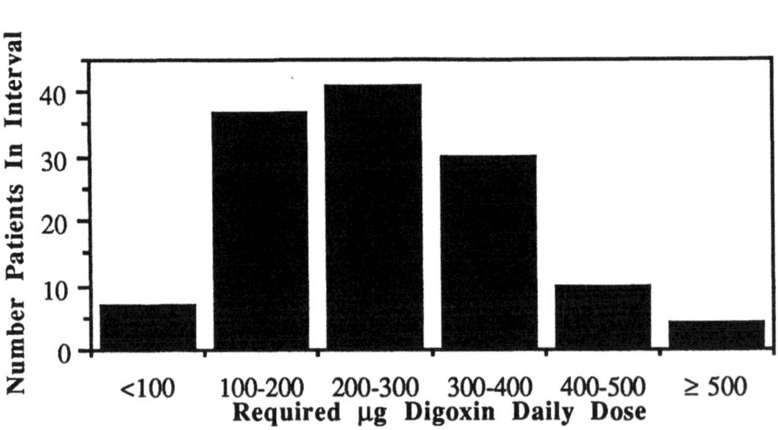

Fig. 3. Estimates by Dobbs et al. (1991) of daily digoxin doses required to keep individual patients at an average steady state serum concentration of 1.6 nmol/liter.

Pharmacokinetic variability is often a prime determinant of drug toxicity in practice. For example, in a classic study of 48 patients who developed toxicity on digoxin therapy and 131 who did not, Smith and Haber (1970) found only a minor difference in prescribed dosage (0.36 ± 0.19 SD mg/day versus 0.31 ± 0.19 SD mg/day) but a significant difference in serum digoxin levels (3.7 ± 1.0 ng/ml for those with toxicity versus 1.4 ± 0.7 ng/ml for those without toxicity--$P < .001$).

More recently, Dobbs et al. (1991) used repeated measurements of digoxin serum levels (on average 4 per patient) to estimate dosages required to produce an average steady-state serum concentration of 1.6 nmol/liter (Figure 3). In this case, the distribution is about equally well fit by either a lognormal or a normal distribution (coefficient of variation = 43%) (plots not shown). There is a strong correlation between this digoxin daily dose requirement and creatinine clearance, a measure of kidney function. Creatinine clearance is approximately normally distributed in this population, with a coefficient of variation of about 42%.

Contributions of Exposure Measures to Elucidating Interindividual Variability in Pharmacodynamics

Clearly, measurements of serum concentration, and other indices of internal exposure are vital for separating pharmacokinetic variability from internal dose response relationships and <u>their</u> interindividual variability. Good examples of these uses can also be found for digoxin.

Buch and Waldorff (1980) found that the effect of digoxin on myocardial contractility was well described by the classic Michaelis-Menten formula which governs most enzyme kinetics:

$$\text{Strength of Effect} = E_{max}\, C/(K_m + C)$$

where is E_{max} is the maximal effect, C is the serum concentration, and K_m is the serum concentration that produces half of the maximal effect. In this case Km was found to be 2.07 nmol/liter, implying that an increase in serum concentration from 2 to 3 nmol/liter would be expected to make only a modest increase in the intensity of the effect (from about 49% to 59% of the maximal effect). Based in part on these results, the usual desirable therapeutic range for digoxin has been defined as 1-2 nmol/liter (Dobbs et al., 1991).

Concentration measurements have also been critical in defining digoxin risk levels. Ordog et al. (1987) recently compiled the results of measurements on over 5,000 patients treated with digoxin and seen in an emergency room for any reason. Total mortality observed over the following four weeks increased progressively from control levels for those found to be below the therapeutic range (2%) to nearly 50% for those showing digoxin levels over 6 ng/ml.

Measuring Drug Use--Implications of Imperfect Patient and Physician "Compliance" with Recommended Dosing

Patients differ not only biologically, but also behaviorally--in how they actually use drugs under different circumstances. Even under the relatively favorable circumstances of clinical studies, patients tend to take less medication than is actually prescribed (Cramer et al., 1989), and the timing of administration often differs substantially from the regular patterns that may be implicitly assumed in the design of dosing regimens (Rudd et al., 1990, 1992). Recently this has been studied with the aid of new microprocessor-based technology in which a special bottle cap automatically records when the bottle is opened. Data obtained using this novel approach for exposure assessment show that (1) patient compliance is greater when drugs can be prescribed to be taken fewer times per day (Table 2), and (2) there is wide interindividual variability in patient compliance (Figure 4).

Table 2. Compliance as a Function of Frequency of Prescribed Dosing--Measurements by Electronic Recording of Bottle-Openings

	% of Days in Compliance (Mean and SD for Different Patients)		
Prescribed Doses/Day	Epileptics (Cramer et al., 1989)	Hypertensives (Eisen et al., 1990)	Hypertensives (Rudd et al., 1990)
1	87 ± 11	84 ± 20	Not done
2	81 ± 17	75 ± 20	63 ± 19
3	77 ± 12	59 ± 30	Not done
4	39 ± 24	Not done	Not done

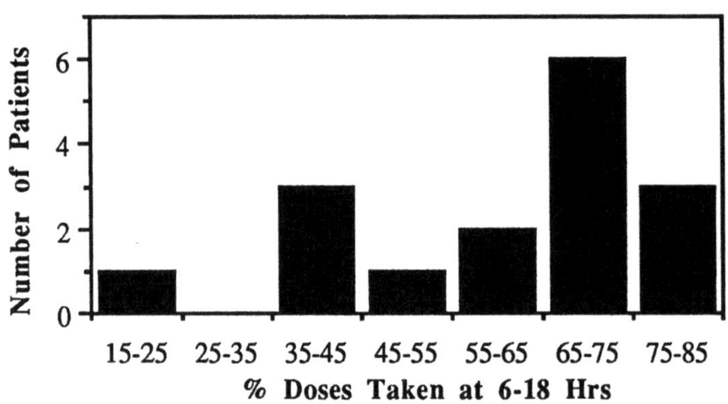

Fig. 4. Rudd et al. (1992) Data on the percentage of doses taken at 6-18 hours after the previous dose (recommended dosing interval was 12 hours).

Monte Carlo Simulation Modeling of the Combined Effects of Multiple Sources of Variability

With quantitative assessment of the basic sources of variability described above, it is possible to use Monte Carlo simulation to analyze likely beneficial and adverse effects of drugs in use in much greater depth than is generally feasible with direct observational studies. In such an analysis the general approach would be to (1) Draw random values from the specified distributions (with correlations, if any) to simulate the practical use of the drug in many simulated "patients". (2) Calculate the frequency and magnitude of expected beneficial and adverse effects for different categories of patients for the "base case" of current usage patterns. (3) Explore the expected effects of changes in recommended protocols for use for different categories of patients., and finally (4) Compare the expected optimized benefits for different therapiess for the same condition. Convenient microcomputer software is now readily available for such Monte Carlo simulation analyses (Hattis and Froines, 1992).

Conclusion

Although there has been a great deal of rhetoric on the balancing of risks and benefits of pharmaceuticals, for the most part this has not taken the form of quantitative modeling of the likely incidence of beneficial and adverse outcomes of drug usage under realistic conditions of practice. Such analyses would need to be based on information about interindividual variability in (A) pharmacokinetics, (B) the pharmacodynamics of beneficial and adverse drug effects, and (C) patient and prescriber drug usage patterns in the field. Inferences about each of these types of variability can be greatly strengthened with the aid of in-use exposure data, designed to reflect as closely as possible the heterogeneity of real patient populations.

Using the three types of variability information mentioned above, Monte Carlo simulation modeling studies have the potential to extend the range of risk and benefit assessments beyond the scope of direct observations in normal clinical trials. This type of analysis can represent an important gain in information particularly in cases where the component distributions are not just empirical formulae, but are based on an understanding of likely causal mechanisms (e.g. physiologically-based pharmacokinetics; receptor-binding as a cause of Michaelis-Menton-like pharmacodynamic relationships). The observations and risk benefit modeling suggested here could help refine recommended dosing amounts, dosing frequency, dosage forms, indications for use, in-use monitoring precautions, and settings for use (e.g., outpatient vs inpatient; prescription vs over-the-counter) for already-marketed drugs for different categories of patients.

References

Bienz-Tadmor B, Dicerbo PA, Tadmor G, Lasagna L (1992) Biopharmaceuticals and conventional drugs: Clinical success rates. Bio/Technology 10:521-525

Catherino WH, Jordan VC (1993) A risk-benefit assessment of tamoxifen therapy. Drug Safety 8:381-397

Cleland PG (1991) Risk-benefit assessment of anticonvulsants in women of child-bearing potential. Drug Safety 6:70-81

Cramer JA, Matton RH, Prevey ML, Scheyer RD, Ouellette VL (1989) How often is medication taken as prescribed? JAMA 261:3273-3277

DiMasi JA, Hansen RW, Grabowski HG, Lasagna L (1991) Cost of innovation in the pharmaceutical industry. J. Health Economics 10:107-142

Dobbs RJ, O'Neill CJA, Deshmukh AA, Nicholson PW, Dobbs SM (1991) Serum concentration monitoring of cardiac glycosides. How helpful is it for adjusting dosage regimens? Clin Pharmacokinet 20:175-193

Duclos P, Bentsi-Enchill A (1993) Current thoughts on the risks and benefits of immunization. Drug Safety 8:404-413

Eisen SA, Miller DK, Woodward RS, Spitznagel E, Pryzbeck TR (1990) The effect of prescribed daily dose frequency on patient medication compliance. Arch. Intern. Med. 150:1881-1884

Finney DJ (1971) Probit Analysis. 3rd Edition, Cambridge University Press, Cambridge.

Harrington R, Ansell, J (1991) Risk-benefit assessment of anticoagulant therapy. Drug Safety 6:54-69

Hattis D, Erdreich L, Ballew M (1987) Human variability in susceptibility to toxic chemicals--A preliminary analysis of pharmacokinetic data from normal volunteers. Risk Analysis 7:415-426

Hattis D, Froines, J. (1992) Uncertainties in risk assessment. In Harvey J. Clewell, IIi, ed., Conference on chemical risk assessment in the DoD: Science, policy, and practice, American Conference of Governmental Industrial Hygienists, Inc., Cincinnati, Ohio, pp. 69-78.

Hattis D, Silver K (1993) Use of biomarkers in risk assessment. Chapter 10 in Schulte PA and Perera FP, eds, Molecular Epidemiology--Principles and Practices, Academic Press, Inc., Boston, London, pp. 251-273

Hattis D., Smith J. (1987) What's wrong with quantitative risk assessment. In J. M. Humber and R. F. Almeder, eds., Quantitative risk assessment. Biomedical Ethics Reviews: 1986, Humana Press, Clifton, New Jersey, pp. 57-105.

Higham JM, Shaw RW (1991) Risk-benefit assessment of drugs used for the treatment of menstrual disorders. Drug Safety 6:183-191

Hughes JR (1993) Risk-benefit assessment of nicotine preparations in smoking cessation. Drug Safety 8:49-56

Kanto J (1991) Risk-benefit assessment of anesthetic agents in the puerperium. Drug Safety 6:285-301

Kyle PM, Redman, CWG (1992) Comparative risk-benefit assessment of drugs used in the management of hypertension in pregnancy. Drug Safety 7:223-234

Lai AA, Fleck RJ, Caplan NB (1988) Clinical pharmacokinetics and the pharmacist's role in drug development and evaluation. Chapter 5 in Cato, AE, ed. Clinical Drug Trials and Tribulations, Marcel Dekker, Inc., New York and Basel, pp. 79-98

Lipworth BJ (1992) Risks versus benefits of inhaled β_2-Agonists in the management of asthma. Drug Safety 7:54-70

Lipworth BJ, Clark RA, Dhillon DP, Brown RA, McDevitt DG (1988) Beta-adrenoceptor responses to high doses of inhaled salbutamol in patients with bronchial asthma. British J. Clin. Pharmacol. 26:527-533

Rudd P, Ahmed S, Zachary V, Barton C, Bonduelle D (1990) Improved compliance measures: Applications in an ambulatory hypertensive drug trial. Clin. Pharmacol. Ther. 48:676-685

Rudd P, Ahmed S, Zachary V, Barton C, Bonduelle D (1992) Compliance with medication timing: Implication from a medication trial for drug development and clinical practice. J. Clinical Research and Pharmacoepidemiology 6:15-27

Peck, CC, Barr WH, Benet LZ, Collins J, Desjardins RE, Furst DE, Harter JG Levy G, Ludden T, Rodman JH, Sanathanan L, Schentag JJ, Shah VP, Sheiner LB, Skelly JP, Stanski DR, Temple RJ, Viswanathan CT, Weissinger J, Yacobi A (1992) Opportunities for integration of pharmacokinetics, pharmacodynamics, and toxicokinetics in rational drug development. J. Pharm. Sci. 81:605-610

Scherokman B, Jabbari B (1993) The risks and benefits of withdrawing antiepileptic therapy. Drug Safety 8:399-403

Schobel HB, Schmieder RE, Messerli FH (1992) Risks versus benefits of withdrawing antihypertensive therapy. Drug Safety 7:395-403

Sheiner LB, Benet LZ (1985) Premarketing observational studies of population pharmacokinetics of new drugs. Clin. Pharmacol. Therap. 38:481-487

Smith TW, Haber E (1970) Digoxin intoxication: The relationship of clinical presentation to serum digoxin concentration. J. Clin. Invest. 49:2377-2386

Thomson AH, Whiting B (1992) Bayesian parameter estimation and population pharmacokinetics. Clin. Pharmacokinet. 22:447-467

Wischnik A (1991) Risk-benefit assessment of tocolytic drugs. Drug Safety 6:371-380

The impact of genetic polymorphisms in risk assessment of drugs

Henrik E. Poulsen and Steffen Loft
Department of Pharmacology, University of Copenhagen, Panum Institute,
3 Blegdamsvej, DK-2200 Copenhagen N, Denmark

Genetic variation is a fundamental characteristic of all living organisms. In humans marked individual variability has been demonstrated at every level of investigation. With regard to drug metabolism variability in responsiveness to a given dose originates from many factors, e.g. environmental influences, alteration of absorption, distribution, excretion and metabolism.

The most important drug metabolizing enzymes are by tradition categorized into two groups, phase I and phase II, according to the chemical reactions they catalyze. Phase I biotransformations include oxygenations, oxidations, reductions and hydrolyses while phase II reactions (sometimes designated synthetic reactions) are conjugation to glucuronic acid, sulfate glutathione, etc. Originally, phase I metabolism followed by phase II metabolism was considered to be the rule. It is now realized that many compounds undergo only one of the phases. From a toxicological point of view there are only very few examples of drugs that are activated into toxic metabolites by phase II metabolism whereas phase I metabolism can frequently yield toxic metabolites. One such example is paracetamol (acetaminophen) to be discussed later in this paper.

In man the CYP (cytochrome P450) isozymes are the major catalysts of drug oxidation, and within the last decade considerable progress has been made in identifying the genetic basis for the huge intraindividual variation in drug metabolism. Although genetic polymorphisms are recognized as major determinants of individual variation in the activities of drug metabolizing enzymes it must be stressed that also with regard to non-polymorphous enzymes large interindividual variability with equally important implications can be found.

In the context of metabolizing chemicals a genetic polymorphism is the presence within a population of at least two groups with distinctly different ability of metabolizing xenobiotics. Presently, a polymorphism is defined by a frequency in the population of at least 1% of at least two phenotypes (Vogel and Motulsky 1982). In this review special focus is on the relevance of genetic polymorphisms in drug metabolism for the risk assessment of drugs.

Genetic polymorphisms in man:

CYP2D6:

CYP2D6 is found in human liver (Gonzalez et al. 1988) and in the brain (Tyndale et al. 1991). A polymorphism in oxidation of the antihypertensive drug debrisoquine was observed (Mahgoub et al. 1977) because of a particular slow elimination and pronounced pharmacodynamic effect in about 8% of the caucasian population. Many drugs are now realized to be oxidized by CYP2D6 (Brøsen and Gram 1989).

The genetic defects in the CYP2D6 gene are well characterized (Kagimoto et al. 1990, Gaedig et al. 1991). Furthermore, the phenotype for debrisoquine hydroxylation can be predicted by DNA-testing (Graf et al. 1992).

Phenotyping: Sparteine, debrisoquine or dextromethorphan (Schmid et al. 1985) can be used as probes for phenotyping individuals. Poor metabolizers (PM) are defined from a metabolic ratio (parent compound/metabolite) in urine of >20 for sparteine (Eichelbaum et al. 1982) , > 12.6 for debrisoquine (Evans et al. 1980) and >0.3 for dextromethorphan (Schmid et al. 1985).

Genotyping: DNA obtained from leucocyte preparation after a conventional venesection can be used for PCR based determination of the CYP2D6 mutations D6-A, D6-B and D6-C (Heim and Meyer 1992).

Example: Codeine is an old analgesic used in many composite analgesic formulations. It is activated to morphine by CYP2D6 (Dayer et al. 1988) and its analgesic action is dependent on biotransformation. It is intriguing that while poor metabolizers have no analgesic effect of codeine fast metabolizers with quinidine induced poor metabolizer phenotype still have the analgesic effect (Weinshilboum and Sladek 1980, Sindrup et al. 1990). Presumably, local bioactivation of codeine in the brain plays a role (Wahlström et al. 1988) and this mechanism may not be inhibited by quinidine.

CYP1A 1/2:

CYP1A1 has been detected in human placenta, skin, lymphocytes and peripheral lung, whereas it is notably absent in human liver (Kadlubar and Guengerich 1992). This enzyme is among the best conserved among xenobiotic-metabolizing enzymes (Nebert 1989). The only available clinically example of a drug that is metabolized by CYP1A1 is coal tar used by dermatologists, the major focus on this enzyme is its role in activation of carcinogens and teratogens (Nebert 1989). In this respect it is noteworthy that the enzyme constitutively has a low expression and that many substances can induce its activity many fold, e.g. cigarette smoke, polyaromatic hydrocarbons etc. In relation to drug metabolism the most interesting feature is whether drugs have CYP1A1 inducing properties that might change the susceptibility of the treated individual to carcinogens or mutagens. In such a case long term treatment might increase the teratogenic or carcinogenic risk.

CYP1A2 has only been identified in the liver and there appears to be a 50 to 100-fold variability in its activity (Guengerich and Shimada 1991). CYP1A2 is important for the metabolism of important drugs, e.g. caffeine, phenacetin and paracetamol, and for a large number of xenobiotics, e.g. heterocyclic amines. A polymorphism has been suggested from a ratio of urinary caffeine metabolites (Butler et al. 1992). However, in studies using the conventional caffeine metabolic ratio on more than 300 persons such a polymorphism could not be detected (Vistisen et al. 1992). The conventional CYP1A2 index that correlates well with the caffeine clearance is between the two step demethylated metabolites of caffeine, i.e. 1-methylxanthine (1X), 1 methyluric acid (1U) and 5-acetyl-6-formylamino-3-methyl uracil (AFMU), and the one step demethylated metabolite 1,7-dimethyluric acid (17U) (Campbell,ME et al. 1987). As stated below genotyping does not suggest any polymorphisms in CYP1A2. Presently it must be concluded that a polymorphism for CYP1A2 still needs to be verified.

Phenotyping: There are no available non-hazardous probes for determination of CYP-1A1 activity *in vivo, in vitro* ethoxyresorufin appears to be a specific probe (Jensen et al. 1993).

CYP1A2 activity can be estimated from dietary caffeine metabolite rations in urine as described above.

Genotyping: DNA from various tissues can, with or without the PCR amplification be used to determine restriction fragment length polymorphisms in CYP1A1 (Kawajiri et al. 1990, Tefre et al. 1991).

CYP1A2 protein analysis of human liver has demonstrated variations, but no liver has been found that lacks this enzyme suggesting that mutant alleles of CYP1A2 are not present in humans (Wrighton et al. 1986, McManus et al. 1988, Shimada et al. 1989).

Example: So far the only example of a drug with CYP1A1 inducing properties is coal tar for topical application whereas the environmentally dominating source of such xenobiotics are tobacco smoke. An association of lung cancer with CYP1A1 polymorphism (RFLP) has been found in a japanese population (Kawajiri et al. 1990) whereas this association could not be found in a Norwegian study (Tefre et al. 1991).

Regarding CYP1A2 a distinct polymorphism remains to be demonstrated, however, the 50-100 fold variability could be equally important for a variety of drug and xenobiotic related effects. A recent example presumably of clinical importance is the inhibitory effect of the antidepressant fluvoxamine as demonstrated by a low K_i for inhibition of CYP1A2 in human microsomes (Brøsen et al. 1993b).

CYP2E1:

In humans CYP2E1 is found in the liver and in peripheral blood lymphocytes. An approximate 10-fold variation in hepatic enzyme activity has been reported (Peter et al. 1990). CYP2E1 was originally called MEOS (microsomal ethanol oxidizing system) and it metabolizes ethanol and other structurally related compounds as for example some of the volatile anesthetics (halothane). The muscle relaxant chlorzoxazone (Peter et al. 1990) is metabolized almost exclusively by CYP2E1 and excreted after glucuronidation into urine, and the analgesic paracetamol (acetaminophen) is activated by CYP2E1 to its arylating metabolite. Presently, there are no systematic investigations available for comparison with the restriction fragment length polymorphisms identified for CYP2E1.

Phenotyping: The only suggested probe for CYP2E1 is chlorzoxazone (Peter et al. 1990). Preliminary investigations in our laboratory show that recovery of the 6-OH metabolite is almost complete in urine and that plasma clearance is about 300 ml/min. It would therefore be possible to use the clearance as a measure of CYP2E1 activity and to develop a simplified one sample method in analogy with the one developed for antipyrine (Poulsen and Loft 1988).

Genotyping: Recently, two reports have described restriction fragment length polymorphisms of CYP2E1. One is based on PstI or RsaI restriction and localized to the transcription regulation region of CYP2E1 (Hayashi et al. 1991). The other polymorphism is localized to entron 6 and revealed by the DraI restriction enzyme (Uematsu et al. 1991).
Example: Paracetamol is activated to an arylating metabolite by CYP2E1, CYP1A2 and CYP3A4 (Thummel et al. 1992, Poulsen et al. 1993) following an overdose that depletes hepatic glutathione. Preliminary investigations in our laboratory points at CYP2E1 as the qualitatively important activating route in mice (unpublished observations), in man, however, another enzyme could be the most important. It is still debated if e.g. alcoholism is a risk factor for development of hepatic injury from paracetamol overdose, since CYP2E1is thought to be induced. However, systematic investigations on this important clinical problem has yet to be done.

CYP2C.
This cytochrome (often called cytochrome $P450_{MF}$) is controlled by an autosomal recessive gene resulting in its absence in approximately 2-3% of the caucasian population as opposed to about 13% in orientals (Bertilsson et al. 1992). An increasing number of drugs are recognized to be metabolized by CYP2C, e.g. diazepam (Bertilsson et al. 1992), omeprazole (Rost et al. 1992) and proguanil (Ward et al. 1991, Funck-Brentano et al. 1992).
Phenotyping: The simplest approach to determine the phenotype appears to be the ratio between the (*R*)- and the (*S*)-enantiomers in urine collected for 8 hours after oral administration of the racemate (Wilkinson et al. 1989).
Genotyping: Presently there are no available methods.
Example: The antimalarial drug proguanil is metabolized by CYP2C to cylcoguanil (Ward et al. 1991), which is responsible for the antimalarial action. Poor metabolizers have only about 25% of the cycloguanil plasma level found in fast metabolizers (Brøsen et al. 1993a), whether this is reflected by a higher risk of therapeutic failure remains to be established.

N-acetyl transferase:
This enzyme is found in many tissues but is most abundant in the liver (Hein et al. 1987). It has been recognized for more than 30 years as genetically determined by two co-dominant alleles (Evans et al. 1960). The acetylation phenotypes, i.e. fast and slow acetylators, predict the rate of elimination of many drugs metabolized by this enzyme, e.g. isoniazid, procaineamide, sulfamethazine, nitrazepam, caffeine (Lou 1990). Both O- and N-acetylation of hydroxylamines are under the same genetic control and the two reactions are important for the metabolic activation and deactivation of environmental carcinogens, e.g. arylamines (Hein 1988).

The genetic defects are now well characterized. The dominant, autosomal allele for fast acetylators and the multiple alleles for slow acetylators have been located to chromosome 8 (Blum et al. 1990, Ohsako and Deguchi 1990).

Phenotype determination: Acetylator phenotype may be determined by probing with caffeine, isoniazid or sulfadimidine (Clark 1985). It should be noted, however, that some drugs are metabolized equally well by acetylation regardless of the phenotype, e.g. *p*-aminosalicylate (Weber and Hein 1985). By means of dietary caffeine intake acetylator phenotype can be determined from the urinary ratio between the acetylated metabolite 5-acetyl-6-formylamino-3-methyluracil (AFMU) and the alternative metabolite 1-methylxanthine(1X), or 1X plus 1-methyluric acid (1U) plus 1,7 demethyluric acid (17U). This index discriminates the two phenotypes in accordance with other indices (Grant et al. 1983, 1984). Whether this probing methodology can distinguish between homozygous and heterozygous fast acetylators is debated (Grant et al. 1983, 1984, Vistisen et al. 1992).

Genotype determination: Genomic DNA obtained from peripheral blood leukocytes can be used for a polymerase chain reaction based test for determination of the wild type (wt) and the mutated genes (M1, M2 and M3) (Graf et al. 1992). By this technique homozygous and heterozygous fast acetylators can be distinguished from slow acetylator genotypes.

Example: Weber et al. (Weber and Hein 1985) concluded that slow acetylators are more likely to develop peripheral neuropathy than fast acetylators from isoniazid, and patients suffering from isoniazid hepatitis have a higher frequency of hepatic disorders if they are fast acetylators. Further supporting this hypothesis is the higher prevalence of isoniazid-induced liver toxicity in orientals (fast acetylators constitute 80 - 90% of the population) than in caucasians (fast acetylators constitute 40 -50% of the population).

Nevertheless, the main importance regarding this polymorphism is probably related to environmental and occupational exposure to aromatic amines and related cancer risk (Hein 1988).

Sulfoxidation.

Sulfoxidation is suggested to be genetically polymorphic regulated. However, solid evidence on the protein and gene level is still not available. A number of sulfur containing drugs are eliminated by sulfoxidation. The formation of a metabolite from the mucolytic agent carbocisteine as assessed by a urinary metabolic ratio has appeared to be under genetic control (Mitchell et al. 1984). Although a low rate of metabolite excretion has been associated with various inflammatory and neurological diseases (Steventon et al. 1990, Gordon et al. 1992), it has only been related to a few reactions to sulfur containing drugs, i.e.D-penicillamine (Emery et al. 1984), and this could be due to an apparently worse clinical course in patients with this trait (Emery et al. 1992). In fact, the claim that the putative carbocisteine metabolite is a product of sulfoxidation and its importance have been disputed by others (Brockmoller et al. 1991, Meese et al. 1991, Gregory et al. 1992). No definite polymorphism with respect to sulfoxidation of drugs has yet been identified.

Methylation.

This reaction lacks the in depth investigations that are available for the CYPs. However, conjugation of methyl groups to thiols, oxygen and nitrogen appear to be under separate genetic control (Weinshilboum 1989, 1992). Thiopurine methyltransferase (TPMT), a cytoplasmatic enzyme expressed in all tissues, catalyzes the S-methylation of aromatic and heterocyclic sulfhydryl compounds, including 6-mercaptopurine and azathioprine. The activity of TPMT may be assessed from red blood cells. From population and family studies the activity appears to be controlled by a single gene with two alleles and gene frequencies for high and low activity of 94% and 6%, respectively (Weinshilboum 1992). Thus, in 298 subjects 88.6%, 11.1% and 0.3% had high, intermediate, and no activity, respectively (Weinshilboum and Sladek 1980). Whereas patients with low TPMT activity may suffer toxicity to the thiopurine drugs (Lennard et al. 1987, 1989), those with particularly high activity may risk underdosing and relaps more frequently when treated for leukemia (Lennard et al. 1990).

Thiolmethyltransferase (TMT), a membrane bound enzyme expressed in most tissues, catalyzes the methylation of aliphatic sulphydryl drugs, such as captopril and D-penicillamine (Weinshilboum 1989, 1992). The activity as assessed from red blood cells appear to be under some genetic control. Although particular TMT activities have been associated with some neurological diseases (Waring et al. 1989), no relationship with any drug reaction has been reported.

Cytoplasmic cathecol O-methyltransferase (COMT) metabolizes some neurotransmitters and catechol drugs (Weinshilboum 1989, 1992). The distribution of COMT activity assessed from red blood cells in a population suggests regulation by a single locus with two alleles and almost equal gene frequencies. Whereas COMT activity correlates with the rate of metabolism of L-dopa and methyldopa (Reilly et al. 1980, Campbell,NRC et al. 1984), the clinical significance has yet to be demonstrated.

From a large family study the activity of N-methyltransferase expressed in red blood cells appears to be under genetic control (Scott et al. 1988). Although histamine is N-methylated by this enzyme, no drug has yet been identified as an important substrate.

Sulfation:

Sulfotransferases belong to a family of enzymes found mainly in the cytosol of hepatocytes but also in platelets. In contrast to the CYPs they are not investigated in the same details regarding the proteins and the genes.

Both thermolabile and thermostable phenol sulfotransferase as assessed from platelets appear to be under genetic controls in population and family studies (Weinshilboum 1990). The extent of paracetamol sulfation correlated with the platelet activity of both forms in one study (Reiter and Weinshilboum 1982) but only with the thermostable form in another study (Bonham-Carter et al. 1983). In subjects supplemented with sodium sulfate the extent of methyldopa sulfation correlated with only the thermolabile sulfotransferase activity of platelets (Campbell,NRC et al. 1985), although this does not predict the activity of this form in other tissues (Weinshilboum 1990). The clinical significance of genetic variation in the sulfation capacity, which may also be dependent on (activated) sulfate availability has yet to be determined.

Miscellaneous reactions:
A variety of other reactions are not reviewed in details here because they either are consdered irrelevant for drug metabolism, because they represent particular problems that should be addressed separately in the risk assessment of drugs (e.g. the hepatic cytosolic alcohol dehydrogenase) or because only very sporadic information is available (e.g. some of the CYPs, the paroxonase/arylesterase polymorphisms) on the proteins and genes.

Phenotyping versus genotyping:

Several consideration on the choice between phenotyping and genotyping can be made although it should be realized that the two methods are supplementary rather that alternatives. Phenotyping has the advantage of continuous scale measurement and can therefore determine potential drug interactions. For example will quinidine administration turn fast CYP2D6 metabolizers phenotype to slow metabolizer phenotype (Sindrup et al. 1992), an effect that cannot be revealed by genotyping. Analogous arguments are valid for inducing effects. Furthermore, most polymorphisms show a considerable variability even in one of the genotypes. Very fast fast-metabolizers for example have a enzyme activity that cannot be distinguished from slow fast-metabolizers by genotyping, whereas phenotyping offers the possibility of revealing reactions related to such a trait. Furthermore, it can be argued that the huge variability that can be found in non-polymorphic enzyme activities can be of equal magnitude and importance for administration of drugs. The draw-back of probing is that it requires a trial that is planned for this purpose and that it requires administration of a test compound and sampling of relevant biological material.

The advantage of analysis on DNA for genotyping is that it only requires minute amounts of genomic DNA from e.g. lymphocytes because of the PCR technique and that it can be done retrospectively if such material is stored. Genotyping has a high potential for screening and the molecular techniques have a high capacity and a price that get cheeper and cheeper. It should be noted that only in case of CYP2D6 and N-acetyltranferase there are solid evidence that the genotyping predict the phenotype with sufficient accuracy. Interestingly, multiple genes are said to be found in fast fast-metabolizers of debrisoquine.

Which polymorphisms are relevant for drug development and registration?

The polymorphisms in CYP2D6, CYP2C and N-acetyltransferase are clearly demonstrated of importance for the metabolism of many drugs. For a safety evaluation it is mandatory to include investigations on whether a new drug is metabolized by these enzymes. If it can be established that the metabolism by either of these pathways only constitute a minor pathway further investigations are not necessary unless in the case(s) where the metabolism leads to conversion of a prodrug to an active metabolite, confer the codeine and proguanil examples mentioned above. The safety evaluation should also include whether a new drug has inhibiting effects on either of these enzymes, e.g. by estimating metabolism of probe drugs in human microsome systems or transfected cellular systems. For example we have recently shown that the antidepressant fluvoxamine is a potent inhibitor of CYP1A2 (Brøsen et al. 1993b). A similar effect has led to withdrawal of the drug furafylline due to its inhibition of caffeine metabolism leading to caffeine intoxication from dietary intake. Transfected cell lines can be used for specific elucidation of analogous effects (Jensen et al. 1993).

Recommendations:

In the safety evaluation of new drugs it should be examined whether or not they are metabolized by the enzymes that exhibit genetic polymorphism, i.e. CYP2D6, CYP2C and N-acetyltransferase. If this is the case further safety studies are required particularly if the drug has a steep dose response curve and/or the pharmacodynamic effects include influence on vital functions, i.e. a narrow therapeutic index. Particular emphasis should be put on drug that are activated by polymorphous enzymes, in which case even a minor metabolic route may be of considerable importance.

Regardless of the drug metabolism route it should be documented that new drugs do not inhibit the enzymes of the major drug metabolism polymorphisms. These recommendations can be extended to existing drugs as well and to other important routes of drug elimination.

If a new drug depends on bioactivation by CYP2D6 or CYP2C about 4 - 10% of the population would have no therapeutic effect while they still may suffer from the side-effects. If a new drug is metabolized by a genetic polymorphous enzyme it should be determined if drug action depends on this and/or if the partial clearance is of a sufficient magnitude to result in interindividual variation in drug elimination .

Acknowledlements:

Supported by the Danish Strategic Environmental Research Programm.

We thank Dr. Kim Brøsen, Odense University, Denmark, for valuable comments.

References:

Bertilsson L, Lou YQ, Du YL, Liu Y, Kuang TY, Liao XM, Wang KY, Reviriego J, Iselius L, Sjoqvist F (1992) Pronounced differences between native Chinese and Swedish populations in the polymorphic hydroxylations of debrisoquin and S-mephenytoin. Clin Pharmacol Ther 51:388-397

Blum M, Grant DM, McBride W, Heim M, Meyer UA (1990) Human arylamine N-Acetyltransferase genes: isolation, chromosomal localization, and functional expression. DNA and Cell Biology 9:193-203

Bonham-Carter SM, Rein G, Glover V, Sandler M, Caldwell J (1983) Human platelet phenol sulphotransferase M and P: substrate specificity and correlation with in vivo sulphoconjugation of paracetamol and salicylamide. Br J Clin Pharmacol 15:323-330

Brockmoller J, Staffeldt B, Roots I (1991) Evaluation of proposed sulphoxidation pathways of carbocisteine in man by HPLC quantification. Eur J Clin Pharmacol 40:387-392

Brøsen K, Gram LF (1989) Clinical significance of the sparteine/debrisoquine oxidation polymorphism. Eur J Clin Pharmacol 36:537-547

Brøsen K, Skjelbo E, Flachs H (1993a) Proguanil metabolism is determined by the mephenytoin oxidation polymorphism in Vietnamise living in Denmark. Br J Clin Pharmacol in press:

Brøsen K, Skjelbo E, Rasmussen BB, Poulsen HE, Loft L (1993b) Fluvoxamine is a potent inhibitor of cytochrome P4501A2. Biochem Pharmacol 45:1211-1214

Butler MA, Lang NP, Young JF, Caporaso N, Vineis P, Hayes RB, Teitel CH, Massengrill JP, Lawsen MF, Kadlubar FF (1992) Determination of CYP1A2 and NAT2 phenotypes in human populations by analysis of caffeine urinary metabolites. Pharmacogenetics 2:116-127

Campbell ME, Spielberg SP, Kalow W (1987) A urinary metabolite ratio that reflects systemic caffeine clearance. Clin Pharmacol Ther 42:157-165

Campbell NRC, Dunette JH, Mwaluko G, Van Loon J, Weinshilboum RM (1984) Platelet phenol sulfotransferase and erythorcyte catechol O-methyltransferase activities: correlation with methyldopa metabolism in man. Clin Pharmacol Ther 35:55-63

Campbell NRC, Sundaram RS, Werness PJ, Van Loon J, Weinshilboum RM (1985) Sulfate and methyldopa metabolism: metabolite patterns and phenol sulfotransferase activity. Clin Pharmacol Ther 37:308-315

Clark DWJ (1985) Genetically determined variability in acetylation and oxidation therapeutic implications. Drugs 29:342-375

Dayer P, Desmeules J, Leemann T, Striberni R (1988) Bioactivation of the narcotic drug codeine in human liver is mediated by the polymorphic monooxygenase catalyzing debrisoquine 4-hydroxylation. Biochem Biophys Res Commun 152:411-416

Eichelbaum M, Bertilsson L, Säve J, Zekorn C (1982) Polymorphic oxidation of sparteine and debrisoquine: Related pharmacogentic entities. Clin Pharmacol Ther 31:184-186

Emery P, Panayi SP, Huston G, Welsh KI, Mitchell SC, Smith RL, Waring RH (1984) D-penicillamine induced toxicity in rheumatoid arthritis: the role of sulphoxidation status and HLA-DR3. J Rheumatol 11:626-632

Emery P, Bradley H, Arthur V, Tunn E, Waring R (1992) Genetic factors influencing the outcome of early arthritis - the role of sulphoxidation status. Br J Rheumatol 31:449-451

Evans DAP, Manley KA, McKusick VA (1960) Genetic control of izoniazid metabolism in man. Br Med J 2:485-491

Evans DAP, Mahgoub A, Sloan TP, Idle JR, Smith RL (1980) A family and population study og genetic polymorphism of debrisoquine oxidation in a white British population. J Med Genet 17:102-105

Funck-Brentano C, Bosco O, Jacqz-Aigrain E, Keundjian A, Jaillon P (1992) Relation between chloroguanide bioactivation to cycloguanil and the genetically determined metabolism of mephenytoin in humans. Clin Pharmacol Ther 51:507-512

Gaedig KA, Blum M, Gaedig KR, Eichelbaum M, Meyer UA (1991) Deletion of the entire cytochrome P450 CYP2D6 geneas a cause of impaired drug metabolism in poor metabolzers of the debrisoquine/sparteine polymorphism. Am J Human Genet 48:943-950

Gonzalez FJ, Skoda RC, Kimura S, Umeno M, Zanger UM, Nebert DW, Gelboin HV, Hardwick JP, Meyer UA (1988) Characterization of the common genetic defects in humans deficient in debrisoquine metabolism. Nature 331:442-446

Gordon C, Bradley H, Waring RH, Emery P (1992) Abnormal sulphur oxidation in systemic lupus erythematosus. Lancet 339:25-26

Graf T, Broly F, Hoffmann F, Probst M, Meyer UA, Howald H (1992) Prediction of the phenotype for acetylation and for debrisoquine hydroxylation by DNA-tests in healthy human volunteers. Eur J Clin Pharmacol 43:399-403

Grant DM, Tang BK, Kalow W (1983) Variability in caffeine metabolism. Clin Pharmacol Ther 33:591-602

--- (1984) A simple test for acetylator phenotype using caffeine. Brit J Clin Pharmacol 17:459-464

Gregory WL, James OF, Idle JR (1992) Carbocisteine polymorphism and disease. Lancet 339:616-617

Guengerich FP, Shimada T (1991) Oxidation of toxic and carcinogenic chemicals by human cytochrome P-450 enzymes. Chemical Research in Toxicology 4:391-407

Hayashi S, Watanabe J, Kawajiri K (1991) Genetic polymorphisms in the 5'-flanking region change transscriptional regulation of the human cytochrome P450IIE1 gene. Biochem J 110:559-565

Heim MH, Meyer UA (1992) Evolution of a highly polymorphic human cytochrome P450 gene cluster: CYP2D6. Genomics 14:49-58

Hein DW (1988) Acetylator genotype and arylamine-induced carcinogenesis. Biochim Biophys Acta 948:37-66

Hein DW, Flammang TJ, Kirlin WG, Trinidad A, Ogolla F (1987) Acetylator genotype-dependent metabolic activation of carcinogenic N-hydroxylaamines by S-acetyl coenzyme A-dependent enzymes of inbred hamster tissue cytosols: relationship to arylamine N-acetyltransferase. Carcinogenesis 8:1767-1774

Jensen KG, Önfelt A, Poulsen HE, Doehmer J, Loft S (1993) Effects of benzo[a]pyrene and ±-trans-7,7-dihydroxy-7,8-dihydro-benzo[a]pyrene on mitosis in Chinese hamster V79 cells with stable expression of rat cytochrome P4501A1 or 1A2. Carcinogenesis in press:

Kadlubar FF, Guengerich PF (1992) Inducibility of human cytochromes P-450 primarely involved in the activation of chemical carcinogens. Chemosphere 25:201-204

Kagimoto M, Heim M, Kagimoto K, Zengin T, Meyer UA (1990) Multiple mutations of human cytochrome P450IID6 gene (CYP2D6) in poor metabolizers of debrisoquine. Study of the functinal significance of individual mutation by expression of chimeric genes. J Biol Chem 265:17209-17214

Kawajiri K, Nakachi K, Imai K, Shinoda N, Watanabe J (1990) Identification of genetically high risk individuals to lung cancer by DNA polymorphisms of the cytochrome P4501A1 gene. FEBS Lett 263:131-133

Lennard L, Van Loon JA, Lilleyman JS, Weinshilboum RM (1987) Thipopurine pharmacogenetics in leukemia: correlation of erythrocyte methyltransferase activity and 6-thioguanine nucleotide concentrations. Clin Pharmacol Ther 41:18-25

Lennard L, Van Loon JA, Weinshilboum RM (1989) Pharmacogenetics of acute azathioprine toxocity: relationship to thiopurine methyltransferase genetic polymorphism. Clin Pharmacol Ther 46:149-154

Lennard L, Lilleyman JS, Van Loon J, Weinshilboum RM (1990) Genetic variation in response to 6-mercaptopurine for childhood acute lymphoblastic leukaemia. Lancet 336:225-229

Lou YC (1990) Differences in drug metabolism polymorphism between orientals and caucasians. Drug Metab Rev 22:451-475

Mahgoub A, Idle JF, Dring LG, Lancaster R, Smith RL (1977) Polymorphic hydroxylation of debrisoquine in man. Lancet 2:584-586

McManus ME, Stupans I, Ioannoni B, Burgess W, Robson RA, Birkett DJ (1988) Identification and quantitation in human liver of cytochromes P-450 analogous to rabbit cytochromes P-450 forms 4 and 6. Xenobiotica 18:207-216

Meese CO, Fischer C, Kupfer A, Wisser H, Eichelbaum M (1991) Identification of the "major" polymorphic carbocysteine metabolite as S-(carboxymethylthio)-L-cysteine. Biochem Pharmacol 42:R13-R16

Mitchell SC, Waring RH, Haley CS, Idle JF, Smith RL (1984) Genetic aspects of the polymodally distributed sulfoxidation of S-carboxymethyl-L-cysteine in man. Brit J Clin Pharmacol 18:507-521

Nebert DW (1989) The Ah locus: genetic differences in toxicity, cancer, mutation, and birth defects. Crit Rev Toxicol 20:153-174

Ohsako S, Deguchi T (1990) Cloning and expression of cDNAsfor polymorphic and monomorphic arylamine N-acetyltransferases from human liver. J Biol Chem 265:4630-4634

Peter R, Böcker R, Beaune PH, Iwasaki M, Guengerich FP, Yang CS (1990) Hydroxylation of chlorzoxazone as a specific probe for human liver cytochrome P-450IIE1. Chem Res Toxicol 3:566-573

Poulsen HE, Loft S (1988) Antipyrine as a model drug to study hepatic drug-metabolizing capacity. J Hepatol 6:374-382

Poulsen HE, Roberts DW, Hinson JA (1993) Covalent binding of acetaminophen estimated from protein adducts. In: Jeffery EH (ed) Human Drug Metabolism. From molecular biology to man. CRC Press, Boca Raton, Ann Arbor, London, Tokyo, pp. 43-52

Reilly DK, Rivera-Calimlim L, Van Dyke D (1980) Cathecol O-methyltransferase activity: a determinant of levodopa response. Clin Pharmacol Ther 28:278-286

Reiter C, Weinshilboum R (1982) Platelet sulfotransferase activity: correlation with sulfate conjugation of acetaminophen in man. Clin Pharmacol Ther 32:612-621

Rost KL, Brosicke H, Brockmoller J, Scheffler M, Helge H, Roots I (1992) Increase of cytochrome P450IA2 activity by omeprazole: evidence by the 13C-[N-3-methyl]-caffeine breath test in poor and extensive metabolizers of S-mephenytoin. Clin Pharmacol Ther 52:170-180

Schmid B, Bircher J, Preisig R, Kupfer A (1985) Polymorphic dextromethorphan metabolism Co-segregation of oxidative O-demethylation with debrisoquine hydroxylation. Clin Pharmacol Ther 38:618-624

Scott MC, Van Loon JA, Weinshilboum RM (1988) Pharmacogenetics of N-methylation: heritability of human erythrocyte N-methyltransferase activity. Clin Pharmacol Ther 43:256-262

Shimada T, Iwasaki M, Martin MV, Guengerich FP (1989) Human liver microsomal cytochrome P-450 enzymes involved in the bioactivation of procarcinogens detected by umu gene response in Salmonella typhimurium TA 1535/pSK1002. Cancer Res 49:3218-3228

Sindrup SH, Brosen K, Bjerring P, Arendt-Nielsen L, Larsen U, Angelo HR, Gram LF (1990) Codeine increases pain thresholds to copper vapor laser stimuli in extensive but not poor metabolizers of sparteine. Clinical Pharmacology & Therapeutics 48:686-693

Sindrup SH, Arendt-Nielsen L, Brosen K, Bjerring P, Angelo HR, Eriksen B, Gram LF (1992) The effect of quinidine on the analgesic effect of codeine. European Journal of Clinical Pharmacology 42:587-591

Steventon GB, Heafield MTE, Sturman S, Waring RH, Williams AC (1990) Xenobiotic metabolism in Alzheimer's disease. Neurology 40:1095-1098

Tefre T, Ryberg D, Haugen A, Nebert DW, Skaug V, Brøgger A, Børresen A-L (1991) Human CYP1A1 (cytochrome P450) gene: lack of association between the Msp I restriction fragment length polymorphism and incidence of lung cancer in a Norwegeian population. Pharmacogenetics 1:20-25

Thummel KE, Lee CA, Kunze KL, Nelson;SD, Slattery JT (1992) Oxidation of acetaminophen to N-acetyl-p-aminobenzoquinone imine by human CYP3A4. Biochem Pharmacol 45:1563-1569

Tyndale RF, Sunahara R, Inaba T, Kalow W, Gonzalez FJ, Niznik HB (1991) Neuronal cytochrome P450IID1 (debrisoquine/sparteine-type): potent inhibition of activity by (-)-cocaine and nucleotide sequence identity to human hepatic P450 gene CYP2D6. Molecular Pharmacology 40:63-68

Uematsu F, Kikuchi H, Motomiya M, Abe T, Sagami I, Ohmachi T, Wakui A, Kanamaru R, Watanabe M (1991) Association between restriction fragment length polymorphism of the human cytochrome P450IIE1 gene and susceptibility to lung cancer. Jpn J Cancer Res 82:254-256

Vistisen K, Loft S, Poulsen HE (1992) Foreign compound metabolism capacity in man measured from metabolites of dietary caffeine. Carcinogenesis 13:1561-1568

Vogel F, Motulsky AG (1982) Human Genetics, problems and approaches. Springer-Verlag,Berlin, Heidelberg, New York

Wahlström A, Winblad B, Bixo M, Rane A (1988) Human brain metabolism of morphine and naloxone. Pain 35:121-127

Ward SA, Helsby NA, Skjelbo E, Brøsen K, Gram LF, Breckenridge AM (1991) The activation of the biguanide antimalarial proguanil co-segregates with the mephenytoin oxidation polymorphism - a panel study. Br J Clin Pharm 31:689-692

Waring RH, Sturman SG, Smith MCG, Steventon GB, Heatfield MTE, Williams AC (1989) S-Methylation in motorneuron disease and Parkinson's disease. Lancet ii:356-357

Weber WW, Hein DW (1985) N-acetylation pharmacogenetics. Pharmacol Rev 37:25-77

Weinshilboum R (1989) Methyltransferase pharmacogenetics. Pharmac Ther 43:77-90

--- (1990) Sulfotransferase pharmacogenetics. Pharmacol Ther 45:93-107

Weinshilboum RM (1992) Methylation pharmacogenetics: thiopurine methyltransferase as a model system. Xenobiotica 22:1055-1071

Weinshilboum RM, Sladek S (1980) Mercaptopurine pharmacogenetics: monogene inheritance of erythrocyte thiopurine methyltransferase activity. Am J Hum Genet 32:651-662

Wilkinson GR, Guengerich FP, Branch RA (1989) Genetic polymorphism of S-mephenytoin hydroxylation. Pharmacol Therap 43:53-76

Wrighton SA, Campanile C, Thomas PE, Maines SL, Watkins PB, Parker G, Mendez-Picon G, Haniu M, Shively JE, Levin W (1986) Identification of a human liver cytochrome P-450 homologous to the major isosafrole-inducible cytochrome P-450 in the rat. Mol Pharmacol 29:405-410

KINETIC ANALYSIS USING POSITRON EMISSION TOMOGRAPHY

Per Hartvig, Joakim Tedroff, Anders Lilja, Karl Johan Lindner and Bengt Långström

Uppsala University PET-centre,University hospital, Uppsala University, S-751 85 Uppsala, Sweden

Positron emission tomography, PET

Positron emission tomography, PET is a noninvasive tracer technique which enables quantitation of the kinetics of a tracer molecule in a physiological process in the tissue of the living animal or man. The method has also been characterized as an *in vivo* autoradiographic technique for assessment of tracer kinetics. This presentation will shortly describe the PET-technique and give an introduction to the principles used in the kinetic analysis of the radioactivity distribution of tracer molecules. Tissue uptake and kinetics of highly toxic substances can be assessed using PET, since very minute quantities, in the submicrogram range of radiolabelled tracer molecule is administered to the experimental animal. However, the number of toxicological studies performed so far with PET is still low. Two different experimental designs in toxicological PET studies have been used. One type of study uses PET as an *in vivo* autoradiographic method for measurement of tissue distribution of a radiolabelled compound. The other type of study quantitates the functional physiologic damage exerted by the toxic compound. Some recent PET studies within the toxicologic field will also be reviewed.

Some examples of physiological process studied with PET examplified in Table 1.

Physiologic process	Radiolabelled tracer
Blood volume	Carbon monoxide
Blood flow	Water, butanol
Tissue pH	Sodium bicarbonate
Energy utilization	Oxygen, glucose, fatty acids
Amino acid transport/proteinsynthesis	Amino acids
Enzyme activity/concentration	Substrate/ enzyme inhibitor
Receptor localization/kinetics	Receptor ligand
Neurotransmitter syntheses	Amino acid precursor

Table 1. Physiological processes studied with PET

The radionuclides most frequently used are 15O;13N;11C and 18F with physical halflives of 2; 10; 20 and 110 minutes, respectively. The radionuclide is incorporated in the tracer molecule by fast and reproducible radioorganic synthesis. Due to the rapid physical decay, the time for radiochemical synthesis of most radiotracers is limited to about three times the half-lives. This includes time for pharmaceutical preparation and control of identity as well as analysis of chemical and radiochemical purity by means of chromatographic techniques. The radiolabelled tracer is injected intravenously to the study object, except for those radionuclides that are inhaled in gaseous form. The radioactivity is collected by external detectors for predetermined intervals in fifteen tomographic slices interspaced with about 7 mm. After reconstruction of images the concentration and kinetics of the tracer molecule within the tissue are calculated and used in further analysis in order to quantify the studied physiological process. The obvious advantage of the method is the possibility of *in vivo* quantitation of physiological processes in the healthy and diseased state. Furthermore, tracer amounts of radiolabelled molecules do not affect normal physiology. The rapid dissappearance of radioactivity make repeated studies possible which further strenghten the validity of data. On the other hand, the limited spatial resolution make analysis difficult in discrete regions of a tissue. Another disadvantage of PET is the co-determination of formed radiolabelled metabolites in the regions of interest, which clearly illuminates the need for supplementary techniques for an accurate assessment of the kinetics of the radioactivity. Chromatographic analysis of tracer and metabolites in plasma or in microdialysates may support the analysis with PET

Kinetic analysis using positron emission tomography.

The radioactivities measured by PET are usually analyzed by using comparmental models which may describe the studied physiological process. Depending on the process under study, different methods and models for the quantitative kinetic analysis for the utilization of radioactivity are employed (Wienhard 1992). A series of compartmental models with increasing complexity together with the quantitative estimate that is calculated. are shown in Fig 1

Tracers or radiolabelled metabolites in blood vessels may contribute to the radioactivity in the tissue. Therefore, the blood volume in the tissue has to be measured in a separate experiment eg using radiolabelled carbon monoxide. The blood volume and the blood concentration of the radiotracer must be taken into account as an additional parameter in the model equations. For a tracer which enters the tissue and is completely and irreversibly captured, only the calculation of an unidirectional rate constant, K1 can be estimated as outlined in II. An equilibrium between tissue and blood radioactivities as in model III will be reached asymptotically and is determined as the ratio of K1/k2 signifying a tracer that is not or insignificantly bound in the tissue and cleared back to the blood. This equilibrium may describe the tissue distribution and kinetics of radiolabelled tracer

No	Compartment model	Quantitated estimate
I	Cp	Vascular concentration/blood volume
II	Cp $\xrightarrow{K1}$ Ct	Irreversible trapping
III	Cp $\underset{k2}{\overset{K1}{\leftrightarrow}}$ Ct	Equilibrium: K1/k2
IV	Cp $\underset{k2}{\overset{K1}{\leftrightarrow}}$ Ct $\xrightarrow{k3}$ Cb	Rate: K1 x k3/k2 + k3
V	Cp $\underset{k2}{\overset{K1}{\leftrightarrow}}$ Ct $\underset{k4}{\overset{k3}{\leftrightarrow}}$ Cb	Equation: K1/k2(1 + k3/k4) = = K1/k2 (1 + Bmax/ kd)

Figure 1. Compartment models of increasing complexity together with the estimate that is asymptotically described.Cp;Ct and Cb denote the radioactivities measured in blood/plasma, tissue and specifically bound

and compares the PET technique with an *in vivo* kinetic autoradiographic technique.

Binding in the tissue with the rates k3 and k4 may require an additional compartment to describe the radioactive distribution as outlined in IV and V. In the case of irreversible binding or trapping of radiotracer the off-rate k4 can be equaled to zero and the accumulated rate will be described as K1 x k3/k2 + k3. The relative order of magnitude between k2 and k3 is important. If k3>>k2, the accumulation rate reduces to K1 (See II above) and nothing can be learned about the rate of the binding process. This model, IV, has been shown very useful for a number of applications with PET. The rate constant for the utilization rate, can be transformed to glucose metabolic rate expressed in micromoles per gram and minute using 18F-fluorodeoxyglucose, 18FDG, as tracer. Other applications using this paradigm are the quantitation of the utilization of radiolabelled amino acid in tumor tissue, synthesis rate of neurotransmitters such as dopamine and serotonin radiolabelled with 11C and for receptor ligands or enzyme inhibitors that are "irreversible bound" to specific binding sites during the course of the PET study.

A reversible binding of the tracer, V, will asymptotically approach an equilibrium between tissue and blood radioactivities which is described by the equation: K1/k2 (1 + k3/k4). It can be shown that the ratio k3/k4, also named binding potential, equals the ratio Bmax/kd. Two main approaches in modelling receptor binding using PET data have been applied: the *kinetic* and the *equilibrium model.* In the *kinetic model* a true equilibrium has not been attained and the model depends on that binding relates to the association and dissociation of the radioligand over time. Only the ratio of number of binding sites, Bmax, over receptor affinity, kd, can be obtained. For a separate estimate of these parameters at least two PET studies must be performed using various concentrations of the radioligand. In the *equilibrium model,* it is assumed that an effective equilibrium is reached. at a certain time after administration.By performing PET studies with at least two specific radioactivities, the Bmax and kd-values can be obtained by Schatchard analysis. Hill plots are applied to determine the number of binding sites. In these respects the equilibrium model closely rembles the ones used for *in vitro* studies. Further compartments can be added to the models which will increase the complexity of calculation. The radioactivities measured with PET over a 1 to 2 hours experiment,however, do not allow for extraction of more than three to four model parameters (Wienhard 1992).

The models heavily rely on the accurracy of the in-put function i.e. the time course of the concentration of the radioactive compound in blood or plasma. In the models, compensations may have to be made for changes in blood flow, formation of radiolabelled metabolites as well as receptor occupancy of metabolites irrespective of whether radiolabelled or not. Non-specific processes must be adequately distinguished from the studied process in the validation of the method. Various approaches have also been done in order to circumvent problems regarding the in-put function. Thus, some difficulties encountered above have been overcome by using a brain reference area in the quantitation of the rate in physiological processes within the brain. It is assumed that the tracer kinetics in this reference area closely resembles the kinetics in the studied area except for the specific process quantitated. This must be proven in prestudies in expermental animals where the physiologic process is specifically perturbated.

Tissue kinetic studies using PET.

Autoradiographic studies in the tissue.

Studies of radiolabelled toxins will not only give information of the distribution and accumulation of radioactivity, but also on the tissue kinetics as well as effect of perturbations that will affect both tissue accumulation and kinetics. A few studies using this approach have appeared in the literature. PET used as an *in vivo* autoradiographic method will give information for prediction of effect as well as the onset and duration of a true pharmacologic action. The elucidation of side-effects from specific tissues can be visualizied by PET due to the accumulation of radioactivity. However, this type of study needs cautious interpretation. The presence of the compound in the tissue of interest is no proof of a toxic effect. Furthermore, saturation of a process is possible for the substance given in high concentrations but not accurately visualized with PET using tracer concentration.

The higher cardiotoxicity of bupivacaine as compared to other local anesthetic amides has been a cause for concern and initiated studies using an array of experimental methods to understand the toxic mechanism (Feldmann et al 1993). The distribution in the pig heart wall of bupivacaine radiolabelled with 11C was similar to that of 11C-ropivacaine but lower than that of 11C-lidocaine. However, the radioactivity remained significantly longer in the cardiac muscle as compared to the two other local anesthetic amides (Feldmann et al 1993). The concept of rapid uptake and slow elimination from the cardiac muscle of 11C-bupivacaine was consonant with the long cardiac rescucitation period observed in man following a bupivacaine overdose. Similarly, the CNS side effects of the cytotoxic compound busulphane was discussed from the results on brain distribution and kinetics in a PET study using 11C-labelled drug. (Hassan et al 1992).

MPTP

The story of the parkinsonian inducing neurotoxin, MPTP, N-methyl-1,2,3,6-tetrahydropyridine, starts with the referral of a young male college student to hospital due to severe motor dysfunction. The student was a drug addict and he even tried to synthetize an analogue of the opioid, pethidine. In order to facilitate the procedure, the reaction temperature was increased and the recrystallization procedure omitted. These changes were detrimental and a side product, MPTP was formed. The student later died from an opioid overdose and at autopsy a selective degeneration of dopaminergic pathways of the pars compacta in the substantia nigra of the brain was revealed. The symptoms and autopsy findings closely correlated with those in Parkinson's disease patients.

Radiolabelling of MPTP with 11C and its use in animal studies with PET opened up the possibility to elucidate the mechanism of action of the neurotoxin. 11C-MPTP rapidly reached high concentrations in the brain and remained constant or almost constant in different brain regions indicating an irreversible trapping of the radioactive compound. (Moerlein et al 1986, Hartvig et al 1986) Pretreatment of the animals with monoamine oxidase inhibitors type B dramatically changed the kinetics of brain radioactivity. The brain radioactivities decreased and a clear elimination phase became obvious. This altered kinetic behaviour could not be seen following pretreatment with a MAO-A inhibitor. The dopaminergic re-uptake blocker, nomifensin somewhat decreased brain radioactivity but a trapping was still obvious. (Hartvig et al 1986). The findings with PET and 11C-MPTP agrees with a transport of the neurotoxin via the dopaminergic uptake system into the terminals. By the action of MAO-type B, MPTP is converted to a metabolite, MPP+ in the terminals. Due to its quarternary ammonium function, MPP+ is captured within the cell where the toxcic action is exerted. Blockade of the uptake system with nomifensin will decrease neuronal deposition of MPTP. Inhibition of MAO-B blocked the conversion to the toxic metabolite and the unmetabolized, lipophilic MPTP was easily eliminated from the brain. Thus, the tissue kinetics of MPTP studied with PET combined with pharmacological perturbations was used in the interpretation of mechanism of action of this parkinsonian inducing neurotoxin.

Metal ions

Various pathological conditions involve toxic cumulation and alteration of metal ion kinetics in the brain. Cumulation of iron in certain brain areas has been suggested in Parkinson's disease and in tardive dyskinesia. Certain brain tumors have been characterized by increased cell-membrane transferrin receptor complex density. The blood to brain barrier transport of 52Fe-citrate has been measured using positron emission tomography following various perturbations of the transport mechanisms (Leenders et al 1993).The iron transport to the brain in the healthy state is very low but increased during continuos haloperidol treatment of monkeys. These results raise questions whether psychiatric patients during pharmacological therapy are subjected to an increased brain iron loading and its significance for the pathophysiology of tardive dyskinesia must be further explored. Interestingly, an enhanced brain transport of positron emitting Fe-ions was detected in patients with Parkinson's disease.

Toxicologically induced functional changes evaluated with PET

Due to the limited mechanistic information from tissue distribution of the radiotracer obtained with PET, the evaluation should be supplemented with functional physiological studies on the damage and toxicological effect caused by the toxic compound. PET offers large possibilities for the quantative assessment of normal as well as altered function of vital physiological processes. Some of these possibilities have also been extended to toxicology.

Neuronal degenerative mechanisms measured with PET.

A number of radiolabelled tracers for the quantitative assessment of energy consumption in the tissue have been used with PET. Of these, 18F-fluorodeoxyglucose, 18FDG, has been used for quantitation of glucose utilization. In fact, 18FDG has a high affinity for the enzyme hexokinase. The quantitated 1FDG binding rate only includes the pocesses of transport and binding to the enzyme. By use of a correction factor this figure under certain conditions can be regarded similar to the glucose metabolic rate in the tissue expressed in micromoles per g and minute. This rate is grossly proportional to neuronal activity in the brain. A regionally decreased 18FDG utilization in brain has been paralleled by functional loss in areas most afflicted in a number of neurodegenerative diseases. Therefore, PET has been used for the assessment of progression of brain neurodegenerative disease, localization of epileptic foci and in differential diagnosis of different brain diseases. The 18FDG method has also been applied to studies in toxicology. Exposure to tetrabromoethane in one individual was diagnosed with PET and 18FDG as a solvent induced encephalopathy (Morrow et al 1990). Similarly the destruction of glutaminergic pathways following intoxication with domoic acid from mussels was visualized PET using 18FDG (Gjedde and Evans, 1992). Although PET with 18FDG is a crude measure on neuronal activity, it might give valuable information on neuronal damage of importance for toxic symptoms.

Dopaminergic function

The midbrain dopaminergic projections are by far the nerve tracts within the brain that has been most frequently studied with PET. This is due to the relative accumulation of dopaminergic pathways within the substantia nigra-striatum area giving a high radioactivity contrast to surrounding brain. However, studies on dopaminergic nerve cell function has also been extended to the heart following chemical sympatecthomy with the toxin 6-hydroxydopamine (Goldstein et al 1991) adiotracers measuring different aspects of dopamine turnover have been explored and validated. Tracers are available for dopamine synthesis rate, for quantifying the number of dopamine receptors of different types, the number of re-uptake sites of monoamine transmitter and quantitation of MAO-B concentration. This battery of tracers has been utilized in toxicologic PET studies beside its use in diseases accompanied by functional changes of the dopaminergic system.

MPTP. The discovery and elucidation of the toxic action by MPTP was followed by an interest to use this neurotoxin for the development of an animal model of Parkinson's disease. Infusion of MPTP into one carotid artery in monkeys resulted in an unilateral lesioning of the dopaminergic pathways and a contralateral hemiparkinsonism (Leenders et al 1988). The specific binding of the tracer, 11C-nomifensin for re-uptake sites grossly equavalent to the number of presynaptic dopaminergic terminals decreased by 80 to 90% within the damaged striatum seven to ten days after lesioning. Two days after MPTP-administration, an up-regulation of the D2-dopamine receptors on the lesioned side became obvious with PET measured with 11C-raclopride, a tracer with high affinity for D2-dopamine receptors. In parallel with the loss of presynaptic terminals, the decarboxylation rate for 11C-DOPA and 18F-DOPA decreased. This lower decarboxylatioon rate persisted several years after lesioning. The changes of the dopaminergic system visualized with PET in the MPTP model in monkeys is similar to the PET changes found in patients with Parkinson's disease. Although, the relevance of the MPTP model to resemble Parkinson's disease is questioned, the model can be used for a general assessment of the damage to different nigro-striatal neurons and is used in the validation of new PET tracers for visualization of different nerve systems within the brain.

Metal intoxication. Intoxication by heavy metals is often followed by emotional and motor disturbances of central origin. Repeated manganese administration in monkeys over four months resulted in the development of unsteady gait and a minor clumsiness of hands and feet combined with an obvious hypoactivity (Eriksson et al 1992). Again, 11C-nomifensin binding decreased in the striatal area with time, but was not paralleled by a lower decarboxylation rate of 11C-DOPA. This finding was surprising since 11C-DOPA was supposed to be decarboxylated within the dopaminergic neurons. However, the enzyme catalyzing the reaction is presumably a modulated enzyme, which may have a great capacity for compensation in a slowly degenerating nigro-striatal projection. It seems reasonable to assume that either the number of nerve terminals has been reduced and the remaing dopamine turnover has been up-regulated or that the number of terminals remained constant but with a reduced amount of uptake sites. Manganese intoxication in man assessed with PET and 18F-DOPA did not show any disturbance in L-DOPA utilization (Wolters et al 1989). A transient decrease of postsynaptic D2-dopamine receptor binding using 11C-raclopride was seen, but at the end of the study after ten months the binding was unchanged. In fact, an unaltered D2-receptor density together with maintained L-DOPA decarboxylation is also the prerequisite for successful attempts of L-DOPA treatment in humans poisoned by manganese.

The changes of the dopaminergic system visualized with PET in manganese intoxication was somewhat at variance with those seen in patients with Wilson's disease (Westermark et al 1993). Wilson's disease is an inheritated disorder characterized by deficiencies in the transport and excretion of copper. This means a cumulation of the metal ions giving raise to liver damage and deposition in neuronal tissue giving symptoms of motor disturbances. The number of presynaptic nerve terminals and postsynaptic D2-receptors decreased in parallel in most patients as visualized with PET. As for manganese intoxication the DOPA decarboxylation activity was unaffected. This different toxic action to the dopaminergic system of manganese and copper, respectively as shown with PET should be of importance for the understanding of the toxic mechanism of the two metal ions. A summary of the changes are seen in Table 2.

Toxic reaction	Dopamine syntheses	Presynaptic terminals	Postsynaptic D2-receptors
MPTP	↓	↓	(↑)
Manganese	↔	↓	↔
Wilson's disease	↔	↓	↓

Table 2. Effect on the dopaminergic system induced by different toxic compounds.

Conclusions

Toxicological studies using PET offer the possibility of examining *in vivo* tissue uptake and kinetics of the compound of interest. For a full understanding of the toxic action the information must be supplemented with functional physiologic measures of neurodegeneration, altered neurotransmitter function or kinetic changes of the enzymatic processes. Studies of toxicological effects and mechanism of action may require a multitracer approach which includes different aspects of functional physiological changes induced by the toxic compound. Irrespective of method, a careful validation is mandatory. Drawbacks of the technique such as limited spatial resolution, a short time period available for kinetic analysis as well as the fact that only total radioactivity is measured must be carefully considered. Accurate methods are developed to compensate for these limitations. In summary, by utilizing the full extent of the possibilities, PET furnishes a method that gives information on both distribution and functional toxicologic effects of a toxic substance.

References

Eriksson H, Tedroff J, Thoumas KÅ, Aquilonius SM, Hartvig P, Fasth KJ, Bjurling P,Långström B, Hedström KG, Heilbronn E (1992) Manganese induced brain lesions in Macaca fascicularis as revealed by positron emission tomography and magnetic imaging.Arch Toxicol. 66:403-407

Feldman HS, Hartvig P, Wiklund L, Antoni G, Gee A, Ulin J, Långström B (1993) Regional distribution of 11C labelled lidocaine, bupivacaine and ropivacaine in the hearts of pigs studied with positron emission tomography. Acta Anaesthesiol Scand. Submitted

Gjedde A, Evans AC (1990) PET studies of domoic acid poisoning in humans:excitotoxic destruction of brain glutaminergic pathways, revealed in measurements of glucose metabolism by positron emission tomography Can Dis Wkly Rep Sept 16, Suppl 1E:105

Goldstein DS, Grossman E, Tamrat M, Chang PC, Eisenhofer G, Bacher J, Kirik KL, Bacharach S, Kopin IJ (1991) Positron emission imaging of cardiac sympathetic innervation and function using 18F-6-fluorodopamine:effects of chemical sympathectomi by 6-hydroxydopamine. J Hypertens 9(5):417-423

Hartvig P, Larsson BS, Lindberg BS, Oreland L, Gullberg P, Långström B, Rimland A, Lundqvist H, Malmborg P, Lindquist NL (1986) Influence of monoamine oxidase inhibitors and a dopamine uptake blocker on the distribution of 11C-N-metyl-4-phenyl-1,2,3,6-tetrahydropyridine, 11C-MPTP in the head of Rhesus monkey. Acta Neurol Scand 74:10-16

Hassan M, Öberg G, Ericsson K, Ehrsson H, Eriksson L, Ingvar M, Stone-Elander S Thorell JO, Smedmyr B, Warne M, Widén L (1992) In vivo distribution of 11C-busulphane distribution in Cynomolgus monkey and the brain of a patient. Cancer Chemother Pharmacol. 30: 81-85

Leenders KL, Aquilonius SM, Bergström K, Bjurling P, Crossman A, Eckernäs SÅ, Gee A, Hartvig P, Lundqvist H, Långström B, Rimland A, Tedroff J (1988) Unilateral MPTP lesion in a rhesus monkeys: effects on the striatal dopaminergic system measured in vivo with PET using various novel tracers. Brain Res. 445:61-67

Leenders KL, Antonini A, Roelcke U, Pellikka R, Reist H, Psylla M, Gunther H (1993) Blood to brain barrier transport in monkeys and man measured using 52Fe-citrate and positron emission tomography. Ann Nucl Med 7 Suppl., S36.

Moerlein SM, Stöcklin G, Pawlik G, Wienhard K, Heiss WD (1986) Regional cerebral pharmacokinetics of the dopaminergic neurotoxin 1-methyl-4-phenyl-1,2,3,6-tetrahydropyridine as examined by positron emission tomography in a baboon is altered by tranylcypromine. Neurosci Lett 15.66(2):20

Morrow LA, Callender T, Lotteberg S, Buchsbaum MS, Hodgson MJ, Robin N (1990) PET and neurobehavioral evidence of tetrabromoethane encephalopathy. J Neuropsychiatry Clin Neurosci 2(4):431-435

Westermark K, Tedroff J, Thoumas KÅ, Hartvig P, Långström B, Andersson Y, Hörnfeldt K, Aquilonius SM (1994) Neurological Wilson's disease: Positron emission tomography studies with the dopaminergic markers 11C-nomifensin, 11C-L-DOPA and 11C-raclopride.J Neurol Neurosurg Psychiatr. Submitted

Wienhard K (1991) Modelisation: Application to the D2 receptors. In: Baron JC, Comar D, Farde L, Martinot JL, Mazoyer B (eds) Brain dopaqminergic systems: Imaging with positron tomography. Developments in Nuclear Medicine, Kluwer Academic Publishers, Vol 20, 85-96

Wolters EC, Huang CC, Clark C, Peppard RF, Okada J, Chu NS, Adam MJ, Ruth TJ ,Li O, Calne DB (1989). Positron emission tomography in manganese intoxication. Ann Neurol 26: 647-651

THE APPLICATION OF MODELLING IN TOXICOKINETICS

ROBERT M.J. INGS

Upjohn Laboratories - Europe, Fleming Way, Crawley, West Sussex, RH10 2LZ.

Considering the relatively short time that toxicokinetics has become part of drug development, it has had a profound impact on, not only the conduct of toxicology studies, but also on the whole approach to nonclinical pharmacokinetics and metabolism. In the past it was considered satisfactory to help establish the validity of a toxicology study by showing that the animal species used had a similar metabolism of the compound to that of man. Thus, the studies undertaken tended to focus on the qualitative metabolic aspects, generally using radiolabelled material and commonly referred to as ADME studies, and not on the quantitative pharmacokinetics where specific drug measurements were required. Thus, toxicokinetics was created when quantitative pharmacokinetic principles were applied to toxicology studies with the emphasis on whether there was sufficient drug in the body for a sufficient length of time relative to the anticipated clinical situation.

As with any new scientific discipline, there has been some confusion as to what toxicokinetics is, with the consequence that many different definitions have been proposed, ranging from those of pharmacokinetics which only directly relate to toxicology studies, to virtually all of nonclinical pharmacokinetics and metabolism (ADME).

When applying models to toxicokinetics, it is important that it is adequately defined since the type of model used will depend on the definition employed. For the purposes of this discussion, toxicokinetics will be defined as that described in Draft 10 of the ICH2 guidelines for toxicokinetics developed from that originally proposed by the EFPIA/CPMP working party in Europe. This states that:

'Toxicokinetics is the generation of pharmacokinetic data as an integral component in the conduct of nonclinical toxicity studies and the use of these data in the interpretation of toxicological findings and their relevance to clinical safety issues'

With this definition, toxicokinetics is restricted to those studies which form part of a toxicological study and excludes all the conventional ADME work. As a consequence, the models which should be considered are primarily those which answer the objectives of toxicokinetics, namely, the interpretation of toxicological findings and the relevance to clinical safety, often referred to as safety margins. However, the application of modelling to the improvement of the experimental design of toxicokinetic studies should also be borne in mind since, at the moment, these studies tend to be very resource demanding both in animals and bioanalysis time.

TRADITIONAL TOXICOKINETIC STUDIES

Although every laboratory has their own individual design for toxicokinetic studies, there does appear to be some sort of commonality of approach being adopted, which in some part may be a result of the draft guidelines now appearing. Also, with traditional techniques there are limitations as to what can be done if the overall objectives of toxicokinetics are not to be compromised. Despite this degree of standardisation, we should not be complacent since the present experimental designs are far from optimal and with the judicious application of the correct modelling techniques, significant improvements can be made.

The study design commonly adopted can only be described as a 'belt and braces' job with, usually, a full plasma profile of approximately 8 or more plasma samples taken at the beginning, midway and at the end of the study. This can result in many hundreds of plasma samples which have to be analysed for the compound of interest using a specific and validated analytical method of sufficient sensitivity. All of this has to be completed in full compliance of Good Laboratory Practise (GLP). Moreover, since the relatively large number of samples taken could affect the integrity of the toxicology study, especially with the rodent species, animals on the main study are rarely used. Instead, a satellite group is employed which is housed alongside the main group under identical conditions, including environmental, dosing and drug formulation.

Once the plasma concentrations are available, the exposure is established by the determination of the area under the curve (AUC), usually by the application of the trapezoidal rule, together with the C_{max} and T_{max}, which are taken directly from the analytical data. These are then analysed by an appropriate statistical test, usually fairly superficially, to examine for dose level dependent pharmacokinetics, time dependent pharmacokinetics and gender differences. It is recognised that the power of these tests tends to be low for a typical toxicokinetic study and much of the interpretation, relating the exposure to toxic events, tends to be descriptive. This is especially true when satellite groups of animals have to be used since, although the conditions are kept as close as possible to those of the main study, the animals used for the toxicokinetics are still different from those used for toxicology. Under these conditions it is impossible to relate any toxic event in an individual animal to the exposure for that animal.

The value of toxicokinetic information for a meaningful interpretation of a toxicological event is illustrated by the work of Ramsey (1982) who compared the low dose extrapolation of carcinogenicity, based on a 'one hit model', when using dose levels or steady-state plasma levels as measures of exposure in a situation when there was dose dependent pharmacokinetics. Two scenarios were considered, the first when the parent drug was the carcinogen and the second when the product of the saturable metabolism was the carcinogen. In the former situation extrapolation using dose level would overestimate the carcinogenic potential whereas with the latter there would be a serious underestimate. Steady-state plasma levels, on the other hand, would result in an acceptable estimate of the carcinogenicity for both situations.

Quantifying the toxicokinetics of a compound with its toxicological response for predictive use is rarely attempted, although, conceptually, it is not difficult and has been used routinely in the clinical situation for cancer chemotherapy. The resultant model is then be used to predict and minimise toxic side effects in cancer patients and can just as easily be applied to the determination of the no effect exposure in toxicology studies. This approach has been described extensively (Egorin *et al*, 1986; Ings *et al*, 1992) and has been proposed for use in toxicokinetics in the past (Ings, 1990).

The problem with modelling the toxicology with the toxicokinetics is that most toxicities from long term toxicology studies are chronic and develop over a relatively long period of time in relation to the changing plasma levels during one dosage interval. Thus, as there is often a very significant delay between the plasma levels and the onset of the toxicological response, equivalent to a pronounced hysteresis, it is not possible to use plasma levels directly. Instead, AUC as a measure of exposure should be employed, substituting it into the standard sigmoid E_{max} equation (1).

$$Response = \frac{E_{max} * AUC^{n}}{EC_{50}^{n} + AUC^{n}} \qquad \text{......... (1)}$$

Where E_{max} is the maximum response, EC_{50} is the AUC that gives 50% of the maximum response and n is the Hill constant.

Despite its obvious utility, this approach is open to criticism, since each point represents a different individual with no account taken of inter- and intraindividual variability, which can lead to biased results. However, newer modelling techniques are now being introduced which overcome this problem.

FUTURE TRENDS

The traditional approach, which is largely empirical, supplies valuable information which will enhance the usefulness of toxicological data. It will, no doubt, be continued to be used for many years in the future but it is not without its limitations, both theoretically and practically.

An aspect of toxicokinetics which is causing immediate concern is the amount of resource it is consuming, both in animals and scientist's time. For a typical NDA it has been estimated that approximately 60% of a Drug Metabolism and Pharmacokinetic Department, including bioanalytical, will be spent on toxicokinetics whereas a few years ago it was virtually none. Interest is, therefore, beginning to focus on how this resource can be reduced without compromising the overall objectives of the study. One possibility is the use of limited sampling models to reduce the number of samples needed for analysis whilst still providing a good estimate of exposure. There are three possibilities:

* Stepwise multiple regression

* D-optimisation

* Baysean feedback (MAP)

The first of these is an empirical approach developed to determine limited sampling regimens for patients receiving cyclosporin (Johnston *et al*, 1990), since there is no assumption of a pharmacokinetic model. It examines the correlation of the plasma levels with AUC, successively removing a plasma time point in a stepwise manner and retaining only those which show the best correlation. A reduced data set of the two to four time points which correlates best with the AUC can then be used for the subsequent toxicokinetic studies. Unfortunately, there is a need for an initial reasonably large data base since a training set of data is required for the stepwise multiple regression plus an independent test set of similar size to ensure that the sampling times selected do provide a good estimate of AUC. Often such a data base is not available until the final long-term studies. Nevertheless, its great advantage is that it can cope with most types of plasma profile, even the erratic profiles, since no pharmacokinetic model is assumed.

The second approach is potentially more useful in that it does not require such a large data base, although a pharmacokinetic model has to be assumed and the data fitted to it. Once the model has been obtained, the optimum sampling times can be determined using D-optimisation with programs such as PCNONLIN (Metzler and Weiner, 1992) and ADAPT II (D'Argenio and Schumitzky, 1992). It must be remembered, however, that with this technique it is impossible to have fewer sampling points than the number of parameters in the model. Thus, the more complicated the model, the more sampling times are needed so that a simple biexponential equation would require at least 4 samples and a triexponential equation, 6 samples.

The last of the limited sampling approaches overcomes this limitation so that fewer sampling times than the number of parameters can be selected. This is available on ADAPT II but still needs to derive a pharmacokinetic model. Other programs such as NONMEM (Beal *et al*, 1989) can also be used, although NONMEM is more commonly associated with mixed effect modelling for population pharmacokinetics in a clinical situation. However, this approach probably offers the most exciting advance in the design and interpretation of toxicokinetic studies, overcoming many of the difficulties experienced with the traditional approach. Not only is it able to use sparse data of only 1 to 4 samples per animal, thereby dramatically reducing the number of plasma samples needed, but it can also separate the inter- and intraindividual variability from the main effect of the population mean of a parameter using a pharmacostatistical approach.

This concept has been reviewed by many workers (Sheiner and Beal, 1980,1981,1983; Steimer *et al*, 1985), examining the merits of the technique in relation to traditional approaches and the various design features necessary to achieve the desired objectives. Most of this has been orientated towards clinical objectives and goes beyond the scope of this present article but, to suffice it to say, preliminary studies have shown its potential utility for use in toxicokinetics. As only a few samples are needed per animal, satellite groups can be omitted and the toxicokinetics of a compound determined in the individual animals of the main toxicology study. Thus, a toxic event in an individual animal can be related to the exposure in that animal.

The value of the mixed effect modelling is demonstrated by Ganes *et al* (1993) when examining drug induced anaemia in a 1 month monkey toxicology trial with a novel agent, 3'-deoxy-3'-fluorothymidine, undergoing development for the treatment of AIDS. The degree of anaemia, measured by the fraction of baseline red blood cell count or haemoglobin, was modelled against exposure (AUC) using the sigmoid E_{max} relationship (equation 1). With the use of NONMEM a description of the exposure, relating it to the toxicological response, was obtained over the duration of the study. In addition, each sex was examined independently and gender differences in the toxicokinetic and toxicological response determined. From this model, the degree of anaemia could be predicted for any exposure of the compound or, conversely, the exposure needed to cause a predetermined degree of anaemia established.

SPECIES SCALING

There are two aspects of scaling that have to be considered for toxicology studies; that of extrapolation from high toxicological doses to relatively low clinical doses within the same species and the extrapolation between animal species to the situation in man. There are many models proposed for the former (Gad and Weil, 1986; Ings, 1990) which adequately describe the observed data, but which, when extrapolated beyond the observed data show large discrepancies. Unfortunately, it is not possible to establish which is the most appropriate model statistically, since a prohibited number of animals would have to be used to obtain the necessary statistical power (Ings, 1990). Therefore, all discussions on the relative merits of the various models for intraspecies scaling must be purely theoretical.

Interspecies scaling does not suffer from these problems since, although it may not seem readily apparent, there is an order to life which can be exploited empirically. As Boxenbaum (1982) noted when reviewing the topic, every mammalian species has 4 heartbeats for every breath and these scale in the same way as lifetime. Thus, most mammals have the same number of breaths and heartbeats before they die, the only difference being that the smaller the animal is, the more rapid the heart rate and breathing and the sooner they die. The relationship with body weight, known as allometry, has been used successfully for many years for scaling physiological parameters between animal species. It is not surprising, therefore, that the pharmacokinetic parameters of drugs which are, in turn, governed by physiological variables such as creatinine clearance or hepatic blood flow, can also be scaled between species using allometry (Boxenbaum, 1982; Mordenti and Chappell, 1989; Ings, 1990).

Most pharmacokinetic parameters are not directly related to body size but are related as described by equation 2.

$$\textit{Pharmacokinetic parameter} = A * \textit{Bodyweight}^{\alpha} \qquad \text{.........} \ (2)$$

Where A is a coefficient and α is a power function determined from the intercept and slope respectively of a regression of the logarithm of the parameter versus the logarithm of the body weight of the animal species.

A word of warning, however, for those about embark on allometric scaling. Beware of neoteny! If your drug is primarily eliminated by metabolism using cytochrome P_{450} and the clearance is low compared to hepatic blood flow, the clearance for man will be significantly different from that predicted from other animals. This is due to neoteny, an evolutionary phenomenon, which has given man an advantage over fellow animals. There are many aspects associated with neoteny (Boxenbaum and D'Souza, 1987; Ings, 1990) but the most important of these for allometric scaling are the increased brain weight, increased longevity and reduced cytochrome P_{450}. All these aspects appear to be co-correlated, enabling the incorporation of either brain weight or maximum expected life span into the allometric scaling equation to compensate for the reduced cytochrome P_{450} activity (equation 3).

$$\textit{Pharmacokinetic parameter} = A * \textit{Body weight}^{\alpha} * \textit{Brain weight}^{\beta} \quad \text{.........(3)}$$

Experience with drugs unaffected by neoteny has shown that the power function for a specific pharmacokinetic parameter in equation 2 tends to be constant, with that for volume of distribution approximating to 1, for clearance approximating to 0.75 and for half-life approximating to 0.25. Thus, only volume of distribution shows an approximately direct relationship with body weight. Clearance, on the other hand, which is the elimination parameter governing AUC, shows a closer direct relationship to body surface area. Why, then, is there an insistence on scaling our dose levels for chronic toxicology studies with a dose per body weight and not with a dose per body surface area? It is noteworthy that in cancer chemotherapy, where the drugs can have a narrow therapeutic window, scaling is often based on dose per body surface area. Moreover, a comparison of scaling from an animal species to man using dose levels based on either body weight or body surface area shows a divergence which becomes greater with increasing weight difference between the animal species (Ings, 1990). For example, when scaling between mouse and man, the dose level based on body weight will be approximately 8 times greater than that based on body surface area. Since from allometric scaling, the power function for clearance bears a closer resemblance to body surface area, then in chronic situations, it is the scaling using body weight that is 8-fold in error.

With the advent of toxicokinetics, where the safety margins are determined solely on exposure, without relating back to dose levels (equation 4), the problems described above should be overcome. Of course, this is provided that the investigator is not tempted to refer the exposure back to dose level rather than the absolute dose used in the clinic.

$$\textit{Safety Margin} = \frac{\textit{Animal no toxic effect exposure}}{\textit{Human clinical exposure}} \quad \text{...........(4)}$$

CONCLUSION

There can be no dispute that toxicokinetics offers considerable advantages over the methods previously available, since it uses exposure instead of dose level to determine safety margins. This was especially true, both on theoretical and practical grounds, when scaling was performed with dose levels based on body weight. Most toxicokinetic studies are designed empirically with little consideration of subsequent modelling. However, with the judicious application of modelling, especially with respect to the relationship between the toxicokinetics and the toxicological response and/or to the use of mixed effect models, the predictive value of the studies can be considerably enhanced. In addition, for the latter models, fewer animals would be needed, since satellite groups would no longer be required, and there would be fewer blood samples.

REFERENCES

Beal, S.L., Boeckmann, A.J. and Sheiner, L.B. (1989) *NONMEM Users guide*, NONMEM Project Group, C255, University of California at San Francisco, San Francisco, CA 94143.

Boxenbaum, H. (1982) Interspecies scaling, allometry, physiological time and the ground plan of pharmacokinetics. *Journal of Pharmacokinetics and Biopharmaceutics*, **10**, 201-227.

Boxenbaum, H. and D'Souza, R. (1987) Physiological models, allometry, neoteny, space-time and pharmacokinetics. In *Pharmacokinetics: Mathematical and Statistical Approaches to Metabolism and Distribution of Chemicals and Drugs*, (A. Percile and A. Rescigno, Eds) pp 191-214. Plenum Press, New York.

D'Argenio, D.Z. and Schumitzky, A. (1992) *ADAPT II User's guide*, Biomedical Simulations Resource, University of Southern California, Los Angeles, California 90089-1451.

Egorin, M.J., Van Echo, D.A., Whitacre, M.Y., Forrest, A., Sigman, L.M., Engisch, K.L. and Aisner, J., (1986) Human pharmacokinetics, excretion and metabolism of the anthracycline antibiotic Menogaril (7-OMEN, NSC 269148) and their correlation with clinical toxicities. *Cancer Research*, **46**, 1513-1520.

Gad, S. and Weil, C.S. (1986) *Statistics and Experimental Design for Toxicologists*, Telford Press, New Jersey.

Ganes, D., Bansal, S., Johnson, D., Leal, M., Nicolau, G. and Yacobi, A. (1993) Population pharmacodynamic modeling of drug-induced anemia. *Journal Pharmaceutical Sciences*, **82**, 546-549.

Ings, R.M.J. (1990) Interspecies scaling and comparisons in drug development and toxicokinetics. *Xenobiotica*, **20**, 1201-1231.

Ings, R.M.J., Lelièvre, E., Ardiet, C., Clavel, M., Leyvraz, S., Minaidis D., Lokiec, F., Turpin, F., Solere, P. and Lucas, C. (1992) Value of early pharmacodynamic and pharmacokinetic investigations with anticancer drugs: data from phase I tolerance studies on a new vinca alkaloid derivative. *Xenobiotica*, **22**, 871-880.

Johnston, A., Sketris, I., Marsden, J.T., Galustian, C.G., Fashola, T., Taube, D., Pepper, J. and Holt, D.W. (1990) A limited sampling strategy for the measurement of cyclosporin AUC. *Transplantation Proceedings*, **22**, 1345-1346.

Metzler, C.M. and Weiner, D.L. (1992) *PCNONLIN User Guide - Version 4.0*, SCI Software, SouthCreek Office Park, 2365 Harrodsburg Road, Suite A-290, Lexington, Kentucky, 40504-3399.

Mordenti, J. and Chappell, W. (1989) The use of interspecies scaling in toxicokinetics. In *Toxicokinetics and New Drug Development*. (A. Yacobi, J.P. Skelly and V.K. Batra Eds) pp 42-96. Pergamon Press, New York.

Ramsey, J.C. (1982) Nonlinear pharmacokinetics relative to toxicity and use of toxicological data. *Drug Metabolism Reviews*, **13**, 779-797.

Sheiner, L.B. and Beal, S.L. (1980) Evaluation of methods for estimating population pharmacokinetic parameters. I. Michaelis-Menten model: routine clinical pharmacokinetic data. *Journal of Pharmacokinetics and Biopharmaceutics*, **8**, 553-571.

Sheiner, L.B. and Beal, S.L. (1981) Evaluation of methods for estimating population pharmacokinetic parameters. II. Biexponential model and experimental pharmacokinetic data. *Journal of Pharmacokinetics and Biopharmaceutics*, **9**, 635-651.

Sheiner, L.B. and Beal, S.L. (1983) Evaluation of methods for estimating population pharmacokinetic parameters. III. Monoexponential model: routine clinical pharmacokinetic data. *Journal of Pharmacokinetics and Biopharmaceutics*, **11**, 303-319.

Steimer, J-L., Mallet, A. and Mentre F. (1985) Estimating interindividual pharmacokinetic variability. In *Variability in Drug Therapy: Description, Estimation and Control*. (M. Rowland, L.B. Sheiner and J-L. Steimer Eds) pp 65-111, Raven Press, New York.

Alterations in Kinetic Parameters Resulting in Modification of Toxicity

C. Charuel
Centre de Recherche, Pfizer, BP 159, F-37401 Amboise Cédex, France

Pharmacokinetic measurements are now routinely integrated within toxicity studies as a means of providing information on drug availability and dose response across test species exposed to different dosage regimens for various periods of treatment (Smith et al, 1990). The quantification of plasma concentrations and areas under plasma concentration against time curves (AUC) has proved very useful to identify changes in toxicity linked to formulation, absorption variability, sex- or dose-dependent kinetics, and induction or inhibition of metabolizing enzymes.

Such alterations in drug disposition can be associated with a reduction or an increase of the toxic response depending on the relationship between concentration and effect (Smith, 1993). Furthermore the administration of relatively high dose levels may produce pharmacological effects (modification of blood flows) or damage to key organs (liver, kidney), which impact on drug kinetics. Age and pregnancy also contribute to modifications of the clearance and/or volume of distribution of drugs. Therefore, the interpretation of pharmacokinetic data in the context of toxicity studies is not always straightforward even when test species are considered separately and when there is a reasonable relationship between the biological response and the concentrations of the test substance in body fluids.

The role and limitations of toxicokinetic data during the development of new pharmaceuticals will be addressed under two major chapters. The first will deal with the characterization of the toxic response at the outset of the toxicology program, and the sensitivity of the response to factors which have an immediate impact on drug concentrations (dose, route, gender). The second chapter will focus on the modifications of the toxic response and drug kinetics with species, duration of treatment (induction, inhibition), age, and pregnancy.

Factors affecting the toxic response at the outset of the toxicology programme

Characterization of the toxic/biological response

Toxicity tests in animals are typically performed on relatively homogeneous populations, using small numbers of young healthy specimens and a limited range of doses. In contrast to the pharmacological and clinical situations, dose levels are generally set as high as possible in order to explore the toxic potential of the test substance. This results in a whole spectrum of biological responses ranging from undetectable to detectable and from harmless to toxic at a given, and *a fortiori* different, levels(s) of exposure. Although the methods of investigation have become more sophisticated with the introduction of new tools to study in detail the functional and morphological state of the organism, there is no guarantee that the most relevant information will be obtained in early animal tests. For example, little attention may be given to the central nervous or endocrine systems (Kroes & Feran, 1990), or perhaps the toxic response is confounded by too many alterations in vital functions (multiple clinical chemistry or histopathological changes). Thus, the objective is to identify threshold doses, adapting, whenever possible, the choice of the test systems according to the history (pharmacology, chemical class) of the compound, the information gathered prior to pre-clinical studies (biopharmaceutics, *in vitro* metabolism) and any additional data suggestive of a mechanism of toxicity.

Once appropriate end points (cardiovascular changes, enzyme activities, clinical chemistry measurements, organ weights, histopathological lesions, teratogenicity) have been determined it is important to estimate how the magnitude of the response varies with dose or, preferably, with concentration of test compound (or active metabolite), ideally at the site of action.

Variation of the toxic response with dose and factors affecting drug concentrations

The dose response relationship is a fundamental concept in toxicology as it provides evidence that the measured effect is quantitatively, if not causally, related to the administration of a new drug (Timbrell, 1982 ; Aldridge, 1990). However, quantitative assessments based on dose levels remain unreliable as long as they are not supported by concentration measurements. When this information becomes available, differences in the toxic response with dose regimen or gender can be put into perspective.

Non linear responses with dose (Smith et al, 1990) may reflect poor absorption (sub-proportionality) or saturation of intrinsic clearances (super-proportionality), whilst gender effects should alert the toxicologist to possible differences in rates of biotransformation (Kato & Yamazoe, 1992).

The following examples describe the behaviour of a dihydropyridine derivative (DHP) tested in our laboratory, whose toxicity in rats showed variation with gender and route of administration. This compound caused haemolytic anaemia in this species, characterized by decreased red blood cell counts (RBC), reduced haemoglobin levels (HGB) and increased plasma bilirubin. This response was specific to the rat, dose-related and, in gavage studies only, more marked in females than in males. In these studies, the gender effect was related to a clear sex difference in the drug kinetics, females exhibiting higher blood concentrations and AUC's than males. Interestingly, however, males and females reacted to the drug in a similar manner when dietary administration was used. They showed haemolytic changes of comparable amplitude while drug concentrations were similar and much lower than those achieved after gavage (fig. 1). Overall, however, the biological response correlated with the concentration of drug present in the blood with a threshold *ca* 10 µg/ml below which no major alteration of red blood cell parameters was found, irrespective of sex or route of administration.

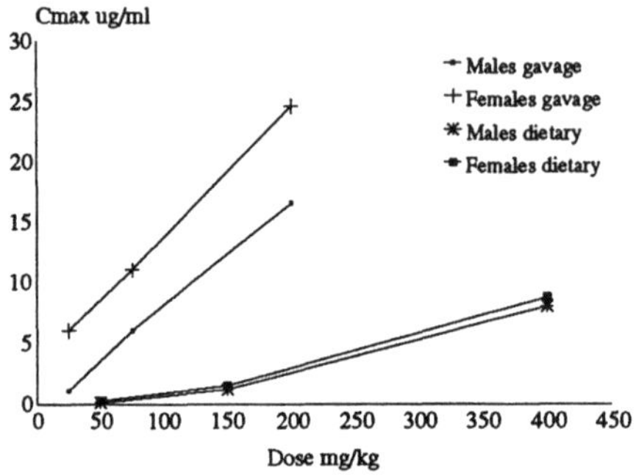

Fig.1 : Toxicokinetics of a dihydro pyridine (DHP) derivative in rats: effect of route of administration on maximal plasma drug concentrations (Cmax)

Therefore, although the unchanged drug *per se* produces no direct lytic effect on isolated red cells, its effects *in vivo* increase with blood concentrations and are very sensitive to modifiers of the drug exposure. The gender effect in particular is consistent with published data on nivaldipine (Niwa et al, 1988), nifedipine and nicardipine (Guengerich et al, 1986). The oxidation of the pyridine ring has been shown to be catalyzed by cytochrome P450 enzymes of the 3A and 2C families which are expressed in adult male rats preferentially to females (Kato & Yamazoe, 1992 ; Guengerich et al, 1986). In contrast, the reason for the lack of sex difference after dietary administration of the drug is unclear. Since this route produced lower peak concentrations and longer exposure than gavage, the rates of the different metabolism processes (food sensitive pathways, low affinity enzymes) in males and females may have been altered (Denk et al, 1989 ; Sipes and Gandolfi, 1991).

Such situations in which the toxic response is confounded by effect of gender or food are not rare (Klamerus et al, 1993), and are difficult to predict using *in vitro* models. One has therefore to rely on data generated *in vivo* within each species and examine whether the biological results are explicable in terms of pharmacokinetics.

Differences in drug kinetics and toxic response with species and duration of dosing

Species differences

There are qualitative and quantitative differences in the way species, or even strains, react to drugs. To some extent such differences are due to differences in bioavailability and metabolism, dependent on the presence of particular enzymes and predominance of certain metabolic pathways (Roe, 1990 ; Gangolli & Phillips, 1990 ; Caldwell, 1992).

Although they are relatively well conserved across species, the cytochrome P450 enzymes (CYP 1A, 2A, 2B, 2C, 2D, 3A) have particular catalytic selectivities

(Smith, 1991 ; Yasumori et al, 1993). This can lead to differences in the rate or route of excretion of drugs between species, and impact on the toxic response. For example, CYP2C in rat and man has now been implicated in the metabolism of acidic drugs such as the anti-inflammatory agent, diclofenac, or the anti-allergy agent, proxicromil, a chromone 2 carboxylic acid. In dogs, this enzyme has poor catalytic activity for these substrates and the hepatotoxic potential of proxicromil correlates with an inability to hydroxylate the drug at an appreciable rate. The dog eliminates the drug virtually unchanged in the faeces by biliary excretion, whilst the rat and monkey metabolize proxicromil readily and excrete the resultant metabolites in both urine and faeces (Smith et al, 1983).

It should be borne in mind, however, that phase II reactions also show species variation (deficiency in N-acetyltransferase in dogs) and may reduce (detoxification) or, on the contrary, further enhance toxicity. When bioactivation occurs leading to covalent binding, the interpretation of the role of simple drug concentrations in the toxic response becomes almost impossible. Therefore caution should be exerted in the way pharmacokinetic data are used and compared across species : the relevance of plasma concentrations depends on the nature and mechanism of the toxicity under investigation (Monro, 1990, 1992).

Induction/Inhibition of drug metabolizing enzymes

Prediction of the toxicological consequences of enzyme induction or inhibition depends on the nature and interplay of the metabolic pathways affected. Originally three different classes of inducers were identified : barbiturates, polycyclic aromatic hydrocarbons and steroids (Snyder et al, 1979). However it is now recognised that the situation is more complex with many drugs being capable of inducing selective cytochromal P450 isozymes and, to a lesser extent, phase II enzymes (Gangolli et al, 1990). This may increase the activity of toxic and/or non-toxic pathways and hence increase or reduce the toxic response on repeated treatment. Similarly, inhibition of drug biotransformation *in vivo* is a complex situation. Inhibitors may directly affect their own metabolism as described with milacemide and monoamine oxidase B (Cavalier, 1988), or produce multiple effects on phase I and phase II reactions, the degree of which can vary with dose and time. Not infrequently a drug may act as both an inhibitor and an inducer.

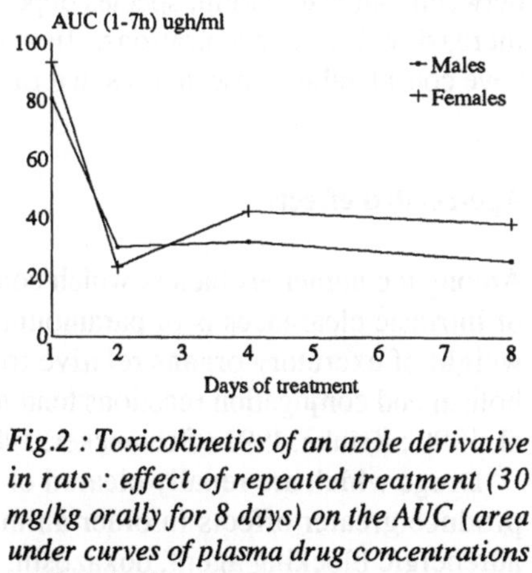

Fig.2 : Toxicokinetics of an azole derivative in rats : effect of repeated treatment (30 mg/kg orally for 8 days) on the AUC (area under curves of plasma drug concentrations against time)

This phenomenon is illustrated in the following example which refers to an azole derivative that showed very peculiar kinetics on repeat administration in our laboratory. After single doses, rats showed dose-dependent kinetics with super-proportional increases of Cmax and AUC resulting from reduced drug clearance as dose increased. However, although significant drug levels were present 24 hours after the high dose, there was no accumulation on repeat administration. On the contrary, the AUC was reduced by 60-70% after the second dose and remained at this low level until the end of an 8-day treatment (fig. 2). Examination of the liver content of cytochrome P450 showed that this reduction of the AUC coincided with an increase in the enzyme activity which persisted throughout the study. At least two isozymes, P450 3A and P450 2B1 were induced as was shown by gel electrophoresis of microsomal proteins and western blotting. It is therefore likely that this compound induced its own metabolism via alterations of liver cytochrome P450's.

In dogs the compound has a long half-life (15-20 hours), and increasing single oral doses produced a reasonably linear increase in the AUC (fig. 3). On repeat treatment, however, autoinduction was seen below 12 mg/kg whilst accumulation prevailed at higher doses. The intersection point (I) corresponded to the threshold dose beyond which a sharp rise in AUC and in the toxic potential of the drug was observed. In males a 15 mg/kg dose produced a 3-fold accumulation of drug accompanied by marked clinical signs, anorexia and body weight loss.

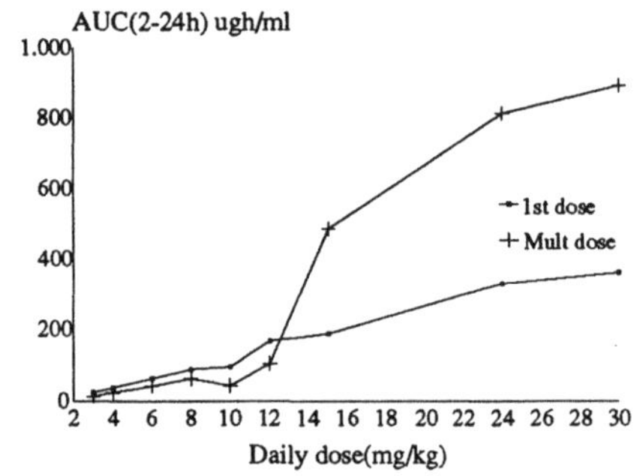

Fig.3 : Toxicokinetics of an azole derivative in dogs : variation of the AUC with dose level on single and multiple dosing

These signs disappeared when the dose was reduced from 15 to 12 mg/kg, ie, when a new steady state with plasma concentrations slightly lower than on day 1 was achieved. One chemical, therefore, may well produce contrasting effects not only between, but also within, species depending on the capacity of the liver to adapt to increasing drug concentrations. In addition, a very small increment in dose may have considerable consequences in terms of AUC and toxic response.

Age-related effects

Among the numerous factors which contribute to age-related changes, modification of intrinsic clearances is of paramount importance as liver and renal blood flows, weight of excretory organs relative to body mass, mixed function oxidase metabolism and conjugation reactions tend to decrease steadily with aging (Hollander et al, 1990 ; Krinke, 1991 ; Schmucker, 1985).

Drugs which are renally cleared or have a high extraction ratio may therefore produce greater effects in older animals. We recently published data on an α-adrenergic blocking agent, doxazosin, whose toxicokinetics in chronic rat studies were dose- and age-dependent (Charuel et al, 1992). The clearance declined and mortality increased, as dose and age increased. Mortality was noted primarily in the heavier animals from the high-dose male group during the second year of the study. As they were progressively eliminated from the calculation of group means, the AUC's became almost proportional to dose suggesting that this particular sub-set of animals was responsible for the non-linear kinetics observed earlier on, and were more sensitive to the drug effects.

These results illustrate the importance of pursuing a systematic toxicokinetic approach in chronic studies in order to try and distinguish whether different responses in older animals are due to the pharmacokinetics or to the pharmacodynamics of a drug, or both.

Toxicity and pharmacokinetics during pregnancy

The way in which chemicals are absorbed during pregnancy, whether or not they reach the conceptus, and in what form, are topics that require consideration along with the physiologic changes occurring during gestation (Juchau & Harris, 1990 ; Manson & Wise, 1991). Pharmacokinetic data contribute to an understanding of fetal exposure and may explain fetal effects due to direct action of drugs, or secondary to maternal toxicity. However, the complexity of the mechanisms involved in fetal development and toxicity is such that the timing of drug administration is sometimes more critical than dose or drug concentrations.

We have shown that daily administration of the imidazole antifungal, tioconazole, to pregnant rats from day 15 post insemination (p.i.) delayed the onset of parturition and suppressed the ante-partum peak in serum estradiol. In contrast, treatment from day 18 p.i. advanced parturition and the pre-partum fall in circulating progesterone (Latrille et al, 1987a).

These results illustrate how dual, apparently paradoxical, effects can be governed almost exclusively by secondary (hormonal) changes, the interplay of which is critically influenced by time of dosing.

Conclusions

Pharmacokinetic investigations in pre-clinical toxicology have been motivated largely by the realization that changes in the time course of drug concentrations in the body elicit changes in the intensity, the duration or even the quality of the toxic response. In many cases this relationship is valid and can be established, quantitated and refined as the toxicology programme unfolds. Early studies provide valuable information on the toxic response *per se* (target organs) and its sensitivity to dose and modifiers of drug concentrations (gender, route, ...). They also provide valuable insights into the design of subsequent or mechanistic studies and the interpretation of data, within a given species. As more information on the nature of the toxic response (time of onset, reversibility) and on the tissue distribution, excretion and metabolism of test agents becomes available, differences in toxicity within or between species can be interpreted with a higher degree of confidence. In particular, the variations of the toxic response secondary to changes in the activity of drug metabolizing enzymes, or ageing can be put into the equation.

However it should be emphasized that blind reliance on plasma concentrations can be misleading since they represent neither the concentration at the site of toxic action, nor -in the case of complex disturbances (teratogenesis, carcinogenesis)- the relevant active molecular entity.

In such cases, the pharmacodynamics of the drug should be explored more systematically, and appropriate indices defined and utilized.

References

Aldridge, WN (1990) The biochemical principles of toxicology. In: Anderson D, Conning IM (eds) Experimental Toxicology, the Basic Principles. Royal Society of Chemistry, Cambridge, pp 56-81

Caldwell J (1992) Problems and opportunities in toxicity testing arising from species differences in xenobiotic metabolism. Toxicology Letters 64/65:651-659

Cavalier RG, de Suray JM, de Varebeke J, Onkelinx CL, Lesne ML (1988) Non linear pharmacokinetics of milacemide, a new CNS drug. In: Bres J, Panis G (eds) Interêt et limites de la pharmacocinétique en recherche et développement. Sauramps Medical, Montpellier, pp 479-484

Charuel C, Comby P, Monro AM (1992) Diurnal exposure profile in rats from dietary administration of a chemical (doxazosin) with a short half-life: interplay of age and diurnal feeding pattern. J Appl Toxicol 12(1):7-11

Denk B, Baumann M, Filser JG (1989) Pharmacokinetics and hepatotoxicity of 2-nitropropane in rats. In: Biological monitoring of exposure and the response at the subcellular level of toxic substances. Arch Toxicol Suppl 13, 330-332

Gangolli SD, Phillips JC (1990) The metabolism and disposition of xenobiotics. In: Anderson D, Conning IM (eds) Experimental Toxicology, the Basic Principles. Royal Society of Chemistry, Cambridge, pp 130-179

Gordon WL, Sherwood OD (1982) Evidence that luteinizing hormone from the maternal pituitary gland may promote ante-partum release of relaxin, luteolysis and birth in rats. Endocrinology, 111:1299-1310

Guengerich FP, Martin MV, Beaune PH, Kremers P, Wolff T, Waxman DJ (1986) Characterization of rat and human liver microsomal cytochrome P450 forms involved in nifedipine oxidation, a prototype for genetic polymorphism in oxidative drug metabolism. J Biol Chem 261:5051-5060

Hollander CF, Delort P, Duprat P (1990) Toxicology in the older age group. In: Clayson DB, Munro JC, Shubik P, Swenberg JA (eds) Progress in predictive toxicology. Elsevier, Amsterdam, pp 113-142

Juchau MR, Harris C (1990) Age: the developing organism as a toxicological target. In: Clayson DB, Munro JC, Shubik P, Swenberg JA (eds) Progress in predictive toxicology. Elsevier, Amsterdam, pp 91-111

Kato R, Yamazoe Y (1992) Sex-specific cytochrome P450 as a cause of sex- and species-related differences in drug toxicity. Toxicology letters 64/65:661-667

Klamerus KJ, Troy SM, Ben Maimon CS, Chiang ST (1993) Effect of age, gender and food on Zalospirone disposition. Clin Pharmacol Ther 53:193

Krinke GJ, Perrin LPA, Hess R (1991) Assessment of toxicopathological effects in ageing laboratory rodents. Recent developments in toxicology trends, methods and problems. Arch Toxicol Suppl 14:43-49

Kroes R, Feron VJ (1990) Toxicity testing: strategies and conduct. In: Clayson DB, Munro IC, Shubik P, Swenberg JA (eds) Progress in predictive toxicology. Elsevier, Amsterdam, pp 15-39

Latrille F, Perraud J, Stadler J, Monro AM, Sutter B (1987a) Effects of tioconazole on parturition and serum levels of 17β-oestradiol, progesterone, LH and PRL in the rat. Biochem Pharmacol 36:1119-1124

Manson JM, Wise LD (1991) Teratogens. In: Amdur, Doull, Klaassen (eds) Casarett and Doull's Toxicology. The basic science of poisons, 4th ed. Pergamon, New York, pp 226-254

Monro A (1992) What is an appropriate measure of exposure when testing drugs for carcinogenicity in rodents ? Toxicol Appl Pharmacol 112:171-181

Monro AM (1990) Interspecies comparisons in toxicology: the utility and futility of plasma concentrations of the test substance. Reg Toxicol Pharmacol 12:137-160

Niwa T, Tokuma Y, Noguchi H (1988) Oxidation of nivaldipine, a new dihydropyridine calcium antagonist, to the corresponding pyridine by rat liver microsomes. Xenobiotica 18:217-224

Roe FJC (1990) Influence of animal species, strain, age, hormonal and nutritional status. In: Anderson D, Conning DM (eds) Experimental Toxicology, the basic principles. pp 23-34

Schmucker DL (1985) Ageing and drug disposition : an update. Pharmacol Rev 37:133-148

Sipes IG, Gandolfi AJ (1991) Biotransformation of toxicants. In: Amdur, Doull, Klaassen (eds) Casarett and Doull's Toxicology. The basic science of poisons, 4th ed. Pergamon, New York, pp 88-126

Smith DA, Neale MG (1983) Metabolism and clearance of proxicromil. Studies in rat, hamster, rabbit, dog, squirrel monkey, cynomolgus monkey, baboon and man. Eur. J. Drug Metab. Pharmacokinet. Vol. 8, 3:225-232

Smith DA, Humphrey MJ, Charuel C (1990) Design of toxicokinetic studies. Xenobiotica 20:1187-1199

Smith DA (1991) Species differences in metabolism and pharmacokinetics: are we close to an understanding. Drug Metabolism Reviews 23:255-274

Smith DA (1993) Integration of animal pharmacokinetic and pharmacodynamic data in drug safety assessment. Eur J. Drug Metab. Pharmacokinet. 1:31-39

Snyder R, Remmer H (1979) Classes of hepatic microsomal mixed function oxidase inducers. Pharmacol. Ther. 7:203-244

Timbrell SA (1982) Biochemical mechanisms of toxicity: specific examples. In: Taylor, Francis (eds) Principles of Biochemical Toxicology. Taylor and Francis, London, pp 177-232

Yasumori T, Chen L, Nagata K, Yamazoe Y, Kato R (1993) Species differences in stereoselective metabolism of mephenytoin by cytochrome P450 (CYP2C and CYP3A). J Pharmacol Exp Therapeut 264:89-94

Mechanisms of Carcinogenicity and Risk Assessment

(Chair: H. Autrup, Denmark)

Importance of Genetic Alterations in Tumour Development

Ken Brown, Chris Kemp, Phil Burns and Allan Balmain

Beatson Institute for Cancer Research, Wolfson Laboratory for Molecular Pathology, Garscube Estate, Switchback Road, Bearsden, Glasgow G61 1BD, Scotland

The wealth of molecular and genetic data over the past decade has resulted in the overwhelming conclusion that cancer is a genetic disease, and that the clonal expansion of a single normal cell to produce a malignant tumour is a complex process which invoves a number of genetic events. One of the most significant advances in cancer research has been the identification of many of the cellular targets which become altered during the development of neoplasia.

Oncogenes and Suppressor Genes

These critical genes can be separated functionally into two classes, namely oncogenes and tumour suppressor genes. Members of the oncogene class, many of which were first identified from studies of tumour viruses (Bishop, 1987), are genes which become activated by mutation, and whose expression now leads to enhanced cell growth. Since the mutational alteration acts in a dominant fashion, it need occur in only one of the two alleles present in the genome. Most of the genes which have been identified to date, as being involved in neoplasia, belong to the oncogene group.

The second group, the tumour suppressor genes, are negative regulators of cell growth, and become functionally inactivated by mutation. This loss of function then results in uncontrolled growth or lack of differentiation of tumour cells. Inactivation of a tumour suppressor gene, *a priori*, requires both alleles to become altered. Inactivation of one copy of the gene can occur, however, in the germ line, resulting in an inherited tumour predisposition. Indeed, this type of gene had been predicted by several independent lines of evidence. Epidemiological studies on inherited and sporadic childhood cancers suggested a two-hit mechanism at a single genetic locus (Knudson, 1973). Somatic cell hybrid experiments also suggested the recessive nature of the genetic changes responsible for tumorigenicity. Hybrid cells generated between tumorigenic and normal cells were non-tumorigenic unless chromosomes were lost from the hybrids to generate tumorigenic variants.

To date, only six tumour suppressor genes have been isolated, each associated with a cancer predisposition syndrome. Several other tumour suppressor genes have been localised to specific chromosomal regions from studies on inherited tumour predisposition syndromes (Stanbridge, 1990). In addition, the analysis of loss of heterozygosity has identified many other candidate loci which are involved in tumour development, and it may be that the majority of changes that transpire in tumour cells occur in genes which belong to this type.

From the wealth of data which has been acquired on the mutation of oncogenes and tumour suppressors, the types of gene alterations are identical to the sorts of genetic changes which are seen seen in tumour cells, and include point mutations, amplification, gene rearrangements, deletions and translocations.

Multiple Genetic Events in Cancers

The development of a tumour arises from multiple genetic hits, each of which results in a selective advantage to the altered cell. This has been termed the clonal evolution model (Nowell, 1989). Thus each genetic change expands the population of altered cells and creates a larger target pool for subsequent alterations. Evidence to support the multiple hit model for carcinogenesis has come from several sources. Epidemiology shows that the incidence of cancer increases dramatically with age. Moreover, mathematical analysis of the data suggests that the number of mutations required to produce an overt malignancy varies from three to twelve depending upon tumour type (Renan, 1993). Morphological data also supports the multistage nature of cancer. In certain human tumours, most notably colon cancer (Fearon and Vogelstein, 1990), and in animal tumour models such as mouse skin (Balmain et al. 1992), it is possible to identify discrete stages of tumour evolution. Some of the molecular events associated with these individual stages have been identified. It has been possible also, by comparing individual areas within the same tumour, to detect mutations associated with the more malignant morphology, in addition to those changes present in all tumour cells.

More direct evidence for the multistage nature of cancer has come from studies of oncogene co-operation both *in vitro* and *in vivo*. The transformation of primary cells cannot be achieved by a single oncogene but requires at least two co-operating oncogenes, as has been shown for *ras* and *myc* (Land et al. 1983). Similar evidence has been obtained from studies using transgenic mice. Mice which carry an oncogenic transgene develop focal tumours, suggesting additional events are required. The latency of tumours, however, can be reduced by crossing transgenics carrying different oncogenes to produce double transgenics (Berns et al. 1989). Similarly, in murine T-cell lymphomas which occur as a result of gene activation by viral insertions, the number of integration events is found to increase as the tumours progress, again indicating multiple events are being selected (Breuer et al. 1989).

What has become apparent from the molecular studies is that some genes are involved in the etiology of many cancers, for example the p53 suppressor gene, mutation of which has been found in most human tumours (Hollstein et al. 1991), whilst many other genes appear to be associated with few or indeed only one tumour type. Additionally, there is some evidence of a temporal order to the genetic alterations. It would appear that only a subset of genetic alterations can initiate the tumorigenic process, with mutations in other genes only being effective at later stages of tumour development. The position of a particular gene in the chain of events may vary depending on the tumour type.

Some mutations may destabilise the genome and increase the frequency of further alterations. Indeed, there is some evidence that progression towards a more malignant phenotype is associated with an increase in the genetic instability of the cell (Meuth, 1990). One gene which may act in this manner has recently been implicated in the evolution of familial colorectal cancer (Peltomäki et al. 1993; Aaltonen et al. 1993; Thibodeau et al. 1993).

Mouse Skin Carcinogenesis Model

Chemical carcinogenesis in mouse skin has been responsible for the development of many of the ideas central to multi-stage carcinogenesis, and continues to be a useful tool for analysing the function of genetic changes during tumour development. Tumour formation in this system proceeds through discrete stages which can easily be identified and isolated for comparitive studies. These stages are illustrated in Figure 1.

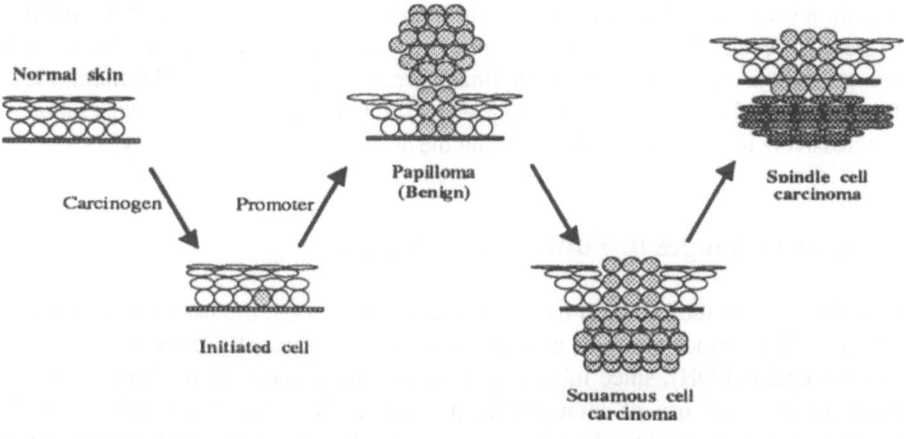

Fig. 1. Multi-Step Carcinogenesis in Mouse Skin. The major stages are indicated.

In this model system, tumours are induced by a single application of a chemical carcinogen. This results in a population of altered *initiated* cells which remain dormant until stimulated to clonally expand upon repeated treatment with a tumour promoter such as 12-O-tetradecanoyl-phorbol-13-acetate (TPA) (Boutwell et al. 1982). This results in the formation of benign well-differentiated papillomas, a few of which undergo further genetic alterations and progress to squamous carcinomas. A further stage involves the the progression of squamous carcinomas to highly invasive undifferentiated spindle carcinomas (Klein-Szanto et al. 1989; Buchmann et al. 1991; Burns et al. 1991). The application of molecular techniques has identified some of the mutational alterations, which occur in proto-oncogenes and tumour suppressor genes, and which are associated with specific stages in the evolution of the tumour (Quintanilla et al. 1986; Bremner, Balmain, 1990; Burns et al. 1991).

In the majority of tumours produced by carcinogen and promoter treatment, the initiating event is the induction of an activating point mutation in the Ha-*ras* gene. This is based on an analysis of the Ha-*ras* gene in tumours produced by a number of carcinogens, in which specific mutation patterns are seen (Bizub et al. 1986; Quintanilla et al. 1986; Brown et al. 1990). The causal role of Ha-*ras* mutation in the initiation step of tumorigenesis has been supported by experiments in which treatment of mouse skin or cells in culture with retroviruses that carry mutant *ras* genes can replace carcinogen treatment (Roop et al. 1986; Brown et al. 1986). More recently, mice containing mutant Ha-*ras* transgenes, which are expressed under the regulation of epidermal-specific keratin promoters, have been produced (Bailleul et al. 1990; Greenhalgh et al. 1993). These mice develop benign tumours upon a promotional stimulus such as wounding or mechanical irritation. One of the effects of mutant *ras* expression is the induction of TGF-α production (Glick et al. 1991), and it is noteworthy that a similar phenotype has been described in mice which contain a TGF-α transgene regulated by a keratin promoter element (Vassar et al. 1992).

The role of mutant *ras* in mouse skin carcinogenesis is not restricted to the initiation stage, however, and changes in mutant gene copy number and the relative expression levels of normal and mutant alleles are frequently detected during both benign tumour development and malinant progression (Bremner, Balmain, 1990; Buchmann et al. 1991). It would appear that mutant *ras* expression levels increase progressively in a stepwise fashion during the development of malignancy.

Recessive Changes in Tumour Progression

One drawback of the standard mouse skin model lies in the the use of inbred mouse strains. This forbids analysis of suppressor gene loss, usually detected by loss of heterozygosity (LOH), since inbred mice strains are by their nature homozygous at each allele. The use of interspecific *M. Spretus/M. Musculus* hybrids in skin carcinogenic experiments, however, overcomes this drawback and enables LOH analysis to be applied to experimental skin tumours. The speed and power of such analysis has been greatly increased by the use of microsatellite polymorphisms which are detected by the polymerase chain reaction (PCR) (Cornall et al. 1991). Using this approach, a large number of skin tumours have been analysed for LOH. Markers on chromosomes 1, 4, 6, 7, 11, and 17 were frequently imbalanced or lost, whereas the majority of markers on chromosomes 2, 8, 9, 12, 14, 16, 18, and 19 retained heterozygosity (C. Kemp *et al*, in preparation). Chromosome 7 imbalance was the most frequent change observed, and is detected in almost all papillomas and carcinomas which contain mutant *ras* genes. The Ha-*ras* allele is located on chromosome 7 and in every case, the chromosome carrying the mutant allele was over-represented. These results confirm previous findings that trisomy of chromosome 7 by non-disjunction is a frequent and early event in the genesis of skin tumours. The tumour promoter TPA has been shown to induce chromosomal changes in cells *in vitro* and in mouse epidermis *in vivo* (Furstenberger et al. 1989), and may be responsible for the induction of chromosome 7 trisomy. Interestingly the non-tumour-promoting hyperplasiogenic agent retinoyl-phorbol-acetate does not induce chromosomal changes and does not promote papilloma development upon treatment of initiated mouse skin (Furstenberger et al. 1989). It is tempting to propose that trisomy of chromosome 7, with duplication of the mutant Ha-*ras* allele, represents the *conversion step* proposed to be an early event in the clonal selection of initiated cells (Slaga, 1985).

The role of p53

Chromosome 11 was also found to show frequent LOH, with the loss being detected in carcinomas but not in papillomas. As the murine p53 gene is located on this chromosome, further study was undertaken to determine whether mutational inactivation of p53 was associated with the progression of mouse skin tumours. Direct sequencing of the p53 gene from a number of carcinomas, homozygous for p53, showed point mutations present in the gene (Burns et al. 1991). Interestingly, sequence analysis of p53 from a tumour, induced by repeated carcinogen treatment, which displayed no LOH, showed both alleles to be independantly mutated. No changes were found in any of the benign papillomas examined. Thus, p53 loss appears to be a late event occuring at the stage of progression from benign to malignant tumours.

These findings have been extended by using p53 *knock-out* mice, developed by Donehower et al (1992), in two-stage skin carcinogenesis experiments (C. Kemp et al, in press). No tumours were induced in homozygous null mice by TPA treatment alone, indicating that absence of p53 cannot function as an initiating event. The number, size, and growth rate of benign papillomas was not increased in the p53 heterozygous mice compared to wild-type, however there was an icrease in progression frequency and this was associated with the loss of the remaining wild-type allele. In contrast, the p53 null mice showed a decrease in papilloma yeild but these underwent rapid progression to produce poorly differentiated carcinomas. All of the tumours produced in the p53 null transgenics showed the same genetic alterations as had previously been seen in wild-type mice. Thus p53 loss, despite being the first genetic change in this system, is functionally dormant and does not lead to a selective advantage in the early steps of tumour formation. Other genetic alterations are required before such an advantage is conferred. In addition, p53 loss cannot function as an initiating event in alternative routes to tumour formation.

Location of the Target Cell

One unresolved problem in the mouse skin system concerns the nature and location of the cell which becomes initiated. The epidermis is comprised of several compartments which differ in the degree of differentiation and the proliferative capacity of the component cells. It is still unclear as to where in the skin the cell, which gives rise to the tumour, is located. The target cell was thought to belong to a stem cell compartment, since initiated cells can remain dormant in the skin for long periods. This would place the initiated cell in the basal layer of the epidermis or in the hair follicle (Sun et al. 1991; Cotsarelis et al. 1990). The studies involving transgenic mice, in which *ras* gene expression is restricted to the differentiating suprabasal layer of the skin (Bailleul et al. 1990) would counter this proposal. Committed cells within the suprabasal compartment can be reprogrammed by the expression of a mutant *ras* gene and can give rise to papillomas. Similar findings have been reported for mutant *ras* expression in both the basal and suprabasal layers of the skin (Greenhalgh et al. 1993). In both studies, no progression to malignancy was observed. There has been some evidence to support the existence of terminally benign papillomas in mouse skin, which cannot progress to squamous carcinomas, together with a subset with increased progression frequency (Burns et al. 1976; Aldaz et al. 1988; Hennings et al. 1985). It is tempting to postulate that terminally benign papillomas arise from the differentiating compartment of the skin, and those papillomas which can convert to squamous carcinomas arise from stem cell compartments, such as the bulb region of the hair follicle.

Conclusions

It is now apparent that tumour evolution proceeds through a series of genetic events which can be correlated with altered biological properties of the developing tumour cells. Much has been achieved in identifying those genes which can function as individual components in the formation of various tumours. It now remains to elucidate the various networks and combinations of alterations which are capable of producing neoplasia.

References

Aaltonen LA, Peltomäki P, Leach FS, Sistonen P, Pylkkänen L, Mecklin J-P, Järvinen H, Powell SM, Jen J, Hamilton SR, Petersen GM, Kinzler KW, Vogelstein B, De la Chapelle A (1993) Clues to the pathogenesis of familial colorectal cancer. Science 260:812-816

Aldaz CM, Conti CJ, Larcher F, Trono D, Roop DR, Chesner J, Whitehead T, Slaga TJ (1988) Sequential development of aneuploidy, keratin modifications, and gamma-glutamyltransferase expression in mouse skin papillomas. Cancer Res 48:3253-3257

Bailleul B, Surani MA, White S, Barton SC, Brown K, Blessing M, Jorcano J, Balmain A (1990) Skin hyperkeratosis and papilloma formation in transgenic mice expressing a *ras* oncogene from a suprabasal keratin promoter. Cell 62:697-708

Balmain A, Kemp CJ, Burns PA, et al (1992) Functional Loss of Tumour Suppressor Genes in Multistage Chemical Carcinogenesis. In: Harris CC, Hirohashi S, Ito N, et al (eds)Multistage Carcinogenesis. Japan Sci.Soc.Press, Tokyo/CRC Press, Boca Raton, pp 97-108

Berns A, Breuer M, Verbeek S, van Lohuizen M (1989) Transgenic mice as a means to study synergism between oncogenes. Int J Cancer Supplement 4:22-25

Bishop JM (1987) The molecular genetics of cancer. Science 235:305-311

Bizub D, Wood AW, Skalka AM (1986) Mutagenesis of the Ha-*ras* oncogene in mouse skin tumors induced by polycyclic aromatic hydrocarbons. Proc Natl Acad Sci USA 83:6048-6052

Boutwell RK, Verma AK, Ashendel CL, Astrup E (1982) Mouse skin: a useful model system for studying the mechanism of chemical carcinogenesis. Carcinog Compr Surv 7:1-12

Bremner R, Balmain A (1990) Genetic changes in skin tumour progression: correlation between presence of a mutant *ras* gene and loss of heterozygosity on mouse chromosome 7. Cell 61:407-417

Breuer ML, Cuypers HT, Berns A (1989) Evidence for the involvement of pim-2, a new common proviral insertion site, in progression of lymphomas. EMBO J 8:743-748

Brown K, Quintanilla M, Ramsden M, Kerr IB, Young S, Balmain A (1986) v-*ras* genes from Harvey and BALB murine sarcoma viruses can act as initiators of two-stage mouse skin carcinogenesis. Cell 46:447-456

Brown K, Buchmann A, Balmain A (1990) Carcinogen-induced mutations in the mouse c-Ha-*ras* gene provide evidence of multiple pathways for tumour progression. Proc Natl Acad Sci USA 87:538-542

Buchmann A, Ruggeri B, Klein-Szanto AJP, Balmain A (1991) Progression of squamous carcinoma cells to spindle carcinomas of mouse skin is associated with an imbalance of H-*ras* alleles on chromosome 7. Cancer Res 51:4097-4101

Burns FJ, Vanderlaan M, Sivak A, Albert RE (1976) Regression kinetics of mouse skin papillomas. Cancer Res 36:1422-1427

Burns PA, Kemp CJ, Gannon JV, Lane DP, Bremner R, Balmain A (1991) Loss of heterozygosity and mutational alterations of the p53 gene in skin tumors of interspecific hybrid mice. Oncogene 6:2363-2369

Cornall RJ, Aitman TJ, Hearne CM, Todd JA (1991) The generation of a library of PCR-analyzed microsatellite variants for genetic mapping of the mouse genome. Genomics 10:874-881

Cotsarelis G, Sun TT, Lavker RM (1990) Label-retaining cells reside in the bulge area of pilosebaceous unit: implications for follicular stem cells, hair cycle, and skin carcinogenesis. Cell 61:1329-1337

Donehower LA, Harvey M, Slagle BL, McArthur MJ, Montgomery CA,Jr., Butel JS, Bradley A (1992) Mice deficient for p53 are developmentally normal but susceptible to spontaneous tumours. Nature 356:215-221

Fearon ER, Vogelstein B (1990) A genetic model for colorectal tumorigenesis. Cell 61:759-767

Furstenberger G, Schurich B, Kaina B, Petrusevska RT, Fusenig NE, Marks F (1989) Tumor induction in initiated mouse skin by phorbol esters and methyl methanesulfonate: correlation between chromosomal damage and conversion ('stage I of tumor promotion') in vivo. Carcinogenesis 10:749-752

Glick AB, Sporn MB, Yuspa SH (1991) Altered regulation of TGF-beta 1 and TGF-alpha in primary keratinocytes and papillomas expressing v-Ha-*ras*. Mol Carcinog 4:210-219

Greenhalgh DA, Rothnagel JA, Quintanilla MI, Orengo CC, Gagne TA, Bundman DS, Longley MA, Roop DR (1993) Induction of Epidermal Hyperplasia, Hyperkeratosis, and Papillomas in Transgenic Mice by a Targeted v-Ha-*ras* Oncogene. Molecular Carcinogenesis 7:99-110

Hennings H, Shores R, Mitchell P, Spangler EF, Yuspa SH (1985) Induction of papillomas with a high probability of conversion to malignancy. Carcinogenesis 6:1607-1610

Hollstein M, Sidransky D, Vogelstein B, Harris CC (1991) p53 mutations in human cancers. Science 253:49-53

Kemp C.J., Donehower L.A., Bradley A., Balmain A Reduction of p53 gene dosage does not increase initiation or promotion but greatly enhances malignant progression of chemically induced skin tumours. Cell, in press.

Klein-Szanto AJP, Larcher F, Bonfil RD, Conti CJ (1989) Multistage chemical carcinogenesis protocols produce spindle cell carcinomas of the mouse skin. Carcinogenesis 10:2169-2172

Knudson AG (1973) Mutation and human cancer. Adv Cancer Res 17:317-352

Land H, Parada LF, Weinberg RA (1983) Tumorigenic conversion of primary embryo fibroblasts requires at least two cooperating oncogenes. Nature 304:596-602

Meuth M (1990) The structure of mutation in mammalian cells. Biochim Biophys Acta 1032:1-17

Nowell PC (1989) The clonal nature of neoplasia. Cancer Cells 1:29-30

Peltomäki P, Aaltonen LA, Sistonen P, Pylkkänen L, Mecklin J-P, Järvinen H, Green JS, Jass JR, Weber JL, Leach FS, Petersen GM, Hamilton SR, De la Chapelle A, Vogelstein B (1993) Genetic mapping of a locus predisposing to human colorectal cancer. Science 260:810-812

Quintanilla M, Brown K, Ramsden M, Balmain A (1986) Carcinogen-specific mutation and amplification of Ha-*ras* during mouse skin carcinogenesis. Nature 322:78-80

Renan MJ (1993) How many mutations are required for tumorigenesis? Implications from human cancer data. Mol Carcinog 7:139-146

Roop DR, Lowy DR, Tambourin PE, Strickland J, Harper JR, Balaschak M, Spangler EF, Yuspa SH (1986) An activated Harvey *ras* oncogene produces benign tumours on mouse epidermal tissue. Nature 323:822-824

Slaga TJ (1985) Mechanisms involved in multistage skin tumorigenesis. Carcinog Compr Surv 10:189-199

Stanbridge EJ (1990) Human tumor suppressor genes. Annu Rev Genet 24:615-657

Sun TT, Cotsarelis G, Lavker RM (1991) Hair follicular stem cells: the bulge-activation hypothesis. J Invest Dermatol 96:77S-78S

Thibodeau SN, Bren G, Schaid D (1993) Microsatellite instability in cancer of the proximal colon. Science 260:816-819

Vassar R, Hutton ME, Fuchs E (1992) Transgenic overexpression of transforming growth factor alpha bypasses the need for c-Ha-*ras* mutations in mouse skin tumorigenesis. Mol Cell Biol 12:4643-4653

Transgenic Mouse Models in Chemical Carcinogenesis Studies

R. Tennant

Laboratory of Environmental Carcinogenesis and Mutagenesis, National Institute of Environmental Health Sciences, 111 Alexander Drive, Research Triangle Park, NC 27709

Introduction

Cancer research today faces some important dilemmas in the search for causes of cancer. It is well known that certain environmental factors, such as particular chemicals can be a cause of cancers in humans, but that results derived from humans, as well as experimental models, have shown that there is an important genetic aspect to the disease. The genetic component involves many different genes that influence susceptibility to carcinogens and tumor progression, and other genes that may be mutated by chemicals which initiate carcinogenesis, or genes lose the capacity to regulate cancer cells.

In order to identify potential carcinogens researchers have relied heavily on rodent bioassays. While such bioassays have detected most of the chemicals or agents known to be associated with human cancers (Ashby and Tennant 1991), another dilemma is faced with chemicals that induce relatively weak or highly specific effects in bioassays. Since rodents are inbred and differ from humans in many genetic traits, there is uncertainty about the actual risk to humans posed by such highly specific carcinogens. Thus, the presumption has long been held that resolution of these dilemmas can occur only by understanding mechanisms by which chemicals induce, promote, or affect the progression of tumors. However, the complexity of the number of events and genes involved in the neoplastic process has made the identification of mechanisms a formidable task. It was not surprising, therefore, that researchers reacted so favorably to the development of transgenic mice.

Methodologies for manipulating genes in bacteria, viruses and other microbes had proven very successful, for example, in the development of the Salmonella strains used in the Ames mutagenicity assay (Ames et al.. 1973). Development of methods by which genes could be transferred into the germ line of mice has opened many avenues of research in carcinogenesis since it is possible to not only evaluate the role of specific genes in neoplasia but to also alter the expression or regulation of specific genes while the genetic background of the animals remains the same. There have been some recent reviews of the applications of transgenic mouse models in cancer research (Adams and Cory 1991; Fowlis and Balmain, 1993; Bradley 1993). This report will deal with two specific applications of transgenic mouse models to better define mechanisms of carcinogenesis and to identify potential carcinogens. First, the effects of selected chemicals in transgenic mice with oncogenes which are targeted for tissue-specific expression will be presented. Second, a model for studying mechanisms of tumor promotion and for the identification of tumor promoters will be discussed. Models developed for identifying tissue-specific mutagenesis will not be discussed in this presentation, but they also offer an important tool in studying carcinogenesis and mutagenesis on a tissue-specific basis (Gossen and Viig 1993).

Chemical Effects in Transgenic Mice with Oncogenes Constitutively Expressed in Specific Tissues

Mice that carried oncogenes that were targeted for expression in specific tissues were among the first transgenic models created (Brinster et al.. 1981; Stewart et al.. 1984; Leder et al.. 1986. Such models potentially can be used to study molecular and cellular pathways involved in the development of specific tumors, and also could possibly be used to aid in identifying potential carcinogens. Since a transgenic oncogene could provide the initiation step for a neoplastic process, it is possible that carcinogenic chemicals could mutate important (tumor suppressor) genes, activate or down regulate other important genes, stimulate cell proliferation or alter cellular differentiation or act through some unknown mechanisms. Such effects could result in either an acceleration of tumor development, or an increase in tumor incidence and thus provide a more sensitive assay for potential carcinogens. The mice chosen for our initial studies were three lines developed by Dr. Philip Leder and his colleagues, and characteristics of these lines has been described elsewhere (Stewart et al.. 1984;Leder et al. 1986;Sinn et al., 1987;Muller et al. 1988). Briefly, the lines each carried the v-Ha-ras, c-myc and activated c-neu oncogenes under the regulatory control of sequences derived from the mouse mammary tumor virus (MMTV). All three of the lines carried the respective transgene on an FVB/N background and were maintained under pathogen-free conditions. The lines carried the following designations; v-Ha-ras (TG.SH), c-myc (TG.M) and c-neu (TG.NK).

A major consideration in our studies involved the selection of chemicals with which to evaluate the response of the mice. The National Toxicology Program has evaluated over 400 chemicals in long term bioassays conducted in mice and rats and over 150 have demonstrated carcinogenic activity. The strategy for choosing chemicals for this study involved two primary questions: a) could a mutagenic carcinogen which was not active in mammary or other glucocorticoid-sensitive tissues induce an altered tumor response in the transgenic lines, and b) would the carcinogenic potential of a known mouse mammary gland carcinogen be reflected in an accelerated or altered pattern of tumor induction? The mutagen selected for these studies was p-Cresidine which had been shown to induce urinary bladder tumors in both mice and rats. The mammary gland carcinogen was reserpine, which is not mutagenic. Doses of the chemicals chosen for administration via feed included those at which tumors were induced during 2 year exposures in $B6C3F_1$ mice (Ashby and Tennant 1991). The duration of exposure of the chemicals was 9 months for the

TG.NK and TG.M lines, but was only 6 months in the TG.SH line due to the earlier development of mammary cancers in this line.

The results of this study have been published (Tennant et al. 1993) and are only briefly summarized in Tables 1 and 2. In all three lines exposed to p-Cresidine there were two discrete patterns of tumors. Transitional cell carcinomas of the urinary bladder were seen in both dose groups of the three transgenic lines and in the FVB/N nontransgenic control mice. No differences in tumor type or rate of tumor development were seen. Only the transgenic mice developed mammary gland adenocarcinomas and p-Cresidine did not affect these tumors since the tumor incidence between 0 dose controls and dosed groups was similar (Table 1). Tumors of the salivary gland developed in some of the transgenic mice but were not affected by p-Cresidine exposure.

The effects of reserpine were seen only in transgenic mice. The duration of exposures (6 and 9 months) in these studies was not sufficient to induce effects in the FVB/N controls similar to those seen in $B6C3F_1$ mice which were exposed for 104 weeks. The high incidences of "spontaneous" mammary tumors in the transgenic lines masked the potential effects of reserpine exposure. The principal effect was an increase in multiple mammary tumors in the TG.M and TG.NK lines (Table 2). There was also a significant increase in the incidence of mammary tumors in the TG.SH females given the 5 ppm dose.

The overall results of these studies show that transgenic models carrying constitutively expressed oncogenes that are targeted for tissue specific expression present both opportunities and problems. Such models may be most useful in studies directed toward identifying events in tumor progression. The time dependent nature of transgene-induced tumors indicates that subsequent epigenetic or genetic events are required in order for the expressed oncogene to manifest its tumorigenic potential. Also, these models may provide important systems with which to identify potential therapeutic agents or intervention treatments. The spontaneous but predictable pattern of tissue specific carcinogenesis may be useful in evaluating therapeutics. The problems presented by these models are most evident in attempts to use them to identify potential carcinogens and to supplement or supplant conventional rodent bioassays. Constitutively expressed oncogenes result in relatively high levels of "spontaneous" tumors often with onset at relatively early age. These tumors present problems in determining the sensitivity and specificity of the model for response to known carcinogens and non carcinogens. In addition, many of the regulatory sequences used to target transcriptional expression of the transgene result in such a high penetrance of the transgene that other chemical effects may be masked.

Inducible Transgene Expression

The most interesting studies that we have conducted thus far involve another line of transgenic mice developed in the Leder laboratory (Leder et al.. 1990) where the v-

Table 1 Tumor Incidence in Mice Treated with p-Cresidine

		Non-Transgenic	Transgenic	
Sex/Strain	Dose (ppm)	Urinary Bladder	Urinary Bladder	Mammary Gland
Male				
	0	0/18	0/11	5/8
TG•SH	2500	8/18	4/11	2/6
	5000	10/13	7/16	3/10
	0	0/11	0/17	0/6
TG•NK	2500	10/14	12/14	0/6
	5000	13/15	13/15	0/4
	0	0/16	0/11	0/5
TG•M	2500	12/15	13/13	1/8
	5000	16/16	14/14	0/5
Female				
	0	0/10	0/17	5/17
TG•SH	2500	6/19	4/9	4/9
	5000	8/11	8/17	4/18
	0	0/11	0/8	7/13
TG•NK	2500	12/17	5/7	6/9
	5000	16/17	8/12	10/13
	0	0/10	0/12	9/19
TG•M	2500	12/16	10/13	7/14
	5000	16/17	6/11	0/12

Table 2 Tumor Incidence in Transgenic Mice Treated with Reserpine

Sex/Strain	Dose (ppm)	Mammary Gland	Multiple Mammary Gland
Male			
	0	3/13	0/13
TG•SH	5	3/10	1/10
	10	5/8	2/8
	0	1/6	0/6
TG•NK	5	1/10	0/10
	10	2/5	0/5
	0	0/6	0/6
TG•M	5	0/6	0/6
	10	1/8	0/8
Female			
	0	2/12	0/12
TG•SH	5	10/15	4/15
	10	5/13	0/13
	0	8/13	6/13
TG•NK	5	14/14	14/14
	10	12/12	12/12
	0	11/17	4/17
TG•M	5	14/14	11/14
	10	13/14	9/14

Ha-ras construct used in the TG.SH line mice was again utilized. However, this time the oncogene was fused to the promoter of a mouse fetal globin gene. The four founder lines derived from this construct designated as TG.AC showed low level fetal hematopoietic expression of the transgene and low level expression in adult bone marrow. However it was the development of papillomas at sites of cutaneous wounds that signaled the phenotype of this line. Subsequently Leder and his colleagues demonstrated progeny of the TG.AC line were sensitive to induction of papillomas by cutaneous application of the phorbol esther TPA. Particularly impressive is the rapid onset of papilloma development in response to relatively low doses of TPA (Table 3) and persistence and even the development of additional papillomas in the absence of further TPA treatments. Thus, in the context of the "two stage" model of carcinogenicity these animals can be categorized as being genetically initiated by the transgene. Therefore, it appeared that they may be useful not only for the study of mechanisms of tumor promotion and progression but also to identify chemicals that may have tumor promoting or carcinogenic potential. Since the development of papillomas is very apparent and not an invasive procedure we initiated studies to explore the potential use and application of the TG.AC transgenic line.

Initial studies evaluated the responsiveness of heterozygous TG.AC mice to the promoting activity of TPA and two other phorbol promoters (2-butanone peroxide and benzol peroxide). Also included in the study were acetone controls, since acetone was used as a solvent for the promoters. In order to conduct studies in mice that were heterozygous for the transgene it was necessary to genotype the animals using DNA extracted from tail biopsies. This then allowed the concurrent nontransgenic controls to be littermates. In none of the these studies did any of the nontransgenic mice respond to chemical exposure, while virtually all of the transgenic littermates responded to all three tumor promoters.

Subsequently similar studies were done with homozygous TG.AC mice (Table 3), (Spalding et al.. 1993) TPA induced the highest incidence of papilloma development by five weeks of exposure but by 20 weeks of exposure the response to 2-butanone peroxide was comparable. No papillomas were induced by exposure to acetone or acidic acid. Thus these results showed that the TG.AC line is highly and specifically sensitive known tumor promoters. Benzol peroxide is a free radical generating chemical that has been shown to have promoting activity in DMBA-initiated SENCAR mice, while 2-butanone peroxide has been described as a weak promoter. Development of TPA-induced papillomas showed that a total dose of 25 to 30 micrograms administered between 3 and 10 applications induced an average of 11 to 15 papillomas per mouse. No quantitative differences were seen between the response of heterozygous and homozygous TG.AC mice in their response to various promoters administered to comparable doses. In order to evaluate the potential systemic effects of the transgene, TG.AC mice were given weekly IP injection of urethane for 20 weeks. Urethane did not induce skin papillomas and the incidence of lung tumors induced both in the TG.AC and FVB/N control mice were comparable. These results suggests that the effects of the transgene are preferentially

Table 3 **Comparison of Tumor Promoting Activity of Different Promoters in Homozygous TG•AC Mice**

Promoter[1]			TTFT[2]	Average Papilloma Incidence/Mouse 20[3]
TPA	2.5 μg	M[4]	3	55
		F	3	60
2-BUP	5.0 mg	M	3	77
		F	5	50
BPO	5.0 mg	M	7	1
		F	7	30
BPO	10.0 mg	M	6	4
		F	7	40
Acetic Acid	30.0 mg	M	4	<1.0
		F	0	0.0
Acetone (control)	200 μl	M	0	0.0
		F	0	0.0

[1] Animals were treated twice weekly for 20 weeks with the indicated promoter in 200 μl of acetone
[2] Time to first tumor in weeks
[3] Number of weeks
[4] Three male (M) and 3 female (F) mice in each dose group

expressed principally in the skin but studies with a broader range of systemic carcinogens will be required to verify this.

When animals are housed separately, or females group housed, where there is no fighting the spontaneous incidence of skin papillomas is near zero. Some small papillomas appear on lips, eye lids and genitalia as the animals age and odontogenic tumors have been identified in the jaws of some animals over 20 weeks of age. Most or all of these neoplasms may be related to tissue injury. Full thickness surgical incisions result in papilloma development along the wound line. The papillomas continue to develop after the wound is fully healed and persist for long periods of time in both surgically treated and chemically treated animals. Malignancies have been observed in association with some papillomas. Results thus far indicate that at least a third of mice that develop papillomas also subsequently develop a malignancy. These skin malignancies have been identified by Cardiff (Cardiff et al. 1993) as spindle carcinomas, which we refer to as fibrosarcomas. These malignancies form an important characteristic of this line since they are in contrast to the epithelial derived squamous cell carcinomas that have been identified to occur most often in conventional 2-stage initiation promotion experiments in SENCAR and other mouse strains. The development of squamous cell carcinomas from the squamous papillomas led to the concept of tumor progression and conversion. Development of the mesenchymally derived fibrosarcomas or spindle cell tumors in the TG.AC line represents malignancies of a different tissue origin. It will be important to understand the relationship between papilloma development, transgene expression in the dermis and the evolution of these malignancies.

Evidence that the induction of papillomas is directly related to expression of the transgene derives from immunoprecipitation and western analysis of homogenates of normal TG.AC skin and epidermal papillomas of dorsal skin and papillomas using an anti-ras (Y13-259 Oncogene Science, Inc.) antibody. Endogenous P21 was distinguished from the transgene product based upon differences in electrophoretic mobility compared to standards derived from a v-Ha-ras overexpressing NIH 3T3 cell line (Hansen and Tennant 1993). These analyses failed to reveal any examples of normal dorsal skin that expressed the transgene product, while abundant a mutated P21 was detectable in samples of different epidermal papillomas and even TPA treated skin that demonstrated epidermal hyperplasia did not express the transgene product. These results therefore suggest that focal induction of transgene expression, either by treatment with tumor promoters or physical wounding, results in the development of papillomas. Studies are currently in progress to analyze the nature of the stimuli that can induce transgene expression and the relationship between transgene expression, epidermal and dermal proliferation and the development of the malignant fibrosarcomas.

Conclusions

These results demonstrate that the TG.AC transgenic line is a valuable model with which to study both genetic and epigenetic aspects of papilloma and malignant tumor development. The rapid onset and high degree of synchrony of tumor development that can be achieved in this line provides a unique opportunity for analyzing early events in both the development of papillomas and in the emergence of the malignancies. In addition, this model shows considerable promise for the study of chemicals that may be tumor promoters or non-mutagenic carcinogens. Studies are currently underway with a number of additional chemicals to determine the range of sensitivity and specificity in the responsiveness of the line. A model such as the TG.AC in which visually observable tumors can be identified on a high sensitive background could provide a useful means for complementing long term rodent bioassays and aid in the recognition of chemicals that may have carcinogenic potential.

References

Adams JM and Cory S (1991) Transgenic models of tumor development. Scienc 254:1161-1166.

Ashby, J and Tennant RW (1991) Definitive relationships among chemical structure, carcinogenicity and mutagenicity for 301 chemicals tested by the US NTP. Mutat. Res. 257:229-306.

Bradley A (1993) Site directed mutagenesis in the mouse. Recent Prog Horm Res 48:237-251.

Brinster RL, Chen HY, Trumbauer M, Senear AW, Warren R and Palmiter RD (1981) Somatic expression of herpes thymidine kinase in mice following injection of a fusion gene into eggs. Cell 27:223-231.

Cardiff RD, Leder A, Kuo A, Pattengale PK and Leder P (1993) Multiple tumor types appear in a transgenic mouse with the RAS oncogene. Am J Path 142:1199-1207.

Fowlis DJ and Balmain A (1993) Oncogenes and tumor suppressor genes in transgenic mouse models of neoplasia. Eur J. Cancer 29A: 638-645.

Gossen J and Viig J (1993) Transgenic mice as model systems for studying gene mutations in vivo. T Pers. Genet 9:27-31.

Hansen L and Tennant R. Focal transgene expression associated with papilloma development in v-Ha-ras transgenic TG.AC mice. Mol Carcinog.

Leder A, Juo A, Cardiff RD, Sinn E and Leder P (1990) v-Ha-ras transgene abrogates the initiation step in mouse skin tumorigenesis: effects of phorbol esters and retinoic acid. Proc. Natl. Acad. Sci. USA 87:9178-9182.

Leder A, Pattengale PK, Kuo A, Stewart T and Leder P (1986) Consequences of widespread deregulation of the c-myc gene in transgenic mice: multiple neoplasms and normal development. Cell 45:485-495.

Muller WJ, Sinn E, Pattengale PK, Wallace R and Leder P (1988) Single-step induction of mammary adenocarcinoma in transgenic mice bearing the activated c-neu oncogene. Cell 54:105-115.

Sinn E, Muller W, Pattengale P, Tepler I, Wallace R and Leder P (1987) Co-expression of MMTV/v-Ha-ras and MMTV/C-myc genes in transgenic mice: synergistic action of oncogenes in vivo. Cell 49:465-475.

Spalding JW, Momma J, Elwell M and Tennant RW (in press) Chemically-induced skin carcinogenesis in a transgenic mouse line (TG.AC) carrying a v-Ha-ras genes. Carcinogenesis, in press, 1993.

Stewart TA, Pattengale PK, and Leder P (1984) Spontaneous mammary adenocarcinomas in transgenic mice that carry and express MTV/Myc fusion genes. Cell 38:627-637.

Tennant RW, Rao GA, Russfield A, Seilkop S. Braun AG (1993) Chemical effects in transgenic mice bearing oncogenes expressed in mammary tissue. Carcinogenesis 14:29-35.

CELL PROLIFERATION AND CELL DEATH IN RAT LIVER CARCINOGENESIS BY CHEMICALS

G.M.Ledda-Columbano, P.Coni and A.Columbano

Istituto di Patologia Sperimentale, University of Cagliari, Via Porcell 4, 09124 Cagliari, Italy

Cell proliferation has often been implicated to play a critical role in the carcinogenic process. In the liver, for example, it is believed that at least one round of cell division is necessary for the accomplishment of the first step of the carcinogenic process, initiation, probably by allowing fixation of carcinogen-induced critical lesions into the newly made DNA (Cayama et al. 1978; Ishikawa et al. 1980; Columbano et al. 1981). In the promotion stage, enhanced cell division may be a stimulus for a clonal expansion of initiated hepatocyte (Pound and McGuire 1978a; Dragani et al. 1986; Farber and Sarma 1987; Columbano et al. 1990). The finding that several non-genotoxic agents which induce hepatic growth also induce liver tumor formation, has renewed interest about the role of cell proliferation in tumor development (Rao and Reddy 1987; Mirsalis 1987; Marsman et al. 1988;). Thus, in the past few years the manner of conducting and interpreting rodent cancer biossays (based on treatment of the animals with the Maximum Tolerated Dose (MTD)), has been criticized because chemically-induced cell proliferation is often seen at the MTD. It has been postulated that this increased proliferation per se plays a major role in tumor formation and that doses that do not induce cell proliferation would not be carcinogenic (Ames and Gold 1990; Cohen and Ellwein 1990).

However, a clear-cut correlation between cell proliferation in a target organ and tumor formation has not yet been established; it is known that while there may be several examples where induced cell proliferation correlates with cancer, there are many where it does not (Columbano et al. 1987; Eldridge et al. 1992; Melnick 1992). In addition, most of the non-genotoxic mitogenic compounds, with the exception of WY,14643 (Marsman et al. 1988), induce only a very transient increase in cell proliferation (Yeldandi et al. 1989; Eacho et al. 1991). This raises the question whether such a transient induction of liver cell proliferation might be related to a late event such as tumor formation. The situation is further complicated by the finding that the nature of the proliferative stimulus may constitute an important factor in the carcinogenic process in the liver. In fact, acute proliferative stimuli provided by primary mitogens, unlike compensatory regeneration such as that induced by partial hepatectomy (PH) or CCl_4, do not support initiation of chemical hepatocarcinogenesis (Columbano et. 1987; Ledda-Columbano et al. 1989). In addition, mitogen-induced liver growth also fails to exert a promoting effect on the growth of carcinogen-initiated cells (Columbano et al. 1990).

Compensatory Regeneration vs Mitogen-Induced Liver Growth

In adult organisms, hepatocytes have a long life span, ranging from 200 to 400 days or more. However, it is well known that under particular conditions liver cells are able to divide. Basically, liver cell proliferation may be induced by two different mechanisms: compensatory regeneration and direct hyperplasia. While a great number of studies have described the sequence of events taking place during compensatory regeneration following PH, very few have been undertaken to characterize at a biochemical and molecular level the liver growth induced by primary mitogens.

Figure 1 schematically illustrates some of the differences exhibited in these two proliferative stimuli. When cell proliferation occurs by a compensatory mode such as that induced by PH or CCl_4, cell loss is the primary event and cell proliferation is triggered in order to replace the cells that were lost. The proliferative process is a very controlled one; once the number of the cells or DNA content is back to normal the process stops.

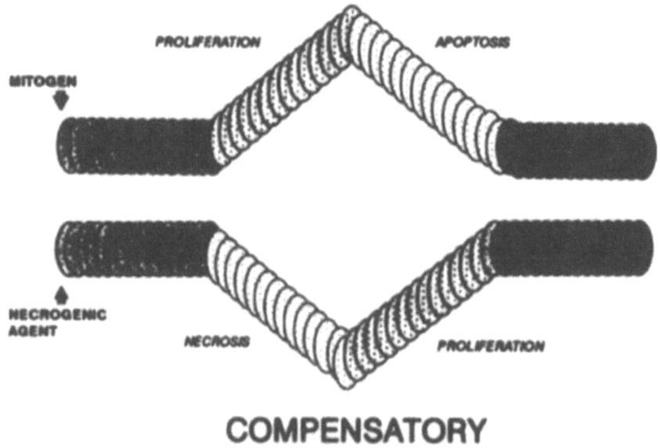

Fig.1. Schematic representation of two different types of cell proliferation

When cell proliferation is induced by primary mitogens, the situation is quite different. In fact, DNA synthesis is induced in the absence of any previous cell loss, suggesting that these chemicals can possibly interfere with growth control regulatory mechanisms normally operating in the organ. The consequence is that the mitogenic stimulus which is now the primary event results in an excess of the liver mass, excess DNA and presumably excess of cells. Interestingly, as soon as the mitogenic stimulus is withdrawn, liver mass and DNA content go back to their original values, indicating that this regression process, as the proliferative process seen during regeneration, is also a well controlled process. The rapid elimination observed during the regression phase appears to be due to a particular type of cell death, namely apoptosis (Wyllie et al. 1980).

Compensatory Regeneration vs Mitogen-Induced Hyperplasia: Growth Factors and Immediate Early Genes

It has been suggested that Hepatocyte Growth Factor/Scatter Factor (HGF/SF) may be a critical hepatotrophic factor for the triggering of liver regeneration (Michalopoulos 1990; Asami et al. 1991). Indeed, HGF/SF mRNA and HGF/SF activities were found to be markedly induced in the liver or plasma of rats after various insults. It was also recently proposed that a protein, injurin, appearing in the blood of rats subjected to injury may be involved in the regeneration of the organ through its ability to stimulate the synthesis of HGF/SF in distant sites such as the lung (Matsumoto et al. 1992). Therefore, it was of interest to us to determine whether HGF mRNA expression could be induced also under conditions where no organ injury was occurring. Preliminary studies indicate that liver growth induced in Wistar rats by the mitogen Lead Nitrate (LN) (Columbano et al. 1985), unlike compensatory regeneration induced by PH or CCl_4, does not induce the expression of hepatic HGFmRNA at any time after treatment (Shinozuka et al. 1993). This is in good agreement with recent data (Masuhara et al. 1992) showing that liver growth induced by the peroxisome proliferator BR931, unlike compensatory regeneration induced by a choline-deficient diet, also fails to induce HGF mRNA in rat liver. These results suggest that different growth factors may be involved in mitogen induced liver growth. Interestingly, a rapid and transient increase in the hepatic expression of tumor necrosis factor-alfa (TNF) was found following treatment with LN (Shinozuka et al. 1993). Whether cell proliferation induced by LN might be mediated by TNF is currently under investigation.

It is known that DNA synthesis induced in hepatocyte culture by addition of growth factors is preceded by an increased expression of some immediate early genes or primary response genes. A similar expression of these genes also occurs during liver regeneration following PH (Goyette et al. 1983; Makino et al. 1984; Morello et al. 1990). In order to study whether the differences seen in the expression of growth factors such as HGF could also result in a different pattern of expression of cell cycle related genes, we have examined the pattern of some genes during liver regeneration induced by PH or mitogen-induced liver growth induced by LN and other liver mitogens, ethylene dibromide (EDB), cyproterone acetate (CPA), nafenopine (NAF).

The results obtained indicate that while an increased expression of these genes is always seen during compensatory regeneration, no such increase is observed during liver growth induced by certain mitogens (Coni et al. 1990; Coni et al. 1993). No differences between the two proliferative stimuli were found when the expression of genes associated to later phases of the cell cycle was analyzed.

Liver response to mitogens thus appears to be different from that evoked by regenerative stimuli not only in terms of growth factors but also in the expression of immediate early genes.

Compensatory Regeneration, Mitogen-Induced Hyperplasia and Liver Ploidy

Recently, we have studied the effect of proliferative stimuli of different nature on the ploidy state of the liver. Normal adult rat liver is constituted predominantly by 4c cells (90%). A small percentage of diploid cells (8-10%) and octaploid cells (1%) is also present. Among the 4c cells, we have a 80% frequency of mononucleate hepatocytes and a 20% frequency of binucleate hepatocytes (BN) which are absent immediately after birth and appear at about 3 weeks after birth, immediately preceeding the occurrence of mononucleate tetraploid (MT) cells (Nadal and Zajedela 1966).

After PH there is a rapid decrease in the number of BN cells with an almost complete disappearance 3 days after operation (see Figure 2). At the same time an increase in the number of MT and MO cells is also observed. On the other hand, the ploidy pattern of the hepatocytes after LN is different. The most striking effect is a doubling in the population of binucleate hepatocytes. Interestingly, the increase is due to 4x2c and 8x2c hepatocytes, the incidence of 2x2c cells being decreased. Thus while liver regeneration is associated to a disappearance of binucleate cells, LN induced hyperplasia is accompanied by an increase in the percentage of these cells (Melchiorri et al.1993).

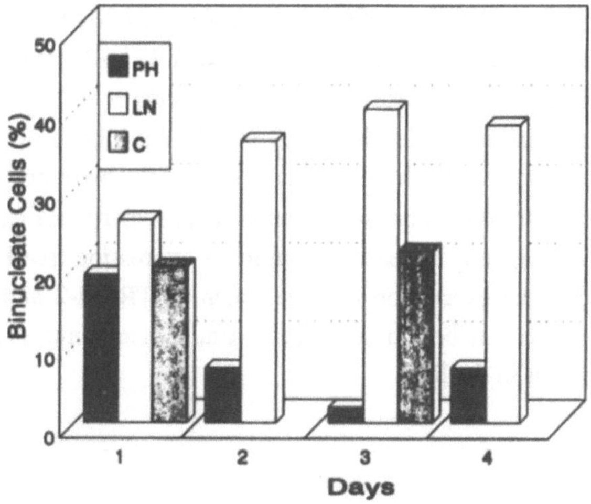

Fig.2. Percentage of binucleate hepatocytes (BN) 3 days after PH or LN

Mitogen-Induced Liver Growth and Apoptosis

Another aspect characteristic of the process of mitogen-induced hyperplasia is the rapid regression of the excess tissue once the mitogenic stimulus is withdrawn. We have shown that cellular elimination occurs via a particular mode of cell death, namely apoptosis, as indicated by the presence of scattered membrane-bounded globules, often containing fragments of nuclear material (Columbano et al. 1985). Besides the morphological features characteristic of apoptotic cell death, it is important to note that the apoptotic process does not occur soon after treatment with LN (therefore is not a consequence of the toxicity of the chemical), but only when the DNA content has reached its maximum and it stops once the DNA content has regained its original value (Columbano et al. 1985). This suggests that under our experimental conditions, apoptosis plays a homeostatic role.

Apoptosis has been proposed to be triggered by the activation of a gene programme; among the possible critical genes, TRPM2 has been associated to programmed cell death in various conditions (Legierm et al. 1987; Buttyan et al. 1989). Therefore, studies were undertaken to determine whether apoptosis occurring during involution of LN-induced hyperplasia could be associated with elevated levels of TRPM-2. However, our data indicate that no increase in the levels of TRPM-2 occurs during regression of liver hyperplasia (Fig 3a), suggesting that this gene is not involved in the apoptotic process taking place in these experimental conditions. A similar conclusion stems from studies by others wherein using the same experimental model the expression of TRPM-2 was measured by a more sensitive technique (in situ hybridization) (Helvering et al. 1993).

Recently, an increase in the levels of p53 mRNA have been observed in association with apoptotic cell death during involution of the prostate (Colombel et al. 1992). However, our studies on involution of liver hyperplasia caused by LN and CPA did not indicate any increased p53 mRNA expression (see Fig. 3b). This finding suggests that the expression of genes such as TRPM-2 and p53 during programmed cell death may be a tissue specific phenomenon and may not be induced during apoptosis in the liver.

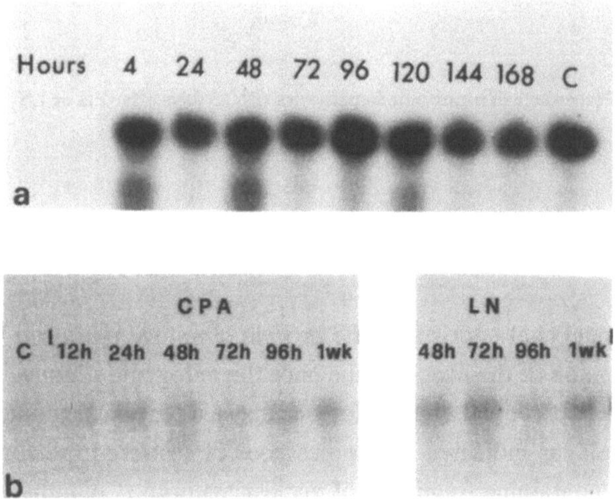

Fig.3. Expression of TRPM-2 (3a) or p53 (3b) mRNAs in rat liver at various time intervals after LN and CPA

CONCLUSIONS

We have shown some of the differences existing between compensatory regeneration and mitogen-induced liver growth. How these differences may affect the carcinogenic process induced in the liver by chemicals is not known. However, previous studies have indicated that acute treatment with primary mitogens, unlike compensatory regeneration, do not support initiation and do not efficiently promote the growth of initiated cells to form preneoplastic foci and/or nodules (Columbano et al. 1987; Ledda-Columbano et al. 1989; Columbano et al. 1990). Although at present we do not understand the mechanism(s) responsible for such a different effect, two possible explanations might be proposed: i) initiated cells while responding to regenerative stimuli, do not efficiently respond to the proliferative stimuli induced by primary mitogens or, alternatively, ii) initiated cells respond to mitogens as efficiently as normal hepatocytes but are eliminated (together with normal excess cells) by apoptosis during the regression phase of liver hyperplasia.

An in-depth study of the mechanisms underlying the differences between these two types of proliferative stimuli may help to clarify the mechanism(s) responsible for the inability of mitogen-induced cell proliferation to support the growth of carcinogen-initiated cells.

ACKNOWLEDGEMENTS

This work was supported by C.N.R (Progetto ACRO), A.I.R.C, and MURST (60%), Italy.

REFERENCES

Ames, B.N. and Gold,L.S. (1990) Too many rodent carcinogens: mitogenesis increases mutagenesis. Science, 249, 970-971.

Asami O, Ihara I, Shimidzu N, Tomita Y, Ichihara A, Nakamura T (1991) Purification and characterization of hepatocytes growth factor from injured liver of carbon tetrachloride-treated rats J Biochem 109:8-13.

Buttyan R, Olsson CA, Pintar J, Chang C, Bandyk M, Ng P-Y, Sawczuk IS (1989) Induction of the TRPM-2 gene in cells undergoing programmed cell death. Mol Cell Biol 9:3473-3481.

Cayama, E., Tsuda, H., Sarma, D.S.R., and Farber E. (1978) Initiation of chemical carcinogenesis requires cell proliferation. Nature, 275, 60-61.

Cohen, S.M. and Ellwein, L.B. (1990) Cell proliferation in carcinogenesis. Science 249, 972-975.

Colombel M, Olsson CA, Ng P-Y, Buttyan (1992) Hormone-regulated apoptosis results from reentry of differentiated prostate cell onto a defective cell cycle. Cancer Res 52:4313-4319.

Columbano A., Ledda-Columbano, G.M., Curto, M., Ennas, G.M., Coni, P., Sarma, D.S.R. and Pani, P. (1990) Cell proliferation and promotion of rat liver carcinogenesis: different effect of hepatic regeneration and mitogen-induced hyperplasia on the development of enzyme altered foci. Carcinogenesis, 11, 771-776.

Columbano, A., Ledda-Columbano, G.M., Coni,P., Faa,G., Liguori, C., Santacruz, G. and Pani,P. (1985) Occurrence of cell death (apoptosis) during the involution of liver hyperplasia. Lab Invest., 52, 670-675.

Columbano, A., Ledda-Columbano, G.M., Lee, G., Rajalakshmi, S. and Sarma, D.S.R. (1987) Inability of mitogen-induced liver hyperplasia to support the induction of enzyme-altered islands induced by liver carcinogens. Cancer Res., 47, 5557-5559.

Columbano, A., Rajalakshmi, S., and Sarma, D.S.R.(1981) Requirement of cell proliferation for the initiation of liver carcinogenesis as assayed by three different procedures. Cancer Res. 41, 2079-2083.

Coni P, Pichiri-Coni G, Ledda-Columbano GM, Rao PM, Rajalakshmi S, Sarma DSR, Columbano A (1990) Liver hyperplasia is not necessarily associated with increased expression of c-fos and c-myc mRNA. Carcinogenesis, 11, 835-839.

Coni P, Simbula G, Carcereri De Prati A., Menegazzi M, Sarma DSR, Ledda-Columbano GM, Columbano A (1993) Differences in the steady state levels of c-fos, c-jun and c-myc mRNA during mitogen induced liver growth and compensatory regeneration. Hepatology (in press).

Dragani, T.A., Manenti, G. and Della Porta G. (1986) Enhancing effect of carbon tetrachloride in mouse hepatocarcinogenesis. Cancer Lett., 31, 171-179.

Eacho, P.I., Lanier, T.L. and Brodhecker, C.A. (1991) Hepatocellular DNA synthesis in rats given peroxisome proliferating agents: comparison of WY-14,643 to clofibric acid, nafenopin and LY171883. Carcinogenesis 12, 1557-1561.

Eldridge, S.R., Goldsworthy, T.L., Popp, J.A., and Butterworth, B.E. (1992) Mitogenic stimulation of hepatocellular proliferation in rodents following 1,4-dichlorobenzene administration. Carcinogenesis, 13, 409-415.

Farber E, Sarma DSR (1987) Hepatocarcinogenesis: A dynamic cellular perspective. Lab.Invest., 56, 4-21.

Goyette M, Petropulos CJ, Shank PR, Fausto N (1983) Expression of cellular oncogene during liver regeneration. Science 219:510-512.

Helvering LM, Richardson KK, Horn DM, Wightman KA, Hall RL, Smith WC, Engelhardt JA, Richardson FC (1993) Expression of TRPM-2 during involution and regeneration of the rat liver. Proc Am Assoc Cancer Res 34:673.

Ishikawa, T., Takayama, S. and Kitagawa, T. (1980) Correlation between time of partial hepatectomy after a single treatment with diethylnitrosamine and induction of adenosine triphosphatase-deficient islands in rat liver. Cancer Res., 40, 4261-4264.

Ledda-Columbano, G.M., Columbano, A., Curto, M., Ennas, M.G., Coni, P., Sarma, D.S.R. and Pani, P. (1989). Further evidence that mitogen-induced cell proliferation does not support the formation of enzyme-altered islands in rat liver by carcinogens. Carcinogenesis, 10, 847-850.

Legerm JG, Montpetit ML, Tenniswood MP (1987) Characterization and cloning of androgen-repressed mRNAs from rat ventral prostate. Biochem Biophys Res Commun 147:196-203.

Makino R, Hayashi K, Sugimura T (1984) C-myc transcript is induced in rat liver at a very early stage of regeneration or by cycloheximide treatment. Nature 310:697-698.

Marsman, D.S., Cattley, R.C., Conway, J.G. and Popp, J.A. (1988). Relationship of hepatic peroxisome proliferation and replicative DNA synthesis to the hepatocarcinogenicity of the peroxisome proliferators di(2-ethylhexyl)phtalate and (4-chloro-6(2,3 xylidino)-2-pyrimidinylthio)acetic acid (Wy-14,643) in rats. Cancer Res. 48, 6739-6744.

Masuhara M, Katyal SK, Nakamura T, Shinozuka H (1992) Differential expression of hepatocyte growth factor, transforming growth factor-alfa and transforming growth factor-beta messenger RNAs in two experimental models of liver cell proliferation. Hepatology 16:1241-1249.

Matsumoto K, Tajima H, Hamanoue M, Kohono S, Kinoshita T, Nakamura T (1992) Identification and characterization of "Injurin", an inducer of expression of the gene for hepatocyte growth factor. Proc Natl Acad Sci USA 89:3800-3804.

Melchiorri C, Chieco P, Zedda AI, Coni P, Ledda-Columbano GM, Columbano A (1993) Ploidy and nuclearity of rat hepatocytes after compensatory regeneration or mitogen-induced liver growth. Carcinogenesis (submitted).

Melnick, R.D. (1992) Does chemically induced hepatocyte proliferation predict liver carcinogenesis? FASEB J.,6, 2698-2706.

Michalopoulos GK (1990) Liver regeneration: Molecular mechanism of growth control. FASEB J 4:240-249.

Mirsalis, J.C. (1987) In vivo measurement of unscheduled DNA synthesis and S-phase synthesis as an indicator of hepatocarcinogenesis in rodents. Cell Biol.Toxicol. 3,165-173.

Morello D, Lavenu A, Bobinet A (1990) Differential regulation and expression of c-jun, c-fos and c-myc protooncogenes during mouse liver regeneration and after inhibition of protein synthesis. Oncogene 5:1511-1519.

Nadal C, Zajedela F (1966) Polyploidie somatique dans le foie de rat. I. Le role des cellules binuclees dans le genese des cellules polyploides. Exp Cell Res 42:99-116.

Pound, A.W. and McGuire, L.J. (1978) Repeated partial hepatectomies as a promoting stimulus for carcinogenic response of liver to nitrosamines in rats. Br.J.Cancer, 37, 585-594.

Rao, M.S. and Reddy, J.K. Peroxisome proliferation and hepatocarcinogenesis. (1987).Carcinogenesis 8,631-636.

Shinozuka H, Kubo Y, Katyal SL, Coni P, Ledda-Columbano, GM, Columbano A (1993) Roles of hepatocyte growth factor and tumor necrosis factor-alfa on liver cell proliferation in rats induced by lead nitrate. FASEB J 7:A429.

Wyllie AH, Kerr JFR, Currie A.R (1980) Cell death: The significance of apoptosis. Int Rev Cytol 68.251-306.

Yeldandi, A.V., Milano, M., Subbarao, V., Reddy, J.K. and Rao, M.S. (1989) Evaluation of liver cell proliferation during ciprofibrate-induced hepatocarcinogenesis. Cancer Lett. 47, 21-27.

Use of Mechanistic and Other Data in Identifying Carcinogens: A Review Based on the *IARC Monographs* Programme

H. Vainio

International Agency for Research on Cancer,
150, Cours Albert-Thomas, 69372 Lyon Cedex 08, France

Introduction

The implementation of effective preventive measures against human cancer depends primarily on knowledge about its causes. The International Agency for Research on Cancer (IARC) has been carrying out a programme on the Evaluation of Carcinogenic Risks to Humans since 1969. The purpose of the programme has been to provide clear, sound evaluations of the degrees of evidence for the carcinogenicity of agents.

The current procedure for evaluating the carcinogenicity of agents, mixtures or exposure circumstances within the *IARC Monographs* programme consists primarily of two steps. First, independent evaluations are made of the degree of evidence for carcinogenicity in humans from epidemiological studies and for carcinogenicity in animals from experimental studies. Subsequently, these two evaluations are combined, with supporting evidence from other relevant data on, for example, genetic effects, structure–activity relationships, mechanisms of action and pharmacokinetics, to make an overall evaluation of carcinogenicity to humans.

Supporting Evidence for Carcinogenicity

Using these three data sets in making overall evaluations of carcinogenicity was first put into practice in 1979, when Supplement 1 (IARC 1979) to the *Monographs* was prepared, and was continued in the preparation of Supplements 4 and 7 (IARC 1982, 1987). Those supplements consisted of new reviews of all agents that had been evaluated within the *Monographs* programme for which some data on carcinogenicity in humans were available. With the introduction in 1987 of a new Preamble to the *Monographs* (which is, except for minor modifications, some of which are discussed below, that which is used currently), the concept of using relevant data other than those on carcinogenicity as supporting evidence in making an overall evaluation of carcinogenicity to humans was formalized. In all subsequent volumes (43 *et seq.*), overall evaluations of carcinogenicity have been made, and working groups have determined whether or not supporting data have a bearing on the overall evaluation. When they have been used to influence an overall evaluation of carcinogenicity, a brief description of the supporting evidence has generally been included in the monograph. It should be stressed that the two evaluations of degrees of evidence of carcinogenicity are made independently of each other and of other relevant data, e.g., on mechanisms of action. It is only when the overall evaluation of carcinogenicity to humans is made that the three sets of data are combined.

Use of Mechanistic Data

In line with the Agency's policy of monitoring advances in basic research and their relevance to evaluations of carcinogenicity, it has convened two meetings to discuss the mechanistic aspects of carcinogenesis and their relevance to the evaluation process. In 1983, an ad-hoc Working Group was convened with the specific purpose of verifying whether chemical carcinogens could be categorized according to their mechanisms of action and the extent to which consideration of mechanisms of action could contribute to a satisfactory categorization of carcinogens. The opinion of that Group was that, in 1983, no classification of carcinogens according to mechanism of action could be 'exhaustive or definitive'; and, therefore, such a classification could not be used within the *IARC Monographs* programme.

Advances made after that meeting in the understanding of the multistage process of carcinogenesis led the IARC to convene another ad-hoc Working Group, in June 1991, to review again whether, and to what extent, mechanistic data could be used in evaluating carcinogenicity. Specifically, the Working Group was to decide whether an agent could be categorized as causing cancer by a defined mechanism, whether the identification of a causal mechanism could be used to influence the overall evaluation of carcinogenicity to humans and whether that mechanism justified extrapolation of the results of experiments in animals to the human situation. The Group recommended that information on possible mechanisms of carcinogenic action be summarized when appropriate data were available.

Mechanisms can be understood at many different levels. The Group proposed four descriptive dimensions for considering such data: (i) evidence of genotoxicity (i.e., structural change at the level of the gene); (ii) evidence of effects on the expression of relevant genes (i.e., functional changes at the intracellular level); (iii) evidence of relevant effects on cell behaviour (i.e., morphological or behavioural changes at the cellular or tissue level); and (iv) evidence of time and dose relationships of carcinogenic effects and interactions between agents. These dimensions are not mutually exclusive and an agent may have effects on several or all dimensions. Thus, the action of an agent on the expression of relevant genes could be summarized under both the first and the second dimension if it were known with reasonable certainty that those effects resulted from genotoxicity. Because different agents act by different mechanisms, the relevance and importance of each of these levels are dependent on the agent and tumour site that are being considered.

The Group proposed that the strength of the evidence that a particular mechanism of carcinogenesis is or is not operating be evaluated, using terms such as 'weak', 'moderate' and 'strong'. Development of operational guidelines for making such conclusions might be difficult because of the diversity of the phenomena involved; however, each working group would assess the available data with regard to the relevance, reproducibility and sensitivity of the methods used.

In assessing the relevance of the available data on mechanisms to the evaluation of the carcinogenic risk of an agent to humans, a number of issues are pertinent: (a) evidence that the effect lies in the chain of events linking the agent with cancer, (b) relevance of the test system to human responses, (c) similarity of the end-point of the test system to the human target cells (although the validity of the end-point can override this consideration), (d) similarity of toxicokinetic variables in the test system and human beings, and (e) specificity of the measured end-point for the mechanism.

The available evidence may show that similar mechanisms act in humans and experimental animals, but situations in which species-specific activity seems possible are of particular concern. The possibility that humans are the more affected or susceptible species, in the absence of epidemiological data, could be assessed on the basis of knowledge of the mechanisms in humans and in animals. The converse possibility—that humans are the putatively unaffected species—can be assessed only after consideration of certain principles: that the mechanism in question is the primary one in inducing the tumour in the species in which it is observed, that the same or a similar mechanism does not operate in humans, and that the agent does not induce other types of tumour in the species under consideration.

The Group suggested a number of situations in which relevant information on mechanisms could be used in evaluating carcinogenic risk to humans: Such information may confirm the level of classification indicated by data from epidemiological and/or experimental carcinogenicity studies. Strong evidence for a mechanism of action that is relevant to carcinogenicity in humans could justify 'upgrading' of the overall evaluation for a particular agent. An overall evaluation of human cancer hazard based on the results of experimental carcinogenicity tests could be 'downgraded' by strong evidence that the mechanism responsible for tumour growth in experimental animals does not occur in humans. In keeping with the goal of public health, priority must be given to the demonstration that the mechanism is not operative in humans. Obviously, the absence of data on mechanisms would not influence evaluations of human carcinogenic hazard based on epidemiological or experimental carcinogenicity data.

Evaluation in a Higher or Lower Category

Of the 768 overall evaluations of carcinogenicity to humans made in the 58 volumes of *Monographs*, 44 were influenced directly by other relevant data (see Table 1). These include evaluations of eight groups of agents for which other relevant data supported an evaluation extending to the whole group.

Thus, for 36/768 evaluations, supporting evidence from other relevant data was used to make a higher overall evaluation of carcinogenicity to humans than would have been the case if the evaluation had been based solely on carcinogenicity data. Supporting data have been used to influence overall evaluations in three ways: (i) from Group 3 to 2B—for example, an agent for which there were no data or inadequate evidence for carcinogenicity in humans but limited evidence in animals supported by other relevant data; (ii) from 2B to 2A—for example, an agent for which there were no data or inadequate evidence of carcinogenicity in humans but sufficient evidence in animals supported by other relevant data; (iii) from Group 3 to 4—the one example being caprolactam, for which there is evidence suggesting lack of carcinogenicity in experimental animals, which, in the absence of similar evidence in humans, can allow an agent to be placed in Group 4 only on the basis of supporting evidence from other relevant data; in this case, consistently negative results were found in a wide range of assays for genetic and related effects *in vitro*.

Of the 36 cases in which a change in category was made, 25 were considered in Supplement 7 and the remainder in the subsequent 16 volumes (43–58) in the *Monographs* series. Of the category changes made in Supplement 7, nine were for agents on which the data had not been updated; thus, the evaluations were based only on data reported in the original volume, and no supporting evidence for the change of the overall evaluation is given in the Supplement. In general, however, the supporting evidence in these cases would have come from studies of genetic and related effects. Only five (atrazine, bleomycins, styrene, marine diesel fuels, gasoline) of the 36 changes were from Group 3 to Group 2B.

The revised Preamble that appeared in Volume 54 of the *IARC Monographs* allows for the possibility that an agent for which epidemiological evidence of cancer is less than sufficient can be placed in Group 1 when 'there is sufficient evidence of carcinogenicity in experimental animals and strong evidence in exposed humans that the agent acts through a relevant mechanism of carcinogenicity'. During 1992, that possibility was considered on two occasions. In the case of ultraviolet (UV) radiation, while the epidemiological evidence for induction of cancer came mainly from studies of solar radiation, a large amount of data was available on the mutagenicity of UVA, UVB and UVC radiation in experimental models and an increasing amount on mutagenicity in exposed humans. Special tandem base substitutions (T:T to C:C) in a tumour suppressor gene, p53, have been found in human skin carcinomas and in melanomas that have developed at sites exposed to the sun. After considerable discussion, the Working Group adopted a conservative attitude and classified UV radiation (UVA, UVB, UVC) into Group 2A, probably carcinogenic to humans. Solar radiation was considered to be carcinogenic to humans (Group 1) (IARC 1992b).

The second case in which the possibility of placing an agent in Group 1 in the absence of sufficient epidemiological evidence was considered was 4,4′-methylene bis(2-chloroaniline) (MOCA). MOCA is carcinogenic in dogs and rodents and is comprehensively genotoxic. It binds to DNA through reaction with *N*-hydroxy MOCA, and the same adducts that are formed in target tissues for carcinogenicity in animals have been found in urothelial cells from a small number of exposed humans. The evidence from epidemiological studies was inadequate. After lengthy discussions, the Working Group again made an overall evaluation of Group 2A, probably carcinogenic to humans (IARC 1993b).

The Preamble in Volume 54 also allows for the possibility that an agent for which there is sufficient carcinogenicity in animals can be placed in Group 3 (instead of Group 2B, in which it would normally be categorized), when 'there is strong evidence that the mechanism of carcinogenicity in animals does not operate in humans'. Like the possible change to Group 1, this possibility has also not yet been used by any working group.

Table 1. Use by *IARC Monographs* working groups of mechanistic and other supporting evidence in making overall evaluations of carcinogenicity

Agent	IARC Monogr. Vol. (year)	Degree of evidence of carcinogenicity		Overall evaluation	Supporting evidence
		Human	Animal		
Adriamycin	S7 (1987)	I	S	2A	*GRE*: criterion (i)
Atrazine	53 (1991b)	I	L	2B	*Other mechanism*: increased risk of tumours known to be associated with hormonal factors observed in both epidemiological and experimental studies are consistent with the effects of atrazine on the hypothalmic-pituitary-gonadal axis
Azacitidine	50 (1990c)	ND	S	2A	*GRE*: active in a broad spectrum of assays including assays in mammalian cells *Other mechanism:* is a pyrimidine analogue and is incorporated into DNA, causing hypomethylation
Benz[*a*]anthracene	S7 (1987)/ 32 (1983)	ND	S	2A	NS [GRE]
Benzidine-based dyes	S7 (1987)	I		2A	Metabolism: metabolic conversion to benzidine, as detected in urine of exposed dye workers *Structure–activity*: related to benzidine (Group 1)
Direct Black 38 (technical-grade)			S		
Direct Blue 6 (technical-grade)			S		
Direct Brown 95 (technical-grade)			S		
Benzo[*a*]pyrene	S7 (1987)/ 32 (1983)	ND	S	2A	NS [GRE]
Beryllium and beryllium compounds	58 (1993c)	S	S	1	*Evaluation as a group*: all the beryllium compounds studied were found to be carcinogenic in experimental animals
Bleomycins	S7 (1987)	I	L	2B	*GRE*: criteria (i) and (ii)
Cadmium and cadmium compounds	58 (1993c)	S		1	*Evaluation as a group*: based on the fact that ionic cadmium causes genotoxic effects in a wide variety of tests, including human cells
Cadmium compounds			S		
Cadmium metal			L		

Table 1 (contd)

Agent	IARC Monogr. Vol. (year)	Degree of evidence of carcinogenicity		Overall evaluation	Supporting evidence
		Human	Animal		
Captafol	53 (1991b)	ND	S	2A	*GRE*: active in a wide range of assays, including dominant lethal mutation *in vivo* – a generally insensitive assay
1-(2-Chloroethyl)-3-cyclohexyl-1-nitrosourea (CCNU)	S7 (1987)	I	S	2A	*GRE*: criteria (i) and (ii)
para-Chloro-*ortho*-toluidine and its strong	48 (1990a)			2A	*Evaluation as a group*: any salt of the parent compound with a strong acid can be expected to behave chemically in a manner similar to that of the hydrochloride salt in solution and *in vivo*
para-Chloro-*ortho*-toluidine		L			
para-Chloro-*ortho*-toluidine hydrochloride			S		
Chlorozotocin	50 (1990c)	ND	S	2A	*Structure–activity*: alkylating agent structurally related to methyl CCNU (Group 1) and BCNU and CCNU (both Group 2A) *GRE*: consistently positive results in a broad spectrum of assays including mammalian cells
Chromium[VI] compounds	49 (1990b)			1	*Evaluation as a group*: based on combined results of epidemiological and experimental carcinogenicity studies and other relevant data that support the concept that chromium[VI] [as chromate] ions generated at critical sites in the target cells are responsible for the carcinogenic action observed
Chromium[VI] compounds as encountered in chromate production, chromate pigment production and chromium plating industries		S			
Barium chromate			I		
Calcium chromate			S		
Chromium trioxide			L		
Lead chromate			S		
Sodium dichromate			L		
Strontium chromate			S		
Zinc chromate			S		
Cisplatin	S7 (1987)	I	S	2A	*GRE*: criteria (i) and (ii)

Table 1 (contd)

Agent	IARC Monogr. Vol. (year)	Degree of evidence of carcinogenicity		Overall evaluation	*Supporting evidence*
		Human	Animal		
Cobalt and cobalt compounds	52 (1991a)	I		2B	*Evaluation as a group*: but no supporting evidence given
Cobalt metal powder			S		
Metal alloys containing cobalt, chromium and molybdenum			L		
Cobalt[II] oxide			S		
Cobalt[II] sulfide			L		
Cobalt[II] chloride			L		
Cobalt-aluminium-chromium spinel			I		
Cobalt[II,III] oxide			I		
Cobalt naphthenate			I		
Cobalt[III]acetate			I		
Dibenz[*a,h*]anthracene	S7 (1987)/ 32 (1983)	ND	S	2A	NS [GRE]
Diethyl sulfate	54 (1992a)	I	S	2A	*Structure–activity*: alkylating agent which ethylates DNA *GRE*: consistently positive results in a broad spectrum of assays including somatic and germ cells of mammals *in vivo*
Dimethylcarbamoyl chloride	S7 (1987)	I	S	2A	*GRE*: criterion (ii)
Dimethyl sulfate	S7 (1987)	I	S	2A	*GRE*: criterion (ii)
Epichlorohydrin	S7 (1987)	I	S	2A	*GRE*: criteria (i) and (ii)
Ethylene dibromide	S7 (1987)	I	S	2A	*GRE*: criterion (ii)
N-*Ethyl*-N-*nitrosourea*	S7 (1987)/ 17 (1978)	ND	S	2A	NS [GRE, possibly structure-activity]
IQ (2-Amino-3-methylimidazo[4,5-*f*]-quinoline)	56 (1993a)	I	S	2A	*GRE*: mutagenic in a broad spectrum of assays *Metabolism*: metabolized by human liver microsomes to DNA-reactive metabolites
5-Methoxypsoralen	S7 (1987)	I	S	2A	*GRE*: criterion (ii)

Table 1 (contd)

Agent	IARC Monogr. Vol. (year)	Degree of evidence of carcinogenicity		Overall evaluation	*Supporting evidence*
		Human	Animal		
4,4′-Methylene bis(2-chloroaniline) (MOCA)	57 (1993b)	I	S	2A	*GRE*: comprehensively genotoxic *Structure–activity*: aromatic amine *Metabolism*: humans metabolize MOCA to *N*-hydroxy MOCA which can bind to DNA in urothelial cells in man
Methylmercury compounds	58 (1993c)	I		2B	*Evaluation as a group*: based on similarities between various methylmercury compounds with regard to kinetics, metabolism and genetic toxicity
Methylmercury chloride			S		
N-Methyl-*N*′-nitro-*N*-nitrosoguanidine (MNNG)	S7 (1987)	I	S	2A	*GRE*: criterion (ii)
N-Methyl-*N*-nitrosourea	S7 (1987)/ 17 (1978)	ND	S	2A	NS [GRE, possibly structure–activity]
Nickel and nickel compounds	49 (1990b)				*Evaluation as a group*: based on combined results of epidemiological and experimental carcinogenicity studies and other relevant data and the underlying concept that nickel compounds can generate nickel ions at critical sites in their target cells
Nickel compounds				1	
Nickel salts			L		
Nickel sulfate		S			
Combination of nickel oxides and sulfides encountered in the nickel refining industry		S			
Nickel monoxides			S		
Nickel trioxide			I		
Nickel sulfide, amorphous			I		
Nickel sulfides, crystalline			S		
Nickel antimonide			L		
Nickel arsenides			L		
Nickel carbonyl			L		
Nickel hydroxides			S		
Nickelocene			L		
Nickel selenides			L		
Nickel telluride			L		
Nickel titanate			I		

Table 1 (contd)

Agent	IARC Monogr. Vol. (year)	Degree of evidence of carcinogenicity		Overall evaluation	*Supporting evidence*
		Human	Animal		
Nickel and nickel compounds (contd)					
Metallic nickel		I	S	2B	
Nickel alloys		I	L		
Nitrilotriacetic acid and its salts	48 (1990a)	ND		2B	*Evaluation as a group*: nitrilotriacetic acid is liberated to some extent from nitrilotriacetate salts in solution
Nitrilotriacetic acid and its sodium salts			S		
N-Nitrosodiethylamine	S7 (1987)/ 17 (1978)	ND	S	2A	NS [GRE, possibly structure–activity]
N-Nitrosodimethylamine	S7 (1987)/ 17 (1978)	ND	S	2A	NS [GRE, possibly structure–activity]
Procarbazine hydrochloride	S7 (1987)	I	S	2A	*GRE*: criterion (ii); in particular, positive *in vivo* *Structure–activity*: generates an alkylating species
Propylene oxide	S7 (1987)	I	S	2A	*GRE*: criterion (ii) *Metabolism*: alkylates haemoglobin of exposed workers *Structure–activity*: chemical structure analogous to ethylene oxide (Group 1)
Styrene	S7 (1987)	I	L	2B	*GRE*: criteria (i) and (ii) *Metabolism*: metabolized to styrene oxide (Group 2A)
Styrene oxide	S7 (1987)	ND	S	2A	NS [GRE]
Tris(2,3-dibromopropyl) phosphate	S7 (1987)	I	S	2A	*GRE*: criterion (ii)
Ultraviolet A, B and C radiation	55 (1992b)		S	2A	*GRE*: mutagenic in a broad spectrum of assays, including human cells *in vivo*
Vinyl bromide	S7 (1987)	ND	S	2A	NS [structure–activity: analogous to vinyl chloride]

Table 1 (contd)

Agent	IARC Monogr. Vol. (year)	Degree of evidence of carcinogenicity		Overall evaluation	*Supporting evidence*
		Human	Animal		
Mixtures					
Diesel fuels	45 (1989)	I			*Evaluation of components*: straight-run kerosin L in experimental animals, light vaccum distillates and cracked residues derived from the refining of crude oil S in experimental animals
Marine diesel fuel			L	2B	
Gasoline	45 (1989)	I		2B	*GRE*: induced unscheduled DNA synthesis in rodents *in vivo* and *in vitro* *Evaluation of components*: light straight-run naphtha and light catalytically cracked naphtha L in experimental animals; benzene, Group 1; 1,3-butadiene S in experimental animals, Group 2A
Unleaded automotive gasoline			L		

S7, Supplement 7; I, inadequate; S, sufficient; GRE, genetic and related effects; L, limited. For full definitions of degrees of evidence and overall evaluations, see IARC (1987)

Criteria: (i), the agent produces genetic or related effects in exposed humans (i.e., indicative of DNA or chromosomal damage) and also gives positive results in a range of other types of assay; (ii), the agent is active in a broad spectrum of assays for genetic and related effects, including those involving mammalian cells, and there is evidence from structure-activity and metabolism studies that the agent itself reacts covalently with DNA or is likely to be converted to a reactive form in humans.

Other mechanism: effect other than a genetic and related effect

NS: supporting evidence not given in Supplement 7 since the data base was not updated at that time; evaluations are based on data presented in earlier published volumes, as indicated

Types of Supporting Evidence

Genetic and Related Effects

Data on genetic and related effects have had the most impact on overall evaluations of carcinogenicity. In only one instance (atrazine) did the change in category not include such data (Table 1). At least half of the category changes were based solely on such effects; several others were based on these effects in combination with data from metabolic studies and structure–activity considerations.

The Working Group that prepared Supplement 7 generated two criteria for using genetic and related effects to change an evaluation. Although the criteria were developed only as a working practice for that particular Group, they have, by and large, been adopted by subsequent groups. The criteria are as follows (IARC 1987):

(i) The agent produces genetic or related effects in exposed humans (i.e., indicative of DNA or chromosomal damage) and also gives positive results in a range of other types of assay;

or

(ii) The agent is active in a broad spectrum of assays for genetic and related effects, including those involving mammalian cells, and there is evidence from structure–activity and metabolism studies that the agent itself reacts covalently with DNA or is likely to be converted to a reactive form in humans.

These are criteria (i) and (ii) indicated in Table 1. It is not stated explicitly in Supplement 7 which of the criteria were used in the case of agent, but the information is implicit in the summaries of other relevant data.

Structure–activity Relationships

Thus far, data on structure–activity relationships alone have not been considered to provide a sufficient weight of supporting evidence to change an overall evaluation. In three cases, the structure of an agent has been related to that of a known or probable carcinogen. Benzidine-based dyes are structurally related to benzidine (Group 1); propylene oxide has an analogous structure to ethylene oxide (Group 2A) (IARC 1987); and chlorozotocin (IARC 1990c) is structurally similar to the chloroethyl nitrosoureas, 1-(2-chloroethyl)-3-(4-methylcyclohexyl)-1-nitrosourea (methyl-CCNU) (Group 1), bischloroethyl nitrosurea (BCNU) and 1-(2-chloroethyl)-3-cyclohexyl-1-nitrosurea (CNNU) (Group 2A), evaluated previously (IARC 1987).

Metabolism

Data data on metabolism alone have also not yet been cosidered to provide adequate supporting evidence of carcinogenicity. Cases in which such data were considered with other information were the following: Workers exposed to benzidine-based dyes were found to have benzidine in their urine, indicating metabolism of the dyes to a known human carcinogen. Propylene oxide alkylates the haemoglobin of exposed workers, and styrene is metabolized to styrene oxide. MOCA is oxidized to *N*-hydroxy MOCA in exposed humans (IARC 1993b).

Mechanisms Other than Genetic and Related Effects

Data on other mechanisms have been used twice as supporting evidence of carcinogenicity, both on recent occasions. In the case of atrazine (IARC 1991b), mechanistic data alone provided supporting evidence for an upgrading from Group 3 to 2B. Thus, the increased risk of tumours known to be associated with hormonal factors observed in both epidemiological studies on triazine herbicides and experimental studies on atrazine are consistent with the effects of atrazine on the hypothalamic-pituitary-gonadal axis. Azacitidine (IARC 1990c) is not only active in tests for genetic activity but is also a pyrimidine analogue which is incorporated into DNA, causing hypomethylation.

Evaluations of Components of Mixtures

In the case of two agents, marine diesel fuel and gasoline (IARC 1989), studies of components of the mixtures provided a similar or a higher degree of evidence for carcinogenicity in experimental systems and humans than did studies of the mixture itself. The combined weight of evidence, therefore, supported a higher evaluation than would have been the case in the absence of data on components.

Groups of Agents

When groups of chemicals have been evaluated in the supplements and in volumes 1–58 of the *Monographs* series, the criteria for grouping evaluations were often implicit in the text of the monograph and were not clearly outlined in a statement of supporting evidence. Quite different approaches could therefore be taken by different working groups. For example, in Volume 2, the carcinogenicity to humans of 'inorganic arsenic' was summarized; in Volume 23, it was concluded that 'inorganic arsenic compounds' are dermal and pulmonary carcinogens in humans; in Supplement 4, the overall evaluation was for the carcinogenicity to humans of 'arsenic and certain arsenic compounds'; and the overall evaluation in Supplement 7 was for 'arsenic and arsenic compounds'.

In practice, evaluations of groups of chemicals have been fairly consistent from one working group to another, although no rationale for grouping agents was given explicitly, notably at the time of the meeting that gave rise to Supplement 7. In 1988, an ad-hoc Working Group made recommendations on how groups of agents should be evaluated and within what limits (IARC 1988). The Group recognized that groups could be evaluated (i) in the case of closely related chemicals, when studies on individual members of the group provide an accurate estimate of the carcinogenic activity of the group as a whole (e.g., salts and esters of aromatic amines, organic acids, stereoisomers, elements and their salts); and (ii) in cases in which epidemiological data relate to exposures to mixtures composed of several members of a group (e.g., aflatoxins, oestrogenic hormones, alkylating antineoplastic agents). That Working Group therefore endorsed the existing practice of evaluating groups of agents when appropriate. They also modified the Preamble to allow the use of supporting data to conclude that agents related to others for which there is direct evidence of their capacity to induce cancer in animals or in humans 'may also be carcinogenic, a statement describing the rationale for this conclusion is added to the evaluation narrative and an additional evaluation may be made for the broader group of compounds if the strength of the evidence warrants it.' This practice has been adopted in volumes 48 *et seq*., and groups evaluated recently are included in Table 1.

In the four recent evaluations of groups of agents, the rationale for grouping was basically similar: that both the parent compound or element and its salts or related compounds liberate what is considered to be the biologically active principle *in vitro* and *in vivo*. Thus, for chromium[VI] and nickel compounds, the chromium[VI] ion and nickel ion, respectively, are considered to be responsible for the observed carcinogenic activity and to be generated *in vivo* from all chromium[VI] and nickel compounds (IARC 1990b). An identical assumption underlies the evaluation of cobalt and cobalt compounds, although no specific supporting evidence was given by the Working Group which evaluated those agents (IARC 1991a). The rationale for the group evaluation of nitrilotriacetic acid and its salts (IARC 1990a) also rested on the principle that it is the nitrilotriacetic acid ion that is responsible for the observed effect and that it is liberated in solution to some extent from both the parent acid and its salts. A similar assumption holds for the evaluation as a group of *para*-chloro-*ortho*-toluidine and its strong acid salts (IARC 1990a), that the parent amine or related ammonium ion is responsible for the observed effects. Recently, beryllium and beryllium compounds and cadmium and cadmium compounds were evaluated as carcinogenic to humans (Group 1) (IARC 1993c).

Conclusion

Data other than those on carcinogenicity from experimental and epidemiological studies have been used in making overall evaluations of carcinogenicity to humans. Of the 768 evaluations made in the first 58 volumes of the *Monographs* series, genetic and related effects provided the most important supporting evidence. Such considerations led to changed overall evaluations, including one from Group 3 to Group 4, five from Group 3 to Group 2B and 29 from Group 2B to Group 2A. In evaluating groups of agents, consideration has been given primarily to determining whether all members of a group liberate the active principle to some extent *in vivo* and to the results of epidemiological studies in which exposure is to several members of the group concomitantly.

Acknowledgements

I am very grateful to Dr Linda Shuker, PhD, and Mr Julian Wilbourn, MSc, for their help in the preparation of the manuscript, to Mrs Elisabeth Heseltine, MSc, for editing the manuscript and to Mrs Jane Mitchell for secretarial assistance.

References

IARC (1978) IARC monographs on the carcinogenic risk of chemicals to humans. Vol 17: Some N-nitroso compounds. IARC, Lyon

IARC (1979) IARC monographs on the carcinogenic risk of chemicals to humans. Suppl 1: Chemicals and industrial processes associated with cancer in humans. IARC monographs, volumes 1 to 20. IARC, Lyon

IARC (1982) IARC monographs on the carcinogenic risk of chemicals to humans. Suppl 4: Chemicals, industrial processes and industries associated with cancer in humans. IARC monographs, volumes 1 to 29. IARC, Lyon
IARC (1983) IARC monographs on the carcinogenic risk of chemicals to humans. Vol 32: Polynuclear aromatic compounds, part 1: chemical, environmental and experimental data. IARC, Lyon
IARC (1985) IARC monographs on the carcinogenic risk of chemicals to humans. Vol 36: Allyl compounds, aldehydes, epoxides and peroxides. IARC, Lyon
IARC (1986) IARC monographs on the carcinogenic risk of chemicals to humans. Vol 39: Some chemicals used in plastics and elastomers. IARC, Lyon
IARC (1987) IARC monographs on the evaluation of carcinogenic risks to humans. Suppl 7: Overall evaluations of carcinogenicity: an updating of IARC monographs volumes 1 to 42. IARC, Lyon
IARC (1988) Report of an IARC working group to review the approaches and processes used to evaluate the carcinogenicity of mixtures and groups of chemicals. IARC Int Tech Rep No 88/02. IARC, Lyon
IARC (1989) IARC monographs on the evaluation of carcinogenic risks to humans. Vol 45: Occupational exposures in petroleum refining; crude oil and major petroleum fuels. IARC, Lyon
IARC (1990a) IARC monographs on the evaluation of carcinogenic risks to humans. Vol 48: Some flame retardants and textile chemicals, and exposures in the textile manufacturing industry. IARC, Lyon
IARC (1990b) IARC monographs on the evaluation of carcinogenic risks to humans. Vol 49: Chromium, nickel and welding. IARC, Lyon
IARC (1990c) IARC monographs on the evaluation of carcinogenic risks to humans. Vol 50: Some pharmaceutical drugs. IARC, Lyon
IARC (1991a) IARC monographs on the evaluation of carcinogenic risks to humans. Vol 52: Chlorinated drinking-water; chlorination by-products; some other halogenated compounds; cobalt and cobalt compounds. IARC, Lyon
IARC (1991b) IARC monographs on the evaluation of carcinogenic risks to humans. Vol 53: Occupational exposures in insecticide application, and some pesticides. IARC, Lyon
IARC (1992a) IARC monographs on the evaluation of carcinogenic risks to humans. Vol 54: Occupational exposures to mists and vapours from strong inorganic acids; and other industrial chemicals. IARC, Lyon
IARC (1992b) IARC monographs on the evaluation of carcinogenic risks to humans. Vol 55: Solar and ultraviolet radiation. IARC, Lyon
IARC (1993a) IARC monographs on the evaluation of carcinogenic risks to humans. Vol 56: Some naturally occurring substances: food items and constituents, heterocyclic aromatic amines and mycotoxins. IARC, Lyon
IARC (1993b) IARC monographs on the evaluation of carcinogenic risks to humans. Vol 57: Occupational exposures of hairdressers and barbers and personal use of hair colourants; some hair dyes, cosmetic colourants, industrial dyestuffs and aromatic amines. IARC, Lyon
IARC (1993c) IARC monographs on the evaluation of carcinogenic risks to humans. Vol 58: Beryllium, cadmium, mercury, and exposures in the glass manufacturing industry. IARC, Lyon

Late Manuscripts

(accepted for publication
after the end of the congress)

Developmental Toxicity - Magnitude of the Problem

Richard Wiger
Department of Environmental Medicine,
National Institute of Public Health, Oslo, Norway

Human reproduction is a complex process, and can be disturbed in many phases by host as well as environmental factors. It has therefore been difficult to distinguish environmental causes of abnormal development from other factors related to the parents' innate properties. Among the environmental factors, chemical insult to the conceptus can result in any one or more of the following manifestations of developmental toxicity: Death, structural abnormality, growth alteration, and/or functional deficit. These manifestations encompass a wide array of adverse developmental endpoints, such as spontaneous abortions, stillbirths, malformations, early postnatal mortality, reduced birth weight, mental retardation, sensory loss, and other adverse functional or physical changes that are manifested postnatally (Fed. Reg. 1991).

Dose-response relationships

The severity of the toxic effect(s) in the embryo/fetus is related to both dose and time of exposure. It is well accepted that developmental toxicity is a threshold phenomenon, i.e. below a certain dose no developmental toxicity occurs. Furthermore, relationships exist between the dose of a developmental toxin and the severity/type of effect (Fig. 1). Lower doses are considered to cause functional impairment whereas the highest doses cause death. The intermediate doses may lead to growth retardation and/or malformations. We will come back to these relationships near the end of this presentation.

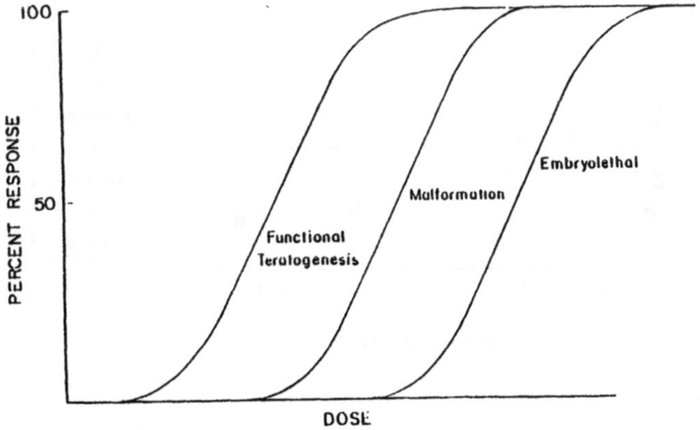

Figure 1. Theoretical set of dose-response curves showing the relationship between three manifestations of developmental toxicity (after Vorhees and Butcher 1982).

In the following information and reflections concerning developmental toxicity will be presented beginning with prenatal and early postnatal death.

Prenatal and Early Postnatal Death

In his 1990 study entitled Survival Probability of Human Conceptions from Fertilization to Term, Boklage began with the following words: "The ultimate outcome of all human conception is death. Less well known, but no less true, the majority of those deaths occur before birth" (Boklage 1990). Although preterm loss of clinically recognized pregnancies is well documented, efforts to quantify losses before clinical recognition have been hampered by uncertain methodology and scarcity of appropriate data. Boklage utilized data from several studies which applied different methodologies in order to estimate the fates of conception products. Although the cited studies have involved considerable variation in sample structure and study design, and in the timing, specificity, and sensitivity of human chorion gonadotropic hormone (hCG) assay methods, hCG elevation remains the best available indicator of pregnancy before clinical recognition. In this context the term implantation refers to the process of trophoblast differentiation and uterine attachment, which occurs about day 4-7 post fertilization in the human. Current understanding is that hCG elevation represents differentiated trophoblast function.

Before examining the fate of human conceptions let us consider the efficiency of the fertilization process in humans. Interpretations of available data indicate that only 60-70% of all oocytes are successfully fertilized given optimum timing of natural insemination in couples with proven fertility.

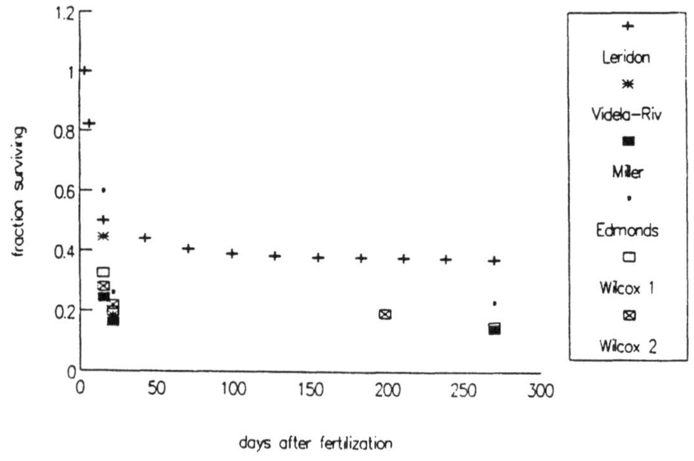

Figure 2. Conception and survival of natural pregnancies plotted as fractions of cycles pregnant versus time after ovulation (after Boklage 1990).

Concerning the extent of death in utero, the results clearly show that preterm death of the human conceptus is common. As can be seen in Fig. 2, there is a consistent biphasic pattern in the rate of loss from biochemical pregnancy detection to term suggesting that most wastage occurs prior to clinical recognition. The data are convincing and indicate that at least 73% of natural single conceptions have no real chance of surviving 6 weeks of gestation. Of the remaining 27%, about 90% will survive to term. What then are the fates of conception products which survive beyond 6 weeks? Some answers to this question are found in a 1991 study by Kalter entitled - Five-decade international trends in the relation of perinatal mortality and congenital malformations: Stillbirths and neonatal death compared.

In the last 50 years the importance of congenital malformations as causes of perinatal death has risen in the western world. This increased rate of congenital malformations is related to the major reduction in pernatal mortality rate in the last few decades in these regions (Fig. 3) and not to any change in their frequency, since the overall frequency of major congenital malformations of about 3% does not appear to have altered. The reduced perinatal mortality rate was largely brought about by progress in managing many of the causes of such deaths, e.g. hemolytic disease of newborn, difficult labor and birth injury, asphyxia and immaturity.

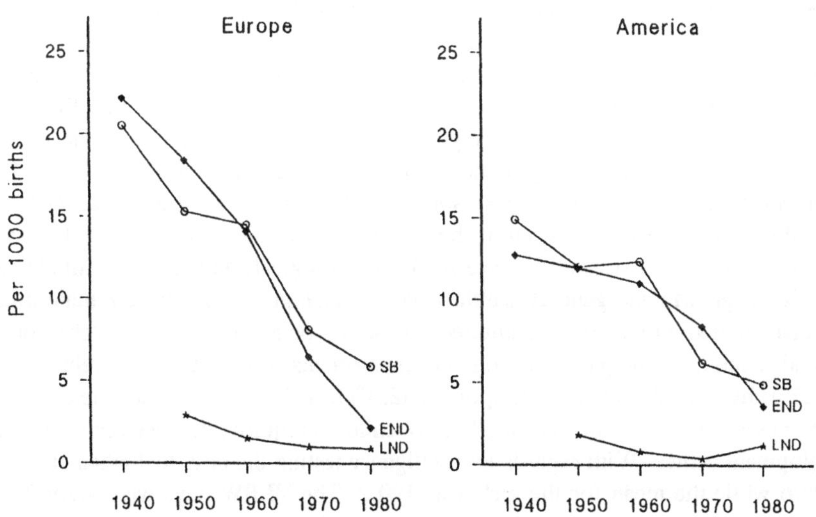

Figure 3. Rates of still birth (SB), early neonatal death (END), and late neonatal death (LND), in offspring >1000 g, from the 1940s to mid-1989, in several European countries (called Europe) and the USA and Canada (called America), derived from hospital reports (after Kalter 1991).

Perinatal mortality generally refers to death over an extended span of time divided into four intervals: early and late fetal death (20-27 and 28 weeks or more of gestational age, respectively) collectively called stillbirth, plus early and late neonatal death, within 7 completed and at eight through 28 days of postnatal age, respectively. The predominant specific defects in stillbirths (75-85%) were those of the central nervous system, especially failure of neural tube closure. In neonatal deaths, on the other hand, the commonest type were cardiovascular defects (38-40%) while those of the central nervous system were a close second (28%) (Kalter 1991). Currently, severe birth defects account for about 20 percent of infant mortality in the USA.

Low birth weight/premature birth

Of those infants born alive, approximately 7.4% are reduced in weight at birth (i.e. below 2500 g), and approximately 3% are found to have one or more congenital malformations at birth, and by the end of the first postnatal year, about 3% more are found to have serious developmental defects (EPA 1991).

In a recent Danish study it was reported that on account of improved management in neonatal intensive care units, the survival rate is presently nearly 50% in babies with birth weights below 1000g and 80% in the group with birth weights from 1000 to 1500 g (Wæver et al. 1992). However, survival at birth does not automatically imply normal patterns of survival and continued growth and development.

What are the prognoses for continued survival and normal development of very low birthweight infants? Roussounis et al. (1993) provide some answers in their Five-year-follow-up study of very low birthweight (VLBW) infants: neurological and psychological outcome. All surviving infants with low birthweight ($\leq$ 1500 g) born in 1982 and 1983 at St. James's University Hospital, Leeds, England were assessed. After five years there were 88 survivors out of the 126 infants (mortality = 32%). In their fifth year the following assessments were made: neurological, audiological, intellectual, behavioral, growth and general health. A comparison group of full-term male infants was also studied with respect to intellectual status, social and emotional behavior and general health. Principal neurological impairments found were: cerebral palsy (10.2%), hydrocephalus (1.1%), epilepsy (2.3%), and sensoneural deafness (2.3%). Furthermore, about one third of the VLBW children required the services of the child development center. With regards to intelligence testing the boys had mean IQ scores of 90.6 while the mean for the girls was 100.2. The VLBW boys were significantly intellectually impaired compared with their peers. However, socially and emotionally they were largely comparable with their full-term peers. The authors concluded that the quality of survival of VLBW children born in the 1980s, despite improvement in perinatal care, remains a major problem.

Premature birth and intrauterine growth retardation have many causes. Some exposures leading to low birth weight include smoking, heavy consumption of ethanol, cocaine addiction, use of anticonvulsants and others. As an example, Medina and Kaempffer (1991) reported that 11% of infant mortality in Chile is attributable to the lower neonatal weight of children of women who smoke. Furthermore, women smokers have twice as many spontaneous abortions as nonsmokers and enter menopause an average of five years earlier. It should be pointed out however, that large variations exist between countries and/or geographical regions with regards to many endpoints of developmental abnormality.

Malformations

Each year more than 250,000 infants are born in the USA with some form of congenital malformation, ranging in severity from extremely mild to life threatening (NICHD 1991). Severe congenital birth defects remain the leading cause of death for infants under the age of 1 year. In order to understand the basis of such congenital malformations, it is essential that we fully understand the mechanisms that underlie normal developmental processes. With an adequate level of understanding, it is conceivable that prophylactic and therapeutic regimens can be designed to reduce the prevalence of developmental abnormalities.

Currently, severe birth defects account for about 20 percent of infant mortality in the USA (NICHD 1991). Babies born with malformations may suffer multiple handicaps, thereby drastically reducing their quality of life. In addition, childhood cancers (caused by uncontrollable growth of embryonic cells) contribute to early death figures in the USA.

During the past 5 years, studies of mammalian embryo development have increased understanding of the genetic and environmental factors responsible for congenital abnormalities and childhood cancer.

It is now well accepted that prenatal mortality, intrauterine growth retardation, behavior alterations, and other postnatal aberrations may accompany structural malformations due to environmental agents or may even occur in the absence of any malformations; therefore, such alterations must also be sought in investigations and included in the study design. A good example of a chemical exposure that produces a wide spectrum of developmental toxic effects is in utero exposure to ethanol.

Fetal Alcohol Syndrome and Fetal Alcohol Effects

The fetal alcohol syndrome (FAS) has been chosen as an example of a developmental toxicant for several reasons:

1) FAS is preventable
2) Ethanol causes all of the endpoints of developmental toxicity, not only malformations
3) There are more than 3,000 publications concerning developmental effects of alcohol

The following statements provide some insight as to the significance of FAS:

> *"Fetal Alcohol Syndrome is the leading cause of mental retardation in the Western world, surpassing Down's syndrome and cerebral palsy" (Luke 1990).*
>
> *The world-wide incidence of the full characteristics of FAS is estimated to be 1.9 per 1000 live births; that of partial characteristics (Fetal Alcohol Effects) is 3-5 per 1000 live births (Spohr et al. 1993).*

As recommended by the Research Society on Alcoholism the diagnosis of FAS (cited from Spohr et al. 1993) requires confirmation of a history of maternal alcohol abuse and that the following criteria are met on examination of the child:

1) Prenatal or postnatal growth retardation (height and weight below the 10th percentile for age or gestational age)
2) CNS dysfunction (any neurological abnormality, developmental delay, or intellectual impairment)
3) Characteristic craniofacial abnormalities, including at least two of microcephaly (head circumference below the 3rd percentile), micropthalmia or short palpebral fissures, and poorly developed filtrum, thin upper lip, and flattening of the maxillary area.

The term Fetal Alcohol Effects (FAE) is used when a child shows two of the three criteria.

Some of the endpoints of developmental toxicity which are associated with FAS include:

Functional Impairment
Decreased sucking behavior, psychomotor developmental delay, cerebral convulsions, muscular hypotonia, distractability, memory deficits, eye abnormalities, reading difficulties, learning problems, lower IQ

Growth Retardation
Intrauterine growth retardation, poor catch-up growth

Malformation
Cardiac defects, skeletal defects, hypospadias, hydronephrosis, cleft palate

Death
Spontaneous abortions and fetal deaths, postnatal mortality

In a recent article by Spohr et al. (1993) some long-term developmental consequences of in utero ethanol exposure were investigated. The most important finding of this long-term follow-up study is that the deleterious effects of intrauterine alcohol exposure on the mental development of children persist for many years. There was little improvement in intelligence even when domestic circumstances changed for the better. These findings suggest that compensatory environmental and educational influences are less important to intellectual outcome of FAS patients in adolescence than expected. Intrauterine alcohol exposure seems to result in teratogenic damage to the developing brain, leading to long-lasting CNS dysfunction. In another study on FAS, Strömland (1992) reported that maternal alcohol abuse during pregnancy causes malformations of the eye with serious consequences to the vision of the affected children. A high percentage (up to 90%) of children suffering from FAS have eye abnormalities of different kinds.

Conjecture on the magnitude of developmental toxicity

To address this issue let us consider the dose-response relationships of developmental toxicity (Fig. 1). If the dose-response relationships for developmental toxicity are applicable to the human situation, then we can use the rate of malformations as a starting point. According to the dose-response relationships, if a particular chemical exposure can lead to malformation, then it is also possible that lower exposures to the same substance could lead to growth retardation and/or functional deficit. Current knowledge links approximately 3-4% of all malformations to exposure to drugs or environmental chemicals. It therefore appears reasonable to assume that the magnitude of developmental toxicity is at least of the same dimension as that for malformations alone, in other words at least 3-4% of live births. However, there is a need for more epidemiological studies which include all the endpoints of developmental toxicity to substantiate this assumption.

What are some of the possiblities for acquiring hard data on human developmental toxicity? It was well known before the iron curtain was lifted that there were a number of regions in the former USSR and in Eastern Europe which were characterized as as ecological catastrophies. Industrial and military activities were conducted with little concern for nature or the people living in the area. There have been a number of reports that these "large scale laboratory experiments" have among other things led to increased cancer, birth defects and poor health in general.

At present there are a number of cooperative efforts between East and West to study health effects of pollution but to date there are no solid data to document developmental toxicity in humans exposed to industrial pollutants. There are many problems to solve in these countries, and how long it will be before such data are generated remains to be seen.

References

EPA (1991) Guidelines for Developmental Toxicity Risk Assessment. Federal Register (Dec. 5, 1991) 63798-63826

Kalter H. (1991) Five-decade international trends in the relation of perinatal mortality and congenital malformations: stillbirths and neonatal death compared. Int J Epidemiol 20:173-179

Luke B. (1990) The metabolic basis of the fetal alcohol syndrome. Int J Fertil 35:333-337

Medina E, Kaempffer AM. (1991) Smoking and health in Chile. Bol Sanit Panam 111:112-121

National Institute of Child Health and Human Development. (1991) An evaluation and assessment of the state of the science: genetics and teratology. U.S. Dept Health Human Service. pp. 1-182

Roussounis SH, Hubley PA, Dear PR. (1993) Five-year-follow-up of very low birthweight infants: neurological and psychological outcome. Child Care Health Dev 19:45-59

Spohr H-L, Willms J, Steinhausen H-C. (1993) Prenatal alcohol exposure and long-term developmental consequences. Lancet 341:907-910

Strömland K. (1992) Malformations of the eyes caused by maternal alcohol abuse during pregnancy. Nordisk Medicin 107:313-315

Vorhees CV, Butcher RE. (1982) Behavioural Teratogenicity. In: Developmental Toxicology (ed.) K Snell. Croom Helm, London

Wæver E, Brandt B, Mortensen T, Jendresen M, Nielsen OH. (1992) The prognoses in neonates with birth weights below 1500 grammes requiring surgery. Ugeskr Læger (Denmark) 154:416-418

Healths problems related to respiratory sensitisation

G. Bylin

Hypersensitivity of the airways has become an important health problem especially in industrialized countries. The hypersensitivity is commonly caused by allergy, and the mechanisms of allergic sensitisation are today beginning to be elucidated However, chemical agents may also induce a state of persistent bronchial hypersensitivity without any detectable signs of allergy.

Allergic rhinitis is characterized by sneezing ,often in series, increased nasal secretion and blockage of the nose. It is often accompanied by eye irritation and conjunctivits. Allergic rhinitis was first describedin 1819 by John Bostock, and reports on separate cases of hay fever were interesting enough to be published in the medical literature during the later half of the nineteenth century . The prevalence of allergic rhinitis in Europe was estimated to about 1 % as late as 1935 (Sahlén and Nordenfors), and now the prevalence is 5-10 % in many industrialized countries (Åberg 1988).

Bronchial asthma gives attacks of dyspnoea due to bronchial constriction , which is reversible either by medical treatment or spontaneously. Asthma is caused by an inflammation of the bronchial mucosa, and treatment by antiinflammatory drugs (mainly inhaled corticosteroids) during the last decades has markedly improved the patients´ control of their symptoms. Nevertheless, the mortality in asthma has not declined and some countries has even experienced an increased asthma mortality during certain periods (Sears 1991).Bronchial asthma was reported to have a prevalence of about 2-3 in Sweden during the nineteensixties (Kiviloog et al 1974). Now the figure has increased to 4-5 % (Lundbäck et al 1988), and a similar increasing trend is reported from many other industrialized countries (Sears 1991).

Improved diagnostic methods may explain a part of the increased prevalence of allergic rhinitis and asthma. It is, however, a generally accepted view among allergologist and pulmonary physicians that we have had a real increase of these respiratory allergic disorders during the last decades.

The first living months seems to be a critical period for allergic sensitisation. It has been shown that children born during spring time run a higher risk of developing a birch pollen allergy (Björkstén 1980), and the same has been shown for exposure to ragweed allergen (Settipane and Hagy 1979).

The rapid increase of allergic respiratory diseases indicates that there are environmental factors contributing to this development . There seems not to have been a significant change of the allergen load during the last decades, which could have explained the increased frequency of allergies. It should be noted that the increased frequency of respiratory sensitisation involves both in-door allergens as mites and pets as well as out-door allergens as pollen. It is likely that other environmental factors act as promoters for the sensitisation process. Increased frequency of upper respiratory infections, changed dietary habits, shorter time of breast feeding and exposure to air pollutants are discussed as possible causes. The hypothesis that air pollutants (e g environmental tobacco smoke, car exhausts) act as adjuvant agents stimulating the sensitisation process has experimental support from animal studies (Muranaka et al 1986) and has attracted great interest during recent years.

References

Björkstén F, Suoniemi I, Koski V. Neonatal birch pollen contact and subsequent allergy to birch pollen.
Clin Allergy 1980, 10, 585-591.

Bostock J. Case of a Periodical Affection of the Eyes and Chest.
Med Chir Trans 1819,10, 61.

Kiviloog J, Irnell L, Eklund G. The prevalence of bronchial asthma and chronic bronchitis in smokers and non-smokers in a representative local Swedish population.
Scand J Resp Dis 1974, 55, 262-276.

Muranaka M, Suzuki S, Koizumi K, Takafuji S , Miyamoto T, Ikemori R, Tokiwa H.
Adjuvant activity of diesel-exhaust particulates for the production of IgE antibody in mice .
J Allergy Clin Immunol 1986, 77, 616-623.

Salén EB, Nordenfors B.Catarrhus aestivus (Bostock´s sommarkatarr, hösnuvan).
Nord Med Tidskrift 1935, 22, 905-910 (in swedish)

Sears M. Epidemiological Trends in Bronchial ASthma. In: Kaliner MA, Barnes PJ, Persson CGA (eds).Asthma- its pathology and treatment, Dekker, New York 1991, p1-49.

SettipaneRJ, Hagy GW. Effect of atmospheric pollen on the newborn.
Rhode Island Med J 1979, 62,477-482.

Åberg N.Allergic diseases in childhood and adolescence.
Dissertation, Gothenburg, Gothenburg University, 1988,160 p.

NEONATAL EXPOSURE TO THE ORGANOPHOSPHATE DFP INDUCES CHANGES IN BEHAVIOUR AND MUSCARINIC RECEPTOR DENSITY IN THE ADULT MOUSE

J. Ahlbom, A. Fredriksson[1], P. Eriksson
Department of Zoophysiology, Uppsala University, Norbyvägen 18a, S-752 36 Uppsala, and [1]Department of Toxicology, Uppsala University, Box 594, S-751 24 Uppsala, Sweden.

Introduction Previous studies have shown an increased vulnerability of the cholinergic system to substances such as diisopropyl fluorophosphate (DFP), paraoxon, DDT, pyrethroids, and PCB´s during rapid development of the neonatal mouse brain, leading to permanent changes in the adult animal.

In order to determine if there is a critical period for these effects of DFP, mice at the ages of 3-days, 10-days and 19-days were given a single sub-symptomal oral dose of DFP (1.5 mg/kg body weight), or the fat emulsion vehicle.

Results and Discussion Behavioral testing performed at the age of four mounths revealed that only mice treated postnatal day 3 or 19 showed significant ($p<0.05$) alterations in spontaneous motor activity.

The density of MAChR of the adult mouse was assayed by measuring [^{3}H]QNB specifically bound in the P2 fraction. A significant ($p<0.01$) decrease in the specific [^{3}H]QNB binding was observed in the cerebral cortex for the mice treated at postnatal day 3 or 10, compared to control.

Conclusion This study provides further support for the hypothesis that the cholinergic system is sensitive to xenobiotics during its rapid development in the neonatal mouse.

Young Scientists Poster Award

The Young Scientists Poster Award is given to the first author of the best poster who must be younger than 35 years.

The winner is announdced at the closing ceremony of the EUROTEX Congress. The abstract is published in the congress proceedings.

The winner 1993 was Jonas Ahlbom, Uppsala.

Subject Index

MIX
Papier aus verantwortungsvollen Quellen
Paper from responsible sources
FSC® C105338

FSC
www.fsc.org

If you have any concerns about our products,
you can contact us on
ProductSafety@springernature.com

In case Publisher is established outside the EU,
the EU authorized representative is:
Springer Nature Customer Service Center GmbH
Europaplatz 3, 69115 Heidelberg, Germany

Printed by Libri Plureos GmbH
in Hamburg, Germany